Ancient Rhetorics and Digital Networks

RHETORIC, CULTURE, AND SOCIAL CRITIQUE

Ancient Rhetorics and Digital Networks

Edited by Michele Kennerly
and Damien Smith Pfister

The University of Alabama Press
Tuscaloosa

The University of Alabama Press
Tuscaloosa, Alabama 35487-0380
uapress.ua.edu

Typeface: Minion and Lithos

Cover image: Adapted from a detail of a Greek "school cup" by the red-figure painter Douris, c. 500 B.C.; courtesy Staatliche Museum, Berlin
Cover design: David Nees

Cataloging-in-Publication data is available from the Library of Congress.
ISBN: 978-0-8173-5904-1
E-ISBN: 978-0-8173-9157-7

To our kids, ingenious combiners of incongruities

Contents

Figures

Preface

Finding Rhetorical *Poroi* in the *Pontos* of Internetworked Media

> Their arms flashed like flame in the sun as the ship / hurried on, and ever their long path gleamed white, visible like a trail through a pale-green plain.
>
> —Apollonius of Rhodes, *Argonautica*, 1.544–46

Seeking concepts to repurpose and metaphors to extend, those who theorize upon communication and its routes often take to the sea. For instance, in the 1940s, Norbert Wiener and his colleagues "decided to call the entire field of control and communication theory, whether in the machine or in the animal, by the name Cybernetics," which they formed from the Greek word for "steersman of a ship."[1] In the early 1980s, the portmanteau "cyberspace" flowed into the popular lexicon after appearing in the sci-fi writings of William Gibson, where he referred to it as a site of "unthinkable complexity."[2] At about the same time, rhetoricians at the University of Iowa and beyond unmoored from the docks of disciplinarity and launched the Project on the Rhetoric of Inquiry (POROI), its acronym spelling the ancient Greek word for paths and passages through the sea.[3] During the next decade, Netscape designers named its internet browser—one of the first—Navigator. Originating from Greek and Latin, respectively, such names reinforce the necessity of directing and orienting oneself and others in conditions of variability and potential vulnerability. A sense and skill of orientation in a vast sea was highly valued in ancient coastal cultures. Assessing many characteristics of ancient Greeks, Edith Hall chose to enumerate ten, listing "seagoing" first. She even framed her entire survey nautically, tracing the Greeks "from Bronze Age seafarers to navigators of the western mind."[4] This volume follows in the wide wake of these earlier pursuits of the novel through the naval.

Pontos was one of many ancient Greek words for the sea. In invoking *pontos*, poets and rhetors summoned an understanding that, "disturbed by the winds which blow across it and by the flux and reflux of the waves, the sea is the most mobile, changeable and polymorphic space."[5] William Thalmann observes that the *pontos* was a "trackless expanse, without spatial coordinates, upon which [the mythological figure] *Poros* performs his work of spatialization, and upon which ships inscribe paths (*poroi*) that create a humanly intelligible space."[6] In the ancient world, the task of the navigator was to find a *poros* through the *pontos*: to avoid crashing upon the Symplegades—Clashing Rocks—or some other nautical impediment, to move to one's destination without the advantage of clear material markers for location, to combine wit and skill in interpreting the ocean's movements in order to safely arrive at a destination.

Likewise, we see our contemporary experience of internetworked media—a term that encompasses the World Wide Web, intranets, the Dark Web, mobile applications, and all similar interlinked digital communication networks—as akin to the ancient sense of *pontos*. Internetworked media are so diverse, so expansive, so sprawling, so fluctuating, and so teeming with both familiar and unfamiliar eddies and currents that navigation presents a constant challenge. Whatever problems we had in keeping track of it all in earlier media ecologies have been intensified with the information abundance, proliferating genres, and new technical innovations afforded by digital media technology. Although rocks that could break a ship are not present literally in digitally networked media ecologies, there are certainly dangers associated with following the wrong path. Yet, for most people living in digitally networked cultures, choosing a *poros* through the internetworked media *pontos* is both necessary and inevitable. However, most of the paths that guide us in our interactions with the digital reflect the preferences created by large media conglomerates. Consider how early dial-up services like America Online taxonomized web content under familiar categories, such as "Today's News," "Sports," or "Reference Desk," to make the fast-growing web intelligible to new users; vestiges of this taxonomy persist in web portals like Yahoo. Google, Facebook, and Twitter, alongside representatives of legacy media like television networks and newspapers, have developed more sophisticated and algorithmically derived ways to organize pathways through the internet. Google's mission, for example, is to "organize the world's information and make it universally accessible and useful." While Google's search results seemingly reflect the way things are, they are really the products of many variables interacting with each other algorithmically. Just as in the ancient world, *poroi* are often naturalized, but no *poros* is natural.

Although these corporate *poroi* are effective at making (profit-minded) sense of the internet's *pontos*, there are many alternative, critically oriented *poroi* that can and ought to be pursued. As Thalmann notes, the ancient concepts of *poros* and *pontos* worked figuratively as well as materially; since every story cuts a path through a broad expanse of phenomena, cleverly selecting characters and episodes to make salient, *poros* could be used to describe the narrativization of experience.[7] Just as a narrative can be understood as a pathway through experience, so can critical perspectives provide *poroi*, as they offer coordinates for understanding complex, expansive phenomena. The premise and the promise of *Ancient Rhetorics and Digital Networks* are one and the same: ancient rhetorical theories offer critical perspectives capable of providing *poroi* through the *pontos* of digital communication networks. In her speech deep within the narrative folds of Plato's *Symposium*, the priestess Diotima presents Poros personified and deified. She activates the meanings of Poros as "resource" and "plenty," and she names Poros the son of Mētis (Cunning).[8] This volume offers to its readers affordances aplenty for avoiding *aporia*—an impasse, literally, a lack of *poroi*—when pondering or discussing the theoretical richness of ancient rhetorics in our digitally networked age. As for the relationships of our contributions to cunning, we leave that to readers to judge.

Notes

1. Norbert Wiener, *Cybernetics, or Control and Communication in the Animal and the Machine*, 2nd ed. (Cambridge, MA: MIT Press, 1962), 11.

2. William Gibson, *Neuromancer* (New York: Ace Books, 1984), 69.

3. John Lyne, personal correspondence with the authors, September 24, 2015.

4. Edith Hall, *Introducing the Ancient Greeks: From Bronze Age Seafarers to Navigators of the Western Mind* (New York: W. W. Norton, 2014), 1.

5. Marcel Detienne and Jean-Pierre Vernant, *Cunning Intelligence in Greek Culture and Society*, trans. Janet Lloyd (Chicago: University of Chicago Press, 1991), 222.

6. William G. Thalmann, *Apollonius of Rhodes and the Spaces of Hellenism* (Oxford: Oxford University Press, 2011), 28.

7. Thalmann, *Apollonius of Rhodes*, 29–30.

8. Plato, *Symposium*, trans. Harold N. Fowler (Cambridge, MA: Harvard University Press 1925), 203b.

Acknowledgments

This volume, about connections, congruous and incongruous, itself came to be through a series of them, which we gratefully acknowledge.

Our professors at the University of Pittsburgh merit pride of place, since it was in their seminars that we shared in the verve and vitality of the intellectual engagement they encouraged—and got to know one another. We recognize in particular John Poulakos, John Lyne, and Gordon Mitchell. John Poulakos did us (and readers, we think) the explicit service of emboldening us in our etymological play.

When we first contemplated this project, during some down time at the Rhetoric Society of America conference in 2012, we knew personally only a very few of those scholars who would eventually contribute to this volume. By the time we started to put it together, we knew a few more. Now, we can unreservedly gush that we consider all the contributors friends. Every one—everyone!—was collegial and inspirational at every step of the process. We have learned a lot from them about scholarly generosity—and about rhetoric, ancient and contemporary.

Dan Waterman and his team at the University of Alabama Press have been enthusiastic stewards of this project since its inventional stage. We appreciate their confident efforts.

We thank also our former and current academic departments. For their sustenance, intellectual and otherwise, Damien thanks his students and colleagues at the University of Nebraska and the University of Maryland. For their curiosity and care, Michele thanks her students and colleagues in the Department of Communication Arts and Sciences at Penn State University.

Finally, for their support and love, we acknowledge our families: Carly Woods, Alexandra Woodsmith, Shaun Mahony, and Naoise Mahony. Because we all are fond of one another, it seems fitting to squeeze together.

Ancient Rhetorics and Digital Networks

Introduction

Michele Kennerly and Damien Smith Pfister

What can ancient rhetorics possibly tell us about new digital media technologies in contemporary public culture? The distance between antiquity and our time seems too far, the cultural conditions too distinctive, the technologies too different; in short, the juxtaposition seems too incongruous to produce meaningful insight. We contend—and the contributors to this volume demonstrate compellingly—that incongruity need not entail incommensurability. Ancient rhetorical theory can illuminate communication phenomena occurring through digital networks. The fundamental communicative issues we deal with today—negotiating information abundance, persuading others in our social network, navigating new media ecologies, shaping broader cultural currents—also pressed upon ancient peoples. Although the ancient-world contexts differ significantly from those of the early-twenty-first century, the image that graces this volume's cover suggests that positing an intersection between ancient rhetorics and digital networks can be more than a fanciful conceit. The original, fifth-century BCE vase painting depicts a student with a wax tablet and stylus.[1] Since the ancient tablet visually resembles the digital tablet, the vase painting could, after a bit of a fashion makeover, pass as a representation of a present-day figure engaging with a screen in a classroom, a coffee shop, or an office. The family resemblance between the wax and electronic tablets has not been lost on present-day observers.[2] *Ancient Rhetorics and Digital Networks* pushes this surface congruity deeper, exploring ancient rhetorical texts, figures, concepts, sensibilities, and practices in light of the glow of digital networks; conversely, this volume surveys the angles and tangles of digital networks from viewpoints afforded by ancient rhetoric. This antimetabole animates the volume.

Historically, major media transitions and ancient rhetorical theory enjoy an iterative relation, linking and looping together at key junctures. It was book rolls made of plant matter that enabled the systematization of rhetorical theory throughout the ancient world, from Egypt to Greece to China. As textual culture unfolded across the Mediterranean Basin, Alexandrians and then Romans turned to the Greeks that preceded them in an effort to contemplate how writing both extends and extenuates memory. The Christian adoption of the codex was differentiated from not only Jewish scrolls but also pagan ones, evidencing a wariness toward classical Greece and Rome even as the writers of the New Testament show themselves to be trained in rhetoric. The European Renaissance witnessed another round of classical revivalism, with the mechanical reproduction of ancient works keeping printing presses churning and Enlightenment mouths yapping. On the Indian subcontinent, media technologies that accompanied colonialism in the early nineteenth century expanded Indian prose production, prompting a recovery of their own ancient rhetorical wisdom. As electronic mass media linked together a global village throughout the twentieth century, Marshall McLuhan predicted the resurgence of oral, rhetorical norms of communication. We make the case in this introductory chapter that the transition from a mass media ecology and toward an internetworked media ecology again amplifies the resonances between ancient rhetorical theory and contemporary communication practices.

Ancient Rhetorics and Digital Networks participates in a larger sounding out of how changing technological and cultural conditions reshape the scope and function of rhetoric as a productive practice and a critical hermeneutic. This is the permanent project of rhetorical theory, a project pursued with urgent purpose since at least the late 1960s. In the wake of those turbulent years, Edward Corbett proposed revisiting ancient rhetoric's primary texts in his contribution to *The Prospect of Rhetoric*: "This is not to recommend an interest in and a respect for the ancient merely because it is old; rather it is to encourage this exploration of the old for its possible bearing on the new."[3] Perhaps the turn to ancient rhetorics at this time was inevitable, since, as Donald Bryant had postulated earlier, "each new generation probably needs to interpret afresh much of the relevant history of thought" to help make sense of new cultural conditions.[4] In effect, the ancients developed ways of thinking about rhetoric that ultimately subtend all communicative activity, including the communication that constitutes digital networks. A strong understanding of those ancient rhetorical theories provides both a necessary and a flexible basis on which to build contemporary rhetorical theory. Yet, cutting and pasting ancient rhetorical theory into our present context is too simple, for the *technē* of rhetoric is necessarily shaped by the technologies that support it. As Thomas O. Sloane observes, "Forty years ago there were no blogs, no hypertexts, no cell phones. As anticipated, technology has had an impact on our thinking about rhetoric and will continue to do so."[5] Returning to ancient texts from new technocultural vantage points shakes up accepted interpretations, produces readings with different nuances, allows old terms to be revivified and reinhabited in new ways, and generates theoretical resources to guide critics, theorists, and publics in negotiating continuity and change.

Five Relations between Ancient Rhetorics and Digital Networks

Ancient Rhetorics and Digital Networks explicitly articulates the value of ancient rhetorical theory for understanding, critiquing, and producing rhetoric in the context of digital networks. Other critical orientations on the connections between rhetoric, internetworked media, and culture are certainly available. One such orientation focuses on the political economy of the internet, tracking how states and corporations influence the development of media infrastructures, enabling or constraining communication and democratic participation.[6] Another perspective examines the internet from the vantage point of critical/cultural studies, examining the politics of representation and popular culture, audience engagement with and resistance to media artifacts, or extension of control in digitally networked environments (an inexhaustive list, to be sure).[7] The various traditions contained by the signifier "media studies" provide yet another pathway to trace how media ecologies shape human communication, consciousness, interdependence, and culture.[8] Contemporary rhetorical studies shapes and is shaped by these other critical currents;[9] nonetheless, it offers its own distinctive perspective on the confluence of the digital and the rhetorical under the rubrics of "digital rhetorics" or "networked rhetorics."[10] Amid the various (postmodernist, posthumanist) twists and (affective, materialist) turns of contemporary rhetorical theory, *this* volume highlights the continuing urgency of engaging with the wisdom of the ancients.

What, then, are the possible connections between the ancient and the internetworked? How might the rhetorical and the digital be meaningfully brought together? While connections between ancient rhetorics and digital networks occasionally appear in both scholarly and public discourses, we organize these episodic linkages into five overarching relations:

(1) *historia* (the antecedent relation): ancient rhetorics are vital historical antecedents to digital communication;
(2) *analogia* (the analogical relation): ancient rhetorics offer rich analogues for practices that appear native to digital media ecologies;
(3) *heuresis* (the heuristic relation): ancient rhetorics serve crucial heuristic functions in theorizing digital and networked rhetorics;

(4) *nomos* (the convention relation): ancient rhetorics cue us to important social customs and moral orientations necessary to surviving and thriving in digital contexts; and
(5) *anakainōsis* (the renewal relation): ancient rhetorics, and the practices and sensibilities accompanying them, are renewed in cultures intensively shaped by digital technologies.

Each of the contributions to this volume deepens, complicates, or intensifies these relations in ways that evidence the potency of reading the ancient and the digital together.[11]

Historia: The Antecedent Relation

Ancient rhetorics antecede rhetorics operating under digitized and networked conditions. Such an obvious claim points in one of four directions. In the first direction, antecedence is taken for anachronism: the ancient world, though it may have some antiquarian appeal, is barely and rarely relevant to today's theorists and critics. It may be important to study the rhetorical imaginaries of the ancient world, but only to understand the early historical conditions that gave rise to later communication practices and cultural formations. This is, for sure, the weakest of all possible relations, identifying the worth of ancient rhetorics only inasmuch as they figure as an early historical link in a long chain of theorizing about communication.

A second direction is bolder and more promising for robust theorizing: ancient rhetorical theories provide guidance for understanding the digital remediation of oral communication. In this view, newer media supplement, but never entirely supplant, older media. The invention of writing, like the invention of print and digital media after it, did not signal the end of oral communication, though it did transform oral norms and many of the practices that accompanied oral culture.[12] However, even in digital contexts, oral communication merits continued attention because, as the scholar of media and rhetoric Marshall McLuhan famously claimed, "the 'content' of any medium is always another medium. The content of writing is speech, just as the written word is the content of print, and print is the content of the telegraph."[13] In McLuhan's terms, the content of digital media is often oratory, of the grand as well as quotidian variety. Despite the early concern that the advent of digital mediation portended further marginalization of public address, digital media technologies appear at least partially responsible for an oratory boom. Millions of people around the world tune in to livestreams of speeches, watch archived versions of speeches on video sharing sites, embed and share speeches through digital social networks, remix and mash-up speeches, and even annotate speeches with their own commentary.[14] In addition to remediating conventional modes of public address, digital communication networks have spawned new genres, among them TED Talks that ask speakers to "communicate your best ideas" in a pop-educational speech. Whether live or recorded, through Skype, FaceTime, Google Hangouts, Facebook Live, or YouTube, we might see oral communication as returning to the center of public life.[15] Following Jay David Bolter and Richard Grusin, the concept of *remediation* is central to understanding orality in the context of digitality, as speech concatenates through the many possible forms and genres of internetworked media.[16] Thus, the need to study orality, which ancient rhetorical theories were largely oriented toward describing and theorizing, is arguably as great today as it ever has been.

A third direction in which to take the antecedent relation is genealogical. Our terms are not our own: they have complex histories, many of which stretch back to antiquity. To provide a reckoning with the history of a concept requires, then, an exploration of the various contexts, materialities, and cultural forces that constituted it, shaped its reception and circulation, and precipitated an evolution in its use. John Durham Peters's critical history of the idea of "information" provides an exemplary model of this kind of genealogical approach. Noting the prevalence of the word "information," Peters tracks the term's rise, diffusion, and shifting meaning from the ancient world onward. Introduced to English through Latin (*informare*, to instruct), the term "information" (idea, instruction, concept, doctrine) took on hylomorphic dimensions in the fourteenth century amid a highly Aristotelian scholarly milieu: "the intelligibility of material objects owes to the forms that *in-form* them, shaping them from within."[17] The seventeenth and eighteenth centuries renovated the medieval metaphysic presupposing some underlying form separate from human perception. Information in this context, following Peters, becomes the "stuff of sense," bodily perceptions capable of empirical analysis.[18] The growth of the state, statistics, and eventually computers produces an understanding of information as rendering large populations into aggregates and making their behaviors visible.[19] Finally, "information" becomes so abstracted that it assumes the role of late twentieth-century metahermeneutic, used to conceptualize the activity of DNA, the processes of cognition, the effects of hormones and enzymes, the processing abilities of the brain, and so on. Such a genealogical approach can expand beyond tracing a single term to apprehend how cultural influences shape rhetorical practices over time. For example, Xiaoye You's study of how ancient Confucian teachings continue to shape the teaching of writing in China, even as they were modified by ever-evolving socialist pedagogies, is a genealogical approach with a wider scope.[20]

One final direction embedded in the antecedent relation draws a parallel between cultures in media transition. In short, the ancient world stages the first documented change in media ecosystems, and contemporary observers might learn about the dynamics of media transition itself from studying an earlier era. As Maryanne Wolf observes, "The historical moment that best approximates the present transition from a literate to a digital culture is found in the ancient Greeks' transition from an oral culture to a literacy-based culture."[21] What kind of insights, then, can be gleaned from examining how the ancients grappled with the cultural and ethical implications implicit in the new medium of alphabetic writing? Plato's *Phaedrus* animates this struggle over the status of writing vis-à-vis rhetoric, an early sally against the convergence of new forms of media and rhetorical study. Rhetoric's resurgence in moments of media transition, as in the shift from manuscript culture to print culture in the European Renaissance, from print culture to electronic broadcast media in the early twentieth century, or the gradual movement beginning in the 1960s from electronic broadcast media to digitally networked media, has an uncanny synchrony with the ancient world. Given the consonance between these four historical-cultural formations undergoing media change, we might posit an "Iron Law of Sophistry": The development of a new medium of communication creates conditions for a rhetorical culture that privileges play, experimentation, argument, contention, and innovation.[22] When the tools change, so does the art of rhetoric, "as entrepreneurial rhetoricians leverage the novel expressive possibilities afforded by a new medium of communication."[23] While rhetorical studies in the United States has historically been focused on the Greek and Roman traditions, we suspect that the Iron Law of Sophistry stretches beyond the Mediterranean whenever different cultural formations grapple with the expressive possibilities generated by the emergence of a new medium.

Analogia: The Analogical Relation

Rhetorical practices in digital networks often inspire analogical comparisons to the ancient world. Everything old, as the commonplace goes, is new again. "Writing on the wall" is not, as Tom Standage observes, a practice native to Facebook, if cave paintings are seen properly as predecessors to digital developments in message-leaving.[24] Emojis are just a digital variation of pictographs, with a history as old as cuneiform writing.[25] YouTube parodies "have ancient roots" in the play prized by the Sophists.[26] Cicero relied on a communication network analogous to our own, though slower and more focused on ruling elites, by circulating papyrus rolls of speeches, treatises, and letters.[27] Neuroscience research that proves how memory is interlinked with spatial awareness is taken as biological proof of the ancient method of loci.[28] Tagging, which is a ubiquitous folksonomic (instead of taxonomic) way of organizing digital content, also has a prehistory in the ancient world: ancient pottery, mosaics, and other fragments of material culture were often inscribed with "tags." Historian of the ancient world Sarah Bond notes of this phenomenon that "much like today, labels functioned in a myriad of ways: to articulate space, to exhibit notions of proper *paideia* (education) and legitimacy, to trigger collective memory, and to provide deeper engagement with an object."[29] The same can be said for tags and hashtags in digital media ecologies.

The culture of likes, favorites, upvotes, and retweets that shapes the digital mediascape is an easy target for some cultural critics, especially as these more democratic signals about the distribution of attention threaten conventionalized tastemakers.[30] Yet, in another example of an analogical path between the ancient and the digital, Rebecca Newberger Goldstein rehabilitates the ancient sense of *kleos* to make sense of these reputational logics of social media:

> *Kleos* means both "glory" or "fame" and also the "song that ensures that glory or fame." The noun is cognate with the Homeric verb *kluō*, meaning "I hear." *Kleos* is sometimes translated as "acoustic renown"—the spreading renown you get from people talking about your exploits. It's a bit like having a large Twitter following. . . . What can people do to withstand time's drowning out the fact that they once had been? The Ethos of the Extraordinary answered that all a person can do is to enlarge that life by the only means we have, striving to make of it a thing worth the telling, a thing that will have impact on other minds, so that, being replicated there, it will take on a moreness. *Kleos*. Live so that others will hear of you. Paltry as it is, it's the only way to beat back uncaring time. Our own culture of Facebook Likes and Twitter followings should put us in a good position to sympathize with an insistence on the social aspect of life-worthiness.[31]

To be fair, to sing a song of my selfie or set the world a-Twitter is not to become Achilles or Homer. Nonetheless, to the extent that internetworked subjects strive for renown and reputation, they participate in an ancient ritual. Small wonder, then, that so much of what we observe in digital discourse communities is wordplay and one-upping, eloquence and novelty. Through such rhetorical practices reputations are made and preserved.

Historical analogy is a refreshing antidote to the ahistorical, often breathless, awe in which "new" digital practices are held. In contrast to the "rhetoric of the technological sublime," the analogical relation relies upon a more pragmatic rhetoric of technological similitude.[32] Analogy, a rhetorical strategy developed across the ancient world, is thus useful in "elucidating obscure phenomena" like the impact of new kinds of mediation.[33] By highlighting similarities, the analogical relation makes an anthropological claim: Humans simply *are* a kind of animal that write on walls, tag objects, rely on communication networks, fold pictographs into otherwise lexical communication, and so on.[34] Yet, as most analogically oriented authors acknowledge, the relation is not fully isomorphic. Analogies often highlight dissimilarities, which prompt richer analyses of cultural changes and media affordances. For example, Goldstein's analogy between *kleos* in the ancient world and *kleos* in the context of digital networks highlights different temporal horizons for acoustic renown. The ancients performed acts of heroism in hopes of being remembered for generations, millennia, even eternity. Now, even Andy Warhol's late 1960s quip about everyone's fifteen minutes of fame seems quaint: in a fast-paced mediascape of digital networks, one would be lucky to get fifteen tweets of fame.

Heuresis: The Heuristic Relation

Ancient rhetorics provide heuristics (from the ancient Greek *heuriskein*, "to find") for understanding digital communication. The evolution of the computer itself can be attributed to the heuristic value of ancient rhetorical theory. Early computers were more like programmable calculators: impressive in their own right but complex to operate and limited in their functionality. Much like the temples of the ancient world, only the anointed were allowed inside rooms that housed computers. In retrospect, we might see interfacing with these early computers as organized around time: The programmer gave their program card to a computer operator, who would run them through the machine and then collect the output. There was no lasting record of the transaction in the computer, no storage memory to speak of that could preserve the program or output. In the 1970s, Nicholas Negroponte and Richard Bolt made a breakthrough in how people could interact with computational technology that ultimately democratized the priesthood of computer operators—a breakthrough they attribute to the Greek poet Simonides of Ceos, known for his astounding memory. His technique for memorization was to tie a section of text to a statue or bust of a deity in a temple. The appropriate text would then spring to mind as Simonides imagined himself walking through the temple's many notable features. This method, known in Latin as the method of *loci*, or place-based mnemonic technique, is still used to aid memorization of texts. Explicitly drawing from the legacy of Simonides, Negroponte and Bolt concluded that memory is intimately tied to spatiality and that enhancing computer memory would require thinking in terms of spatialization. In the wake of this insight was born MIT's Spatial Data Management System, a predecessor to the desktop interface and file structure that spatialized computer functions and consequently made computers the powerful memory machines that they are today.[35] Thus, the very genesis of computing technology's material interface lies in the ancient world.

What Douglas Ehninger identified as the "grammar" of rhetoric—the parts of a speech, canons, genres, proofs, audiences, occasions, and styles—continues to play a central, if modified, role as a heuristic for producing and interpreting communication through computational devices.[36] Understanding strategies for invention, or generic conventions of a particular platform, is a prerequisite for effective communication in digital contexts. The question, then, is about the nature and extent of modification needed to make sense for new media environments: how might ancient rhetorical vocabularies be adapted to account for screen cultures? For audiences radically distributed in space and dispersed in time? For digital media technologies? For publics premised on cultural pluralism rather than on cultural homogeneity? For a capitalist political economy? Kathleen Welch answers these questions by turning to the work of Isocrates, which inspires what she calls an "electric rhetoric" appropriate for the screen.[37] Similarly, Collin G. Brooke recognizes the underlying utility of the rhetorical canons of invention, arrangement, style, memory, and delivery but updates them to account for digital media ecologies.[38] His five *P*s of digital rhetoric—*proairesis*, pattern, perspective, persistence, and performance—are each tied to one of the original canons but transformed because of the conditions of digitality.

These radical and systemic revisions are dramatic examples of ancient rhetoric's continuing heuristic value. The more modest but ubiquitous use of rhetoric's grammar, however, illustrates the resonant heuristic value of ancient terms, figures, and sensibilities.[39] This is certainly true within the interdiscipline of rhetorical studies (note the continuing salience of terms like invention, audience, and style), but we would argue that rhetoric's grammar (often coded, veiled, ahistorical, and thus impoverished) runs through the linguistic, poststructural, postmodern, and affective turns across the humanities and social sciences. In other words, rhetoric is a heuristic on the ascent (again). We see this as intimately tied to cultural conditions of mediation, as the advent of any new medium produces new challenges in negotiating communicative abundance. Following Richard Lanham, rhetoric produces but, importantly, also manages this communicative abundance—*copia*—by providing resources for getting, sustaining, and transforming attention.[40] Rhetoric thus becomes an especially useful cultural technology capable of working through the attention challenges posed by information abundance. The blogosphere, for example, often looks like a hyperagonistic mess. However, if that agonism is seen as a method for proliferating and then sifting through arguments, a defense of agonism marshaled by many a Sophist, then the b*logos*phere might be redeemed as a site of deliberation.[41] Similarly, microblogging platforms like Twitter might be perceived as the ultimate in inanity, unless one sees how character limits goad microbloggers to eloquence in the name of concision. Indeed, the best tweeters are masters of fundamentally rhetorical forms: aphorism, ironic juxtaposition, and quick quip.[42] The vocabulary of rhetoric provides a way to understand communicative practices in digital networks that enrich our appreciation of them.

Nomos: The Convention Relation

Rhetoricians have often (justifiably) been accused of providing a "Greek Solution" for every problem.[43] This tendency signals for Jim Aune an antimodern strain in contemporary rhetorical theory that copes with "the central role of moral autonomy in the emerging view of the liberal self" by returning to some version of premodern *nomos* (roughly: law, convention, custom, or social norm).[44] Antimodern rhetorical theorists such as Wayne Booth, Richard Weaver, Richard McKeon, and Thomas B. Farrell claim that reviving elements of the Greek rhetorical *nomos* provide a credible alternative to the strange polarities of modern culture constituted by "a scientific and technological worldview reluctant to discuss problems of ethics and value *and* a romantic worldview in which individual self-expression seems to undercut the possibility of rational public argument."[45] Farrell, for example, rehabilitates Aristotle's ethico-political vision of a rhetorical culture, "an institutional formation in which motives of competing parties are intelligible, audiences available, expressions reciprocal, norms translatable, and silences noticeable."[46] While the Greco-Roman rhetorical tradition often functions as a font of preferred *nomoi* in twentieth-century US rhetorical theory, scholarship in comparative rhetorics is excavating similar values from other ancient cultures that might provide normative resources for navigating persistent problems that are necessarily becoming globally salient.

Surely the most obvious invocation of an ancient *nomos* to guide current rhetorical practice is the constant call for civility in public discourse. Although these calls are not often accompanied by an explicit hope for a return to the discourse norms of the ancient world, figuring civility as a necessary convention of public argument draws from the rhetorical concept of *to prepon*, the appropriate. Although some degree of civility is required to coordinate social actors, popular calls for civility generally underestimate the disciplinary function of decorum by ignoring how claims of "wrong forum" or "unreasonability" historically marginalize dissent.[47] Rhetorically informed analyses of civility attempt to balance this tension, often reaching back for complementary ancient concepts that can thicken the concept of decorum beyond playing nice in order not to rock the boat.[48]

More broadly, new technologies introduce new ethical challenges that invigorate conversations about *nomoi* appropriate to guiding their use. As Lee Rainie and Barry Wellman note, "the norms of networked individualism have not caught up to the practice of networked individualism."[49] To help norms catch up with practices, critics often turn to the ancient world with an eye toward revitalizing archaic *nomoi* capable of managing digital dramas. William Powers, for example, recasts Plato and Seneca as philosophers of screens that provide a practical philosophy to navigate everyday problems.[50] In Powers's reading of Plato's *Phaedrus*, the value of distance is posited as a countervailing *nomos* that provides an alternative to the immediacy and vibrancy of Athens. Socrates, usually satisfied to engage with the "screen" of Athens, meets Phaedrus beyond the city walls to get yet another perspective on the emergence of rhetoric and writing. Similarly, Powers takes Seneca's prescriptions to find inner peace by cultivating arts of concentration a soothing tonic capable of counterbalancing link hopping through digital networks. Most cultural critics, unlike Powers, do not explicitly attribute their analysis to ancient sources of *nomos*. Nonetheless, these ancient resources subtend—and, if mined for insight, can substantially enrich—the conventions required to enhance rhetorical cultures.

One ancient resource capable of improving the quality of internetworked rhetorical cultures is the concept of *dissoi logoi*, or countervailing arguments. The theory of *dissoi logoi* assumes that pitting opposing arguments against each other is a useful way to test the strength of competing convictions. Such an ancient principle could address one of the most compelling criticisms of internetworked media: that citizens respond to information abundance by attending primarily to opinions that already comport with their own beliefs. In the context of the early internet, Cass Sunstein memorably dubbed this phenomenon the "echo chamber" to capture how like-minded interlocutors tended to parrot each other's perspectives and gradually push each other to more extreme positions.[51] Eli Pariser updated the echo chamber by coining the term "filter bubble" as a way to reflect how algorithms increasingly connect internet users to content deemed more relevant based on personal preferences.[52] Both Sunstein and Pariser claim that another disposition is necessary to preserve the long-term health of democratic cultures; namely, that citizens must be exposed to unpredictable, competing perspectives. Although neither uses the term *dissoi logoi*, the utility of examining countervailing arguments in order to clarify one's own opinion and invent arguments is not lost on any cultural formation influenced by the ancient Greek Sophists. Although no simple "Greek Solution" will ameliorate current problems, revitalizing *dissoi logoi* may well be a precondition for beginning to address the seemingly polarized dramas of present public life.

Anakainōsis: The Renewal Relation

Kainos (new), the word at the core of *anakainōsis*, first appeared in fifth-century Athenian texts, reflecting ancient Athenians' notorious zeal for novelty.[53] The sense of the new inherent to *anakainōsis* is cyclical or seasonal: what was old is new, what was gone has returned afresh. The renewal relation hints at the reemergence of ancient patterns of communication and culture in contemporary cultures shaped by digital communication networks. Thomas Pettitt encapsulates the renewal relation by noting how the "ongoing media revolution was/is in the process of restoring, at a higher level of technology, significant aspects of the way things were before some earlier media revolution of similar magnitude."[54] In short, digital cultures renew some of the communicative patterns of oral cultures; thus, the renewal relation might be seen as a strong or synthesized version of the analogical, heuristic, and *nomos* relations.[55] If the renewal thesis is correct, then the utility of ancient rhetorics as analogy, inspiration, and moral compass is necessarily resurgent, because the practices and sensibilities of digitally networked rhetorical cultures map on more neatly to the premodern than to the modern. This restoration of premodern norms comes on the heels of what Pettitt identifies as the Gutenberg Parenthesis, "a period of four to five centuries in which the mediation of verbal culture has been dominated by print technology in general and the printed book in particular."[56] When conceived as a parenthesis—a "putting (*thesis*) away (*par*) inside (*en*)," which signifies a specific interval or interruption—the rhetorical worlds supported by print media technology appear to be more exception than rule.[57] Seeds for the renewal relation were planted in the pioneering work of Walter Ong and Marshall McLuhan, both of whom were well versed in the history of rhetoric. Of course, neither could have foreseen the ways in which digital technologies extended the sensory, epistemological, and cultural logics of electronic mass media. Yet, McLuhan's *The Gutenberg Galaxy* presaged the development of the Gutenberg Parenthesis by questioning the normalization of print culture. Ong, in his explanation of a "secondary orality," vivified by electronic mass media, noted presciently that the computer "has striking resemblances to the old [orality] in its participatory mystique, its fostering of a communal sense, its concentration on the present moment, and even its use of formulas."[58]

The renewal *topos* signals the rejuvenation of deep communication norms that connect the ancient oral tradition and new digital communication practices. For example, John Miles Foley identifies "the fundamental similarities and correspondences between humankind's oldest and newest thought-technologies: oral tradition and the Internet."[59] As he notes, "despite superficial differences, both technologies are radically alike in depending not on static products but rather on continuous processes, not on 'What?' but on 'How do I get there?' In contrast to the fixed spatial organization of the page and book, the technologies of oral tradition and the internet mime the way we think by processing along pathways within a network. In both media it's pathways—not things—that matter."[60] Jay David Bolter echoes the point, noting that hypertext "more closely resembles oral discourse than it does conventional printing or handwriting."[61] For example, based on hypertextual linking, blogs are "closer to the reciprocity of oral communication that Plato appreciates than to the unresponsiveness of writing."[62] Likewise, according to George Landow, "the reader-author [of hypertext] inevitably has more in common with the bard who constructed meaning and narrative from fragments provided by someone else" than do audiences of the print era.[63] This correspondence is evident in oral metaphors commonly used to describe sites of digital communication, as in the terms "chat room" and "discussion forum," or in the claim that the latest new digital genre is the "new agora."[64] Chat rooms and discussion forums are exemplars of how oral communication norms migrate to hypertext-driven, digital environments. The same could be said for texting, which, according to Pettitt, "is more plausibly a lettered quasi-orality—speech in a textual mode but not a simple transcript of speech: a speaking through the fingers, which is digital in more than one sense."[65] Digital technologies may invite hyperlexicality through text-based genres, but the form of these exchanges mirrors oral discourse more than it does the printed word. "Digital orality," Zizi Papacharissi proposes, is the term that best captures the blending of new modes of data capture and information analysis with traditional narrative and storytelling conventions.[66]

Digital composition troubles the notion, sacrosanct in the dominant understanding of the print tradition, of a "lone genius who writes with intention and autonomy," making social authorship a richer (and much older) conceptualization of the inventive process.[67] Such a renewal signals the larger cultural ramifications of digital technology, of which the shifting fate of rhetorical theory is but one marker. Tom Boellstorff predicts that the advent of digital media technology will usher in an Age of *Technē* that will supersede the Age of *Epistēmē*, which has dominated since Plato.[68] The fast growth of the maker movement lends some credence to such claims, returning rhetoric as an art of design to its ancient role as an architectonic productive art of creative invention. Such a *technē*-centered culture would not fetishize makers in the same way that print culture fetishized the solitary authors, for invention in oral and digital cultures leverages imitation and sociality as much as novelty and individuality. Nowhere is this dance between imitation and sociality more obvious than in remix culture. Digital remix culture, Scott Haden Church claims, restores the "classical culture of collectivism, re-creation, and performance in cultural production."[69] *Imitatio*, seen within the Gutenberg Parenthesis as a guarantee of merely derivative work, is now often conceived (at least implicitly) as the engine driving all digital communication. Audio and video remixes, retweets and shares, memes and virals: these cultural productions thrive and mimetically propagate in collaborative digital environments that specialize in making minor modifications and improvements to the "original." The imitative registers of digital culture invite a rehabilitation of ancient rhetorical theories of *mimēsis*, like those found in the work of Isocrates and Quintilian.

Organizing Principles, Chapter Summaries, and Connective Threads

This volume shows how ancient rhetorics provide unique and currently undertheorized resources useful in apprehending communication practices as varied as remix, image circulation, social networking, identity presentation, public shaming, viral videos, big data, virtual reality technology, social protest, and memes. This diversity of digitally networked sites is purposefully complemented by a critical engagement with a range of rhetorical figures. Many of the contributions draw from key figures in ancient Greek rhetoric, including Protagoras, Gorgias, Plato, Isocrates, and Aristotle. As a result, familiar terms from the rhetorical tradition, like the *homo mensura* fragment, *kairos*, decorum, *imitatio*, genre, *mimēsis*, and *tropos*, play central roles as analytical levers. However, this volume also responds to the turn to comparative rhetorics by including contributions from Buddhist, Confucian, and Jaina rhetorics that add terms like *śūnyatā* (essenceless), *jiàn* (remonstration), and *ahimsā* (nonviolence) to the rhetorical lexicon.[70] Our goal was to produce a volume that establishes the range of insights possible through engagement with ancient rhetorical theory. The chapters are organized chronologically, based on the primary figure or text(s) at the center of each contribution.

Mari Lee Mifsud's "On Network" traces the lexical, mythical, and tropical senses of network in texts ranging from the archaic period to the classical. Ancient senses of net-work often figure women's material net-work as entrapping men; almost always, that kind of networking is marginalized in favor of the more immaterial networking performed by men in pursuit of political and cultural power. This same dynamic is at play in the context of digital networks: Women are often to be found making microchips, the very foundation of digital communication networks, in the sweatshops of countries with poor labor and environmental regulations. Meanwhile, powerful networks predominantly constituted by men in Silicon Valley and elsewhere shape the contours of digital networks and profit handsomely. Rather than the conventional account of networks that promises better living through digitality, Mifsud convincingly shows that *amphibole* should be the dominant trope of the network form. An amphibolic account understands the work of networks to be essentially ambivalent, capable of both taking and giving freedom. Such a conception of net-work is required to face the ongoing challenges of democratic togetherness in digital media ecologies.

Yet, democratic togetherness, as Arabella Lyon reminds us in "Imagining Confucian Audiences: Tactical Media and the Umbrella Movement," is a culturally specific affair. Lyon investigates how three images that circulated during the 2014 prodemocracy movement in Hong Kong envisioned possible relationships and utopic imaginings within the constraints of the centralized Chinese state. The Confucian sense of remonstration (*jiàn*), with its attendant concepts of *shù* (personalized understanding of others, or, as Lyon understands it, the recognition of oneself in others) and *zhōng* (acts of duty to others), contrasts sharply with a Greco-Roman, oratorcentric vision of a rhetor persuading an objectified audience. These images, Lyon argues, work to establish respectful and ethical relations grounded in political community and thus provide a robust rhetorical alternative to the agonism and adversarialism associated with both the ancient rhetorical tradition and modern liberalism.

E. Johanna Hartelius, in "Big Data and Global Knowledge: A Protagorean Analysis of the United Nations' Global Pulse," turns to that ancient maven of measurement, Protagoras, to probe the promise of new digital metrics for improving public welfare. As she details, digital technologies introduce new mechanisms of measurement and new measurers that claim mastery over them. Proponents of big data analytics claim that newly trackable views, hits, clicks, and impressions, among many other metrics, provide a window into human behavior that improve on prior modes of analysis. Yet, as Hartelius compellingly maps out in the context of a United Nations' "Big Data" initiative, the enthusiasm for new digital tools often elides important distinctions between and among data, information, knowledge, and judgment. Drawing on Protagorean concepts of *dissoi logoi*, *homo mensura*, and *orthoepeia*, Hartelius underlines how big-data discourses, in promising more objective analysis of actual rather than reported or observed human behavior, obfuscate how measurement and the concomitant constitution of knowledge are deeply rhetorical processes.

In "On Fear and Longing: Gorgias and the *Phobos* and *Erōs* of Visual

Rhetoric," Nathan Crick examines Gorgias's *Encomium of Helen* as an exploration of the pre-cognitive, affective force of the visual. As Crick notes, interpreters of Gorgias's *Encomium* rightly emphasize the power of *logos* as a *dunastēs megas* ("powerful master"); yet, in so doing, they overlook Gorgias as a theorist of sensation and affect. *Opsis*, or sight, is, as Crick notes, positioned by Gorgias as a force capable of moving people just as much as *logos*. Gorgias figures *opsis* as stimulating fear (*phobos*) or love (*erōs*), affects that are worked over and complicated by *logos*. Crick examines the entanglement of *opsis* and *logos* in the context of the image of three-year-old Aylan Kurdi, a young Syrian refugee found drowned on a Mediterranean beach on September 2, 2015, which became an iconic distillation of an unfolding refugee crisis in southern and eastern Europe. Reworking the relationship between sensation, affect, and *logos* from the heuristic provided by Gorgias, Crick meditates on the impact of a digital mediascape that trades in the constant circulation of images.

The metaphor of "viral media" should attune us to epidemiological dimensions of digital culture, argues Christopher J. Gilbert in "Impure Imaginations: The Rhetorical Humors of Digital Virology." Yet, conceiving of media artifacts with high levels of circulation as a "virus" or "contagion" focuses on the vector of movement rather than the underlying dis-ease that a particular "viral" represents. Gilbert recuperates ancient humoral theory, drawing primarily from ancient Greek myths and Hippocrates, as a way to connect the activities of individual bodies engaging digital artifacts with the collective dispositions and health of a networked body politic. "Festering," he contends, is a more apt metaphor to analyze media content like Chipotle's "Back to the Start" and "The Scarecrow," two advertisements for the fast-food chain that received public attention and industry acclaim but critical skepticism. Although humorism as a theory of medicine has been eclipsed, the vocabulary of humorism—including *stasis*, temperament, disposition, and purification—offers cultural critics a lexicon to further theorize the potent connections between media circulation and public culture.

As Rosa A. Eberly and Jeremy David Johnson explicate in "Isocratean *Tropos* and Mediated Multiplicity," Isocrates's notion of *tropos* provides unique resources to navigate the complexities of digitally networked identity. Isocrates's translators have usually used the term "character" in translating *tropos*; however, in contrast to *ēthos*, which connotes and is often translated as a stable or true character, *tropos* implies that identity is necessarily multiple, shifting, and turning. Eberly and Johnson's adaptation of *tropos* to the play of identity in digital communication networks provides a convincing rejoinder to Facebook founder and CEO Mark Zuckerberg, who, in 2010, asserted that "the days of you having a different image for your work friends or co-workers and for the other people you know are coming to an end pretty quickly . . . having two identities for yourself is an example of a lack of integrity."[71] *Tropos* points us to intraindividual variability; we are many, not one. The action of rhetoric is to unite or disperse these multiple selves—as demonstrated by case studies of the performer Harry Shearer and Reddit user Violentacrez, the so-called biggest troll on the internet.

In "Plato's *Phaedrus* and the Ideology of Immersion," Ekaterina V. Haskins and Gaines Hubbell critically examine the promise of immersion through new virtual reality applications being designed for educational purposes. "Immersion" is a promise that every new media technology makes, and digital media technologies like video games, IMAX theaters, smartphones, and other internetcentric devices certainly make claims to our senses and our attention in immersive ways. The holy grail of digital immersion, of course, is virtual reality—thematized by cyberpunk science fiction and breathless technology prognosticators. Offering a novel rereading of Plato's *Phaedrus* by attending to how the natural world and the world of speech provide immersive environments in this dialogue, Haskins and Hubbell explain how immersion shapes sensory perception, rhetorical performance, and the very possibility of educational exchange. Far from being a screed against the immersive qualities of any particular media form, Haskins and Hubbell maintain that a rhetorical pedagogy for our times must insist upon a constant toggling between immersive activities and critical reflection on the infrastructure—the forms, genres, interests, and other rhetorical scaffoldings—of the immersive media environment.

Carolyn R. Miller extends her landmark work on genre with "Genre in Ancient and Networked Media." Miller identifies the rapid ascent of "genre-talk" in the context of networked media: Not only is there a heady proliferation of genres stimulated by the new publication opportunities of the internet (like mobile phone novels, aristocrunk, dirtbag sitcoms, and shreds videos), but public talk features an increasing amount of metareflection about generic conventions. Genre-talk refers to how publics name genres, how they organize rhetorical artifacts into genres, and how they argue about the conventions appropriate to a genre. Miller situates the proliferation of internetworked genres and the genre-talk about them in the context of *inventio* and *imitatio*, what she calls "rhetoric's drivers of change and stabilization." Tracing ancient genre theory through a range of ancient sources, Miller takes a closer look at Aristotle's work on biology to draw out subtle distinctions between *eidos* and *genos*. The term *eidos*, meaning form or pattern, and the term *genos*, meaning function or family resemblance, are equally important in classifying biological forms of life; by extension, Miller argues that functional-formal approaches to genre in contemporary rhetorical theory are likely to be more useful than approaches that privilege either form or function.

In "*Poiēsis, Genesis, Mimēsis*: Toward a Less Selfish Genealogy of Memes," we (Michele Kennerly and Damien Smith Pfister) evaluate and find wanting the sociobiological focus of "the *selfish* meme," a quality imputed to the meme by Richard Dawkins when he coined the term in 1976 and currently conferred on the meme in analyses of its digital instantiations. Uninspired by the exclusion of human agency from accounts of selfish-meme generativity but inspired by Dawkins's nod to *mimēsis* (representation) when coining "meme," we (Kennerly and Pfister) undertake a new, rhetorical genealogy of memes. Using the three concepts from ancient rhetorical culture featured in their chapter title, we argue that memes are made and done, generative, and mimetic. We consider memes, especially hashtags, a variant of what Seyla Benhabib calls "democratic iterations," public acts that refigure a community or polity's norms of self-definition and self-understanding by invoking or revoking those norms. Our less selfish genealogy of memes divests memes of their sociobiologically tinted origins without discounting their creative and circulatory energy.

"Remix, *Śūnyatā*, and *Prosōpopoeia*: Projecting Voice in the Digital Age," by Scott Haden Church, mashes up two ancient rhetorical traditions to theorize digital remix. *Prosōpopoeia*, often understood as personification in the Greco-Roman tradition but literally translated as "face-making," is often used, as Church notes, to imply the making, restoring, or projecting of voice for those not present. Since remixes call forth voices from the past to forge a new sonic assemblage, *prosōpopoeia* is one of the master tropes of remix. If the meteoric growth of remix culture over the past fifty years teaches nothing else it is that these symbolic fragments are unbelievably malleable, capable of inventive reconfigurations and dramatic transformations. This is so, as Church explains, because there is no *essence* to any symbolic arrangement. Church teases out the essencelessness that is remix's condition of possibility through the Buddhist philosopher Nāgārjuna's conception of *śūnyatā*. The mash-up of *prosōpopoeia* and *śūnyatā* shows the promise of reading ancient rhetorical traditions with each other, since each concept supplements the other in a way that, combined, provides a useful frame through which to apprehend the dynamics of cultures shaped by digital networks.

In "The Jaina Rhetoric of Nonviolence and the Culture of Online Shaming," Scott R. Stroud analyzes the phenomenon of public shaming through digital networks from a perspective rooted in ancient Jainism. Anonymous and pseudonymous individuals often take advantage of networked communication architectures to swarm perceived wrongdoers with criticism, symbolic violence, and physical threats. For Jainism, passions are inherently self-harming; thus, inflaming the passions of the self through public shaming, no matter how righteous the shamer may feel, results in a kind of ego-bolstering that is itself a kind of violence. Stroud examines the Tumblr blog *Racists Getting Fired* to show how, even in instances where moral consensus about the wrongness of racism exists, public shaming of racists exacts its own harmful and ineffective symbolic violence. Stroud's chapter is an excellent example of how moving beyond the Greco-Roman rhetorical tradition can provide critical purchase on the new ethical quandaries raised by digital communication networks.[72]

As those summaries suggest, these essays draw heavily on *analogia*, *heuresis*, or *nomos* as ways to connect the ancient and the digital, though careful readers will also detect hints of both the *historia* and the *anakainōsis* relations. Several crosscutting themes traverse the essays, giving the collection cohesion beyond mere assemblage under the general thematic of ancient rhetorics and digital networks. A concern with the relationship between rhetoric and materiality suffuses the collection but is especially salient in Mifsud's account of the material basis of net-work, in Hartelius's exposition of the materiality of "data exhaust" in big data, and in Crick's and Haskins and Hubbell's emphases on sensation. Circulation, too, is a prevailing theme across many of the essays, but especially in Lyon's description of memetic appropriations of the yellow umbrella, Gilbert's analysis of viral videos, Kennerly and Pfister's rhetorical genealogy of memes, and Stroud's analysis of *Racists Getting Fired*. Many of the contributions underline the continuing centrality of how rhetoric interacts with the kindred terms "multiplicity," "imitation," and "flux": for Eberly and Johnson, *tropos* unlocks the potential to think the self as a formation of multiple selves, while Miller's essay on networked genres ponders the constant multiplication of digital genres by thinking through *imitatio* and Hartelius's contribution identifies the rapid fluctuations of a digitally networked world as a rationale for big data. Digital inflections of identity are yet another theme emergent from these essays, from Stroud's Jaina consideration of the self, passions, and public shaming and Eberly and Johnson's interpretation of *tropos* to Lyon's read of remonstration through visual rhetorics.

The eleven contributions can also be read in clusters that engage similar digital phenomena. Memes are scrutinized in Lyon's analysis of appropriated yellow umbrellas, in Gilbert's take on Chipotle's viral marketing campaign, in Miller's discussion of native digital genres, and in Kennerly and Pfister's critique of the genetic metaphor that underpins most memetic theory. Visual images are taken up in Mifsud's contrast between Lewis Hine's photograph of a spinner girl and a stock photo of male Silicon Valley executives, in Crick's analysis of the iconic image of Aylan Kurdi, and in Gilbert's critical assessment of the Chipotle campaign. New genres afforded by digital media technology are taken up broadly by Miller and more specifically in Church's essay on remix. Techno-utopian rhetoric (or the rhetoric of the technological sublime) can be located in the non-amphibolic rhetorics of network identified by Mifsud, in the optimistic rhetoric of Global Pulse, as analyzed by Hartelius, and in the uncritical embrace of immersion as a pedagogical good detailed by Haskins and Hubbell.[73] The "dark side" of internet culture is reflected in the boundary-pushing postings of Violentacrez on Reddit, assessed in the Eberly and Johnson piece, and in the symbolic violence created by racists and those looking to get them fired in Stroud's essay.

One final thread connects these essays: the (perhaps surprising) suppleness and vivaciousness of ancient rhetorical terms as vehicles for cultural critique of digital communication networks. "Perhaps surprising" is in parenthesis because, we readily admit, even *we* found ourselves surprised at how generative the juxtaposition of ancient rhetorics and digital networks could be as effective tools of cultural criticism. The topics our contributors thematize span issues of social and symbolic justice, equality and discrimination, political economy and power relations, free speech and hate speech. Each contribution shows how ancient rhetorical theory is not just capable of being cleverly (or, worse, mechanically) applied in relation to some emergent digital phenomenon; rather, each essay shows how it is vital and even necessary to situate, understand, analyze, and critique communication practices supported by internetworked media with the furnishings of ancient rhetorical theory (after some burnishings by our authors). Reanimating ancient texts, figures, concepts, sensibilities, and practices in the context of the digital, contributors to *Ancient Rhetorics and Digital Networks* adapt some of rhetoric's richest traditions to account for and give an account of changing media ecologies and public cultures.

Notes

1. See the original image and commentary at "Berlin F 2285 (Vase)," *Perseus*, http://www.perseus.tufts.edu/hopper/artifact?name=Berlin+F+2285&object=vase.

2. See, for example, Lindybeige, "Laptops of the Ancient World," *YouTube*, May 16, 2013, https://www.youtube.com/watch?v=_RwR_nQ6SS4.

3. Edward Corbett, "Rhetoric in Search of a Past, Present, and Future," in *The Prospect of Rhetoric: Report of the National Developmental Project*, ed. Edwin Black and Lloyd Bitzer (Englewood Cliffs, NJ: Prentice-Hall, 1971), 170.

4. Donald Bryant, "Rhetoric: Its Function and Scope," *Quarterly Journal of Speech* 39, no. 4 (1953): 422.

5. Thomas O. Sloane, "Prologue: *The Prospect* as Prospectus," in *Reengaging the Prospect of Rhetoric: Current Conversations and Contemporary Challenges*, ed. Mark J. Porrovecchio (New York: Routledge, 2010), 4.

6. See, for example, Siva Vaidhyanathan, *The Googlization of Everything (And Why We Should Worry)* (Berkeley: University of California Press, 2011); Robert McChesney, *Digital Disconnect: How Capitalism Is Turning the Internet Against Democracy* (New York: New Press, 2013); and Shawn Powers and Michael Jablonski, *The Real Cyber War: The Political Economy of Internet Freedom* (Urbana: University of Illinois Press, 2015).

7. Christine Harold, *OurSpace: Resisting the Corporate Control of Culture* (Minneapolis: University of Minnesota Press, 2007); Jeremy Packer and Stephen B. Crofts Wiley, eds., *Communication Matters: Materialist Approaches to Media, Mobility, and Networks* (New York: Routledge, 2012); Stephanie Schulte, *Cached: Decoding the Internet in Global Popular Culture* (New York: New York University Press, 2013); David M. Berry, *Critical Theory and the Digital* (New York: Bloomsbury, 2014).

8. Lance Strate, *Amazing Ourselves to Death: Neil Postman's Brave New World Revisited* (New York: Peter Lang, 2014); John Durham Peters, *The Marvelous Clouds: Toward a Philosophy of Elemental Media* (Chicago: University of Chicago Press, 2015).

9. Of course, we should acknowledge that rhetoric's status as an architectonic art means that it undergirds all other critical perspectives; see Richard McKeon, "The Uses of Rhetoric in a Technological Age: Architectonic Productive Arts," *The Prospect of Rhetoric* (Englewood Cliffs, NJ: Prentice-Hall, 1971): 44–63. If rhetoric is an architectonic art, then digital networks themselves are products of rhetoric. In other words, both the technical infrastructure of internetworked media and the symbolic activity that circulates through digital networks are, first and foremost, rhetorical accomplishments.

10. Barbara Warnick, *Rhetoric Online: Persuasion and Politics on the World Wide Web* (New York: Peter Lang, 2007); Jeff Rice, *Digital Detroit: Rhetoric and Space in the Age of the Network* (Carbondale: Southern Illinois Press, 2012); Sarah Arroyo, *Participatory Composition: Video Culture, Writing, and Electracy* (Carbondale: Southern Illinois Press, 2013); Jeff Pruchnic, *Rhetoric and Ethics in the Cybernetic Age: The Transhuman Condition* (New York: Routledge, 2014); Damien Smith Pfister, *Networked Media, Networked Rhetorics: Attention and Deliberation in the Early Blogosphere* (University Park: Pennsylvania State University Press, 2014); Douglas Eyman, *Digital Rhetoric: Theory, Method, Practice* (Ann Arbor: University of Michigan Press, 2015).

11. Despite the risks of typologies overschematizing practices and thus eliding their fluidity, we take our cue here from Jeff Pruchnic: "Structurally, [typologies] facilitate not so much epistemological resolution ('so that's what it is!') and the end of investigation but a dynamic potentiality for functional application ('so that's what it

does . . . ') . . . and the beginning of practice . . . the point of such an operation is not the exposure of a certain mechanism but the potential that this mechanism might be manipulated through practice"; in "Rhetoric, Cybernetics, and the Work of the Body in Burke's Body of Work," *Rhetoric Review* 25, no. 3 (2006): 279.

12. See Eric Havelock, *The Muse Learns to Write: Reflections on Orality and Literacy from Antiquity to the Present* (New Haven, CT: Yale University Press, 1986), and Walter Ong, *Orality and Literacy: The Technologizing of the Word* (New York: Methuen, 1982).

13. Marshall McLuhan, *Understanding Media: The Extensions of Man; Critical Edition*, ed. W. Terrence Gordon (Corte Madera, CA: Gingko Press, 2003), 19.

14. Laura Gurak and Smiljana Antonijevic, "Digital Rhetoric and Public Discourse," in *The SAGE Handbook of Rhetorical Studies*, ed. Andrea Lunsford, Kirt Wilson, and Rosa Eberly (Thousand Oaks, CA: Sage Publications, 2009), 497.

15. Matthew Johnson, "Rekindling Rhetoric: Oratory and Marketplace Culture in *Guild Wars*," in *Writing and the Digital Generation: Essays on New Media Rhetoric*, ed. Heather Urbanski (Jefferson, NC: McFarland, 2010), 162–73.

16. Jay David Bolter and Richard Grusin, *Remediation: Understanding New Media* (Cambridge, MA: MIT Press, 1998).

17. John Durham Peters, "Information: Notes Toward a Critical History," *Journal of Communication Inquiry* 12, no. 2 (1988): 10–11. Hylomorphism refers to theories and doctrines about the relationship between matter and form. Ben McCorkle, in *Rhetorical Delivery as Technological Discourse: A Cross-historical Study* (Carbondale: Southern Illinois University Press, 2012), similarly traces how the concept of "delivery" evolves in tandem with the successive development of alphabetic writing, print, electronic media, and digital media; each technological regime "changes the speed, reach, form, and purpose of the communication practices for those people who have lived through such moments of technological transition" (1–2). To think in a complex way about delivery now means thinking through the history of delivery. Grant Bollmer's analysis of the term "connection" in *Inhuman Networks: Social Media and the Archaeology of Connection* (New York: Bloomsbury, 2016) offers yet another outstanding case of a genealogical approach to terms that appear to have digital provenance but actually have lengthier and more complicated histories.

18. Peters, "Information," 13.

19. Peters, "Information," 15.

20. Xiaoye You, *The Devil's Tongue: A History of English Composition in China* (Carbondale: Southern Illinois University Press, 2010).

21. Maryanne Wolf, "Our 'Deep Reading' Brain: Its Digital Evolution Poses Questions," *Nieman Reports* (Summer 2010): http://niemanreports.org/articles/our-deep-reading-brain-its-digital-evolution-poses-questions/.

22. For these characteristics of the Sophists, see John Poulakos, *Sophistical Rhetoric in Ancient Greece* (Columbia: University of South Carolina Press, 1995); Susan Jarratt, *Rereading the Sophists: Classical Rhetoric Refigured* (Carbondale: Southern Illinois University Press, 1998); and Nathan Crick, "The Sophistical Attitude and the Invention of Rhetoric," *Quarterly Journal of Speech* 96, no. 1 (2010): 25–45.

23. Pfister, *Networked Media*, 9.

24. Tom Standage, *Writing on the Wall: Social Media-The First 2,000 Years* (New York: Bloomsbury, 2013), 14–15.

25. Lisa Lebduska, "Emoji, Emoji, What for Art Thou?" *Harlot* 12 (2014): http://harlotofthearts.org/index.php/harlot/article/view/186/157.

26. Christopher J. Gilbert, "Playing with Hitler: Downfall and Its Ludic Uptake," *Critical Studies in Media Communication* 30, no. 5 (2013): 410.

27. Standage, *Writing on the Wall*, 1–3.

28. Jessica Hughes, "Ancient Memory Palace Provides Classical Connection to 2014 Nobel Prize in Medicine," *The Conversation*, October 14, 2014, https://theconversation.com/ancient-memory-palace-provides-classical-connection-to-2014-nobel-prize-in-medicine-32776.

29. Sarah Bond, "Facebook Before Facebook: Tagging in Antiquity," *medievalbooks*, November 21, 2014, http://medievalbooks.nl/2014/11/21/facebook-before-facebook-tagging-in-antiquity/.

30. See, for example, a series of books issuing blunt, catastrophic warnings in their titles. Andrew Keen, *Cult of the Amateur: How Blogs, MySpace, YouTube, and the Rest of Today's User-generated Media are Destroying Our Economy, Our Culture, and Our Values* (New York: Doubleday, 2007); Mark Bauerlein, *The Dumbest Generation. How the Digital Age Stupefies Young Americans and Jeopardizes Our Future (Or, Don't Trust Anyone Under 30)* (New York: Penguin, 2008); Maggie Jackson, *Distracted: The Erosion of Attention and the Coming Dark Age* (Amherst, NY: Prometheus Books, 2008).

31. Rebecca Newberger Goldstein, *Plato at the Googleplex: Why Philosophy Won't Go Away* (New York: Pantheon Books, 2014), 127, 129.

32. James W. Carey and John Quirk, "The Mythos of the Electronic Revolution," *Communication as Culture: Essays on Media and Society* (Cambridge, MA: Unwin Hyman, 1988): 113–41. Such analogical thinking is evidence of pragmatic approaches to theorizing about internetworked media, as recommended by James W. Carey, "Historical Pragmatism and the Internet," *New Media and Society* 7, no. 4 (2005): 443–55.

33. G. E. R. Lloyd, *Polarity and Analogy: Two Types of Argumentation in Early Greek Thought* (Cambridge, UK: Cambridge University Press, 1966), 417, and Xing Lu, *Rhetoric in Ancient China, Fifth to Third Century B.C.E.: A Comparison with Classical Greek Rhetoric* (Columbia: University of South Carolina Press, 1998).

34. Hans Blumenberg, "An Anthropological Approach to the Contemporary Significance of Rhetoric," in *After Philosophy: End or Transformation?* ed. Kenneth Baynes, James Bohman, and Thomas McCarthy (Cambridge, MA: MIT Press, 1987).

35. See Nicholas Negroponte and Richard A. Bolt, *Spatial Data Management* (Cambridge, MA: MIT Press, 1979); Stewart Brand, *The Media Lab: Inventing the Future at MIT* (New York: Viking, 1987), 138–40; Richard A. Bolt, "Spatial Data-Management," in *Multimedia: From Wagner to Virtual Reality*, ed. Randall Packer and Ken Jordan (New York: Norton, 2001), 187–89. The tale of Simonides was circulated anew in this period as a result of the magisterial work of Frances A. Yates's

The Art of Memory (Chicago: University of Chicago Press, 1966). Richard A. Bolt, in *The Human Interface: Where People and Computers Meet* (Belmont, CA: Wadsworth, 1984), 4fn4, cites Yates on this historical connection between Simonides, space, and memory.

36. Douglas Ehninger, "On Systems of Rhetoric," *Philosophy & Rhetoric* 1, no. 3 (1968): 131–44.

37. Kathleen Welch, *Electric Rhetoric: Classical Rhetoric, Oralism, and a New Literacy* (Cambridge, MA: MIT Press, 1999).

38. Collin G. Brooke, *Lingua Fracta: Toward a Rhetoric of New Media* (New York, Hampton Press, 2009).

39. The list here could, indeed, be long. We note just a few exemplars of how scholars draw on the vocabulary of ancient rhetorics to make sense of specifically digital phenomena: Carolyn R. Miller and Dawn Shepard, "Blogging as Social Action: A Genre Analysis of Weblogs," in *Into the Blogosphere: Rhetoric, Community, and the Culture of Weblogs*, ed. Laura Gurak, Smiljana Antonijevic, Laurie Johnson, Clancy Ratliff, and Jessica Reyman (Minneapolis: University of Minnesota Digital Conservancy, 2004): http://hdl.handle.net/11299/172818; James Porter, "Recovering Delivery for Digital Rhetoric and Human-Computer Interaction," *Computers & Composition* 26, no. 4 (2009): 207–24; Jeremiah Dyehouse, "The Cyberspace Incrementum: Technology Development for Communicative Abundance," *Rhetoric Society Quarterly* 39, no. 3 (2009): 281–302; E. Johanna Hartelius's chapter on disposition and Wikipedia in her *The Rhetoric of Expertise* (Lanham, MD: Lexington Books, 2011); and Jeff Pruchnic and Kim Lacey, "The Future of Forgetting: Rhetoric, Memory, Affect," *Rhetoric Society Quarterly* 41, no. 5 (2011): 472–94. In a more pedagogical vein, rhetoric continues to be taught in courses like public speaking, first-year writing, and digital composition. Here, too, the centrality of ancient rhetorics as a heuristic is being revivified, as in Jay Heinrichs's *Thank You for Arguing: What Aristotle, Lincoln, and Homer Simpson Can Teach Us about the Art of Persuasion* (New York: Three Rivers Press, 2007).

40. Richard Lanham, *The Electronic Word: Democracy, Technology, and the Arts* (Chicago: University of Chicago Press, 1993) and *The Economics of Attention: Style and Substance in the Information Age* (Chicago: University of Chicago Press, 2006). See also Pfister, *Networked Media,* 29–33.

41. Damien Smith Pfister, "The *Logos* of the *Blogosphere*: Flooding the Zone, Invention, and Attention in the Lott Imbroglio," *Argumentation and Advocacy* 47, no. 3 (2011): 141–67.

42. Michele Kennerly, "Micrologia." Paper presented at the annual meeting of the National Communication Association, Nov. 19–23, 2014, Chicago.

43. Borrowed from Hannah Arendt, "The Greek Solution," in *The Human Condition*, 2nd ed. (Chicago: University of Chicago Press, [1958] 1998), 192–97.

44. James Arnt Aune, "Coping with Modernity: Strategies of 20th-Century Rhetorical Theory," in *The SAGE Handbook of Rhetorical Studies*, ed. Andrea A. Lunsford, Kirt H. Wilson, and Rosa A. Eberly (Thousand Oaks, CA: Sage, 2009), 85.

45. Ibid., 89, emphasis added.

46. Thomas B. Farrell, *Norms of Rhetorical Culture* (New Haven, CT: Yale University Press, 1993), 1.

47. Nina M. Lozano-Reich and Dana L. Cloud, "The Uncivil Tongue: Invitational Rhetoric and the Problem of Inequality," *Western Journal of Communication* 73, no. 2 (2009): 220–26, and Dana Cloud, "'Civility' as a Threat to Academic Freedom," *First Amendment Studies* 49, no. 1 (2015): 13–17.

48. For examples of rhetorically informed analyses of civility, see Theresa Enos, "A Call for Comity," in *Beyond Postprocess and Postmodernism: Essays on the Spaciousness of Rhetoric*, ed. Theresa Enos, Keith D. Miller, and Jill McCracken (New York: Routledge, 2002): 131–57; Sharon Crowley, *Toward a Civil Discourse: Rhetoric and Fundamentalism* (Pittsburgh, PA: University of Pittsburgh Press, 2006); Jennifer Emerling Bone, Cindy L. M. Griffin, and T. M. Linda Scholz, "Beyond Traditional Conceptions of Rhetoric: Invitational Rhetoric and a Move Toward Civility," *Western Journal of Communication* 72, no. 4 (2008): 434–62; Roderick P. Hart, Jay P. Childers, and Colene J. Lind, "Partisanship and Balanced Tone," in their *Political Tone: How Leaders Talk and Why* (Chicago: University of Chicago Press, 2013): 59–88; the special issue on civility in *Rhetoric & Public Affairs* 17, no. 4 (2014) ed. Thomas A. Hollihan and Francesca Marie Smith, especially Mary E. Stuckey and Sean Patrick O'Rourke, "Civility, Democracy, and National Politics," *Rhetoric & Public Affairs* 17, no. 4 (2014): 711–36.

49. In their *Networked: The New Social Operating System* (Cambridge, MA: MIT Press, 2012), 105.

50. William Powers, *Hamlet's Blackberry: Building a Good Life in the Digital Age* (New York: HarperCollins, 2010).

51. An argument that Sunstein makes originally in *Republic.com* (Princeton, NJ: Princeton University Press, 2001).

52. Eli Pariser, *The Filter Bubble: How the New Personalized Web is Changing What We Read and How We Think* (New York: Penguin Books, 2012).

53. See Armand D'Angour, *The Greeks and the New* (New York: Cambridge, 2011), 72–73.

54. Thomas Pettitt, "Bracketing the Gutenberg Parenthesis," *Explorations in Media Ecology* 11, no. 2 (2012): 95–96.

55. Thanks to Jonathan S. Carter for this insight.

56. Thomas Pettitt, "Media Dynamics and the Lessons of History: The 'Gutenberg Parenthesis' as Restoration Topos," in *Companion to New Media Dynamics*, ed. John Hartley, Jean Burgess, and Axel Bruns (Somerset, NJ: John Wiley and Sons, 2013), 55.

57. Pettitt, "Bracketing," 101. Eric McLuhan uses a rougher figure than the parenthesis to communicate a similar point: "Plato stood on the dividing-line, surrounded by pre- and proto-Literates. We too stand on a dividing line, actually the same one, though facing the other way, and surrounded by proto-Literates and post-Literates. (A post-Literate is someone who knows how to read, but prefers not to, finding it

too slow, too laborious, or simply distasteful—or even pointless.)" In "Literacy in a New Key," *Proceedings of the Media Ecology Association*, vol. 10, ed. Paul A. Soukup, http://www.media-ecology.org/publications/MEA_proceedings/v10/3_literacies.pdf.

58. Ong, *Orality and Literacy*, 136.

59. John Miles Foley, *Oral Tradition and the Internet* (Urbana: University of Illinois Press, 2012), 5.

60. John Miles Foley, *The Pathways Project*, accessed September 28, 2015, http://www.pathwaysproject.org/.

61. Jay David Bolter, *Writing Space: Computers, Hypertext, and the Remediation of Print* (Mahwah, NJ: Lawrence Erlbaum Associates, 1991), 58–59.

62. Jill Rettberg, *Blogging* (Cambridge, UK: Polity Press, 2008), 34.

63. George Landow, *Hypertext: The Convergence of Contemporary Critical Theory and Technology* (Baltimore, MD: Johns Hopkins University Press, 1992), 117.

64. The claim that internetworked media can reproduce the conditions and functions of the ancient Greek agora is one of the most popular scholarly invocations of a term related to ancient rhetoric. A smattering of examples, with varying degrees of subtly, are included here: Niklas Damiris and Helga Wild, "The Internet: A New Agora?," in *An Ethical Global Information Society: Culture and Democracy Revisited*, ed. Jacques J. Berleur and Diane Whitehouse (New York: Chapman and Hall, 1997): 307–17; Mark Poster, *What's the Matter with the Internet* (Minneapolis: University of Minnesota Press, 2001); Colin Sparks, "The Internet and the Public Sphere," in *Mediated Politics: Communication in the Future of Democracy*, ed. W. Lance Bennett and Robert M. Entman (Cambridge, UK: Cambridge University Press, 2001); and Rita Kirk and Dan Schill, "A Digital Agora: Citizen Participation in the 2008 Presidential Debates," *American Behavioral Scientist* 55, no. 3 (2011): 325–47.

65. Pettitt, "Media Dynamics," 63.

66. Zizi Papacharissi, "The Unbearable Lightness of Information and the Impossible Gravitas of Knowledge: Big Data and the Makings of Digital Orality," *Media, Culture, and Society* (online first, 2015).

67. E. Johanna Hartelius, "'Leave a Message of Hope or Tribute': Digital Memorializing as Public Deliberation," *Argumentation and Advocacy* 47, no. 3 (2010): 74; see also Margaret Ezell, *Social Authorship and the Advent of Print* (Baltimore, MD: Johns Hopkins University Press, 2003).

68. Tom Boellstorff, *Coming of Age in Second Life: An Anthropologist Explores the Virtually Human* (Princeton, NJ: University of Princeton Press, 2008).

69. Scott H. Church, "A Rhetoric of Remix," in *The Routledge Companion to Remix Studies*, ed. Eduardo Navas, Owen Gallagher, and xtine burrough (New York: Routledge, 2014), 43.

70. Despite this inclusion, we concede that this volume reflects the bias of rhetorical studies in the United States toward the Greco-Roman tradition. With Arabella Lyon, we agree that "comparative," like "rhetorics," is a sticky term that risks recentering the established tradition as the norm; yet, we believe that the included essays "challenge, critique, and [offer] thick description" that resists as much as it reinscribes. See her "Tricky Words: *Rhetoric* and *Comparative*," *Rhetoric Review* 34,

no. 3 (2015): 245. We also concur with Susan Jarratt's warning that hypostasizing "Greco-Roman rhetoric" as a singular tradition risks concealing how "rhetoric has been multiple, heterogenous, and comparative from its earliest beginnings"; "Recontextualizing Ancient Rhetoric," *Rhetoric Review* 34, no. 3 (2015): 251. Although we ought not use the diversity of Greek and Roman rhetorics to delay an exploration of other traditions, a volume such as this one that appreciates the range of thought in the ancient world can only help to appreciate the distinctiveness of rhetorical theory emerging from different cultural formations over time.

71. Quoted in David Fitzpatrick, *The Facebook Effect: The Inside Story of the Company that Is Connecting the World* (New York: Simon and Schuster, 2010), 199. Both Facebook and Google tried in the first several years of the 2010s to require users to use their "real" names for marketing and advertising purposes, sparking "nymwars" (a portmanteau of pseudonym and wars) about naming practices on digital sites.

72. Also see Scott R. Stroud, "'Be a Bully to Beat a Bully': Twitter Ethics, Online Identity, and the Culture of Quick Revenge," in *Controversies in Digital Ethics*, ed. Amber Davisson and Paul Booth (New York: Bloomsbury Press, 2016): 264–78; Scott R. Stroud, "The Dark Side of the Online Self: A Pragmatist Critique of the Growing Plague of Revenge Porn," *Journal of Mass Media Ethics* 29, no. 3 (2014): 168–83.

73. Evgeny Morozov, *The Net Delusion: The Dark Side of Internet Freedom* (London: Penguin, 2011); James Carey and John Quirk, "The Mythos of the Electronic Revolution," *Communication as Culture: Essays on Media and Society* (Cambridge, MA: Unwin Hyman, 1988): 113–41.

1

On Network

Mari Lee Mifsud

What does it mean in this digital age to network and to be networked? One way to make sense is to trace the histories of the idea and practice. Gregory Flaxman identifies *ergō diktuōtō* as the ancient Greek antecedent of "network."[1] From this phrase, meaning "work with nets," as in fishing nets or hunting nets, Flaxman begins his genealogy of the ancient origin of "network" in fabric and fabrication. Tracing the development of network to its eventual immaterial form in the World Wide Web, Flaxman cautions that the material promises of networked culture must be explored in the context of the immaterial network encroachments into our lives: "In an age when our purchases and predilections are tracked, we are quickly becoming algorithmic functions within a massive information network: political campaigns now calculate and tailor their appeals for contributions on the same basis that Amazon personalizes recommendations, or Google personalizes advertisements, or Target customizes coupons by deducing out when a shopper is pregnant. . . . Perhaps the final step in the genealogy . . . a step already pioneered by the pharmacolonization of the brain, consists in networking our synapses."[2] The dangers of idealizing digital network call for being as circumspect about network as possible. Such circumspection issues from (at least in part) genealogical work in the ancient words, ideas, and cultural practices of network. Networks are not a New Age phenomenon. They were not born in digital form. They precede even the industrial revolution, though this epoch accelerated networks via roads and railways and communication systems, of the telegraph, the telephone, and television.[3] Networks are archaic in and originary to human society. As far back as can be seen, for example, in the ancient Greek past, networks are woven into as well as out of words, myths, and tropes of human knowing, being, doing, relating, and meaning.

In this essay, I card the archaic rhetorics of "network," separating the lexical and mythical threads to prepare for weaving. Then, weaving with these words and myths, a story-cloth emerges, one revealing network in the tropical pattern of amphibolia, an ancient rhetorical term and trope of network. Amphibolia works in and reveals the ambiguity of network originary to society. This ambiguity tells of network as taking, as being taken, and as being free from being taken. In the elaborately woven threads of ancient Greek rhetorics emerge a complex and striking story-cloth of network, which tells as well of being human.

Lexical Network

From Homeric to Hellenistic cultures, we are given a robust vocabulary of networking. We have terms for "nets," for "work," and for "network."[4] Each term gives rise to yet another nuance of the role network plays in being human. In this section on lexical network, I present these terms in a catalog form as an homage to archaic Homeric rhetoric.[5] Homer's catalogs are plentiful in the epics, his catalog of ships being particularly well known.[6] Homeric catalogs call attention to their items. Catalogs circulate well and are an aid to remembering the past, as ever-present. The catalog of "network" I offer takes the form of a list, a sparse accounting of items, in this case related words for "net," "work," and "network." Lists are, as Benjamin Sammons describes them, "inelaborate catalogues (i.e., easily worked up into the fuller form through the addition of elaboration)."[7] From this list, we will begin to see distinctive lexical dimensions of network; elaboration of the list will issue from mythical and tropical dimensions of network.

NET:

agreuma: that which is taken in hunting, booty, prey, spoil, a means of catching, being caught in the net of fate

airein: to take, take away, take by force, take a city, to overpower, kill, often of passions, to seize, conquer, catch, in a good sense, to win over, to catch or detect one doing a thing, to win, gain, in law it means to convict a person of a thing, to prove, as in "*o logos aireei*," reason *proves*, to take for oneself, to take one's supper, to choose, to take in preference, prefer one thing to another, to choose by vote, to elect, to take up, raise, lift up, to step, walk, to hoist, to sail, to get under way, start, set out, raise up, exalt, to excite, to raise by words, to extol, exaggerate, to lift, take away, remove, take off, kill, take up for oneself, to carry off, win, gain, to receive, to take upon oneself, undergo, carry, bear, to undertake, begin, as in war

arkus: a net, hunter's net, the toils and perils of the sword

atelēs: free of end, without end, not brought to an end or issue, unaccomplished, incomplete, not accomplishing one's purpose, without deduction, net pay, pay without state intervention in form of taxes, free from tax or tribute, free from all other taxes, (of sums) without deduction, net, clear, clear gain, uninitiated in mysteries, endless, eternal

broxos: a noose or slip knot for hanging or strangling, a snare for birds, the mesh of a net

diktuon: net for fishing or hunting

sumpodidzein: to entangle in a net

WORK:

ergon: as in works of industry, or occupation: tillage, weaving, feasting, fishing; as in deed or action, primarily war; as in matter, taking something up as a matter, as an issue

NETWORK:

amphiballein: to throw upon, or put around both sides; of clothes, to put them on a person or oneself; to throw the arms around, to embrace; to create doubt or to be doubtful

amphiblēstra; *amphiblēston*: anything thrown around, a casting net, the garment thrown over like a net, a fetter, bond

periballein: to throw around, as in an embrace, to throw around oneself or put on, to throw around oneself for defense, to build a wall around, to have a thing put around one; to involve one in calamities, evils, to surround or enclose for oneself, to put around a person, invest a person with it, to bring into one's power, aim at, to cloak or veil in words, to throw beyond, to excel, surpass

peribolos; *peribolai*; *peribolē*: anything that is thrown around, a covering, associated with the grave—like death thrown around you, a wall around a town, a space enclosed, house of large compass, a circumference, circuit; to make a circuit (*peribolos poiēsthai*); a compassing, endeavoring after (the *archē*); the whole compass of the matter, the long and short of the matter, the spires or coils of a serpent

periptuxai: a network of

peristixidzein: to surround with a net

plegma: a network of

plekō: a plait, twine, twist, braid, weave, as in ropes or nets, or baskets, to plan, devise, contrive, like mostly of tortuous means, of poets weaving words, to twist oneself around

Each lexical thread in this list helps us to see network's character and quality, and each offers a generative energy of sayables (and unsayables) on network, giving patterns and possibilities of meaning-making for network.

Each thread becomes a rhetorical *topos* for weaving network's story. A most colorful, distinctive thread shows network as a making or creating of a circuit (*peribolos poiesthai*).[8] Identifying lexically this creative, endeavoring dimension of network invites query of the character and quality of network's circuit making. Again, we can see in the list that the terms signify network as all-encompassing and ambiguous: as taking and being taken (*airein*; *agreuma*, *arkus*) and as free, of end or issue, or deduction, as in "being net," or being free from being taken (*atelēs*).

What do we do with these ambiguous characters and qualities of network? Turning again to the terms for network, we see another distinctive thread: that of *amphiballein*, from which issues the rhetorical trope amphibolia. If network is amphibolic, then it makes, creates, and endeavors for ambiguity. An appropriate or fitting way to tell network's story is through ambiguation, not idealization. Amphibolia is a rhetorical trope for creating doubt, but at its base means "on both sides" (*amphi*) and "to throw" (*ballein*). We see *amphiballein* has a partner in the list *periballein*, throwing around not just "both sides" but "all sides" to make a circuit encompassing all. These terms suggest a rhetorical turn that could mean to throw one's arms around in an embrace, or to throw doubt around on both, or all, sides of a matter. Amphibolia and peribolia as network's tropes turn us in meaning-making between embrace and doubt and hold us accountable for attending to the ambiguity of being human in and through network.

The ambiguity of network—of endeavoring to take and to be free of being taken, to embrace and to doubt—shows forth in network's ancient Greek lexical range. Ancient Greek myths help us weave even more of network's story.

Mythical Network

The myths of network come to us from various ancient Greek texts, whether of poetry, tragedy, comedy, or philosophy. These myths are archaic. By "archaic" I mean more than that time well before the classical era. Rather, I mean originary. Whereas these myths take the mark of words, they are themselves deeds, acts of material, creative life, not only in the mythmaking and telling but also in the myth-keeping, in the form of material texts from an oral tradition, whether by monastic record-keeping or some other act of recording history. Both words and deeds become our originary material for exploring ancient Greek rhetorics of networks. With such an archaic rhetorical materialist sensibility, we can explore how these words and deeds of ancient Greek network are rhetorical threads weaving a story-cloth of being human.

We see in this list of network terms the range of life through network. The base of these terms relates to the very making of the net and the network itself, the plaiting, twisting, twining, weaving (*plekō*). In myths told with such terms, we find women doing this base material work. As Evy Johanne Håland puts the matter, "In ancient Greek tradition, mostly sources written by men, the sign of the female, first and foremost, is weaving, since women do not speak, they weave."[9] Håland gathers the scenes of female speech, of weaving and planning, connecting both a female way of handling things and female cunning: "Lysistrata applies terms used in wool-working when describing her plan of how she will unite Greece in peace (Ar. *Lys*. 567 ff.). Homer (*Il*. 3.125–8) describes Helen weaving battle scenes and Andromakhe weaving talismans (*Il*., 22.440 . . .), and shows Penelope, holding her suitors at bay for more than three years while she wove a figured funerary cloth, unraveling it every night (*Od*. 2.94–110, 19.139–51, 24.139 f) . . . women's weaving implies a 'writing,' or graphic art, a silent material representation of audible, immaterial speech."[10] Myth tells us that weaving belonged to women. Their networking was an art showing forth their various roles "as mother, provider, worker, entrepreneur and artist."[11] And women's art issues from ambiguity, of "silent material representation of audible, immaterial speech."

Weaving belonged to men, too, as myths do tell. In particular regard to women, myths show men taking women's bodies, identities, and authority. Well known is the myth of Aphrodite caught in Hephaestus's net. When Hephaestus learns about the infidelity of his wife, Aphrodite, with Ares, he forges a hunting net, as fine as gossamer but unbreakable, which he secretly attaches to the bedposts of his marriage bed with Aphrodite. The trap works, and Aphrodite and her lover, Ares, are caught in the net and put on display for the gods to shame. Male gods laugh and jeer at her and devise their plan for controlling her. Only a gift exchange among the men, to force Ares to pay Hephaestus for the marriage gifts and to release Aphrodite for their own services, frees her from the net. Her freedom, though, is just another form of being taken.[12]

This episode of masculine network and the negative consequences for women pales in comparison to a lesser-known, more insidious, myth of Melanion in Aristophanes's *Lysistrata*. This myth reveals that masculine network enacts the political means by which to live in hatred and rejection of women. Here is the scene of the telling of the myth: In a revolutionary act, Lysistrata and her Panhellenic band of women occupy the Acropolis to force the men to end their perpetual warring. In the transition between a scene featuring the women reading prophecy to fortify their courage in staying the course of the occupation and a scene wherein the men are readying to confront the women, we are visited by the men's chorus, which tells this myth of Melanion. There once lived, in olden times, a young man named

Melanion. In flight from marriage, he went off into the wilderness. He lived in the mountains with his dog. There he wove nets. And he hunted. And he never went home again because of his hatred for women. That's how much he loathed women. And, the men's chorus concludes, they, being wise, hate women just as much as Melanion did.[13]

To loathe women is to steal their work, to turn it against them, to structure society via political network in such a way as to trap women, whether by devising through network perpetual war (*ergon*) that unravels the feminine economy of home and peacefulness, as Lysistrata frames the matter, or by trapping and torturing women for men's pleasure and power, like Aphrodite caught in Hephaestus's net. The men's chorus singing support for Melanion's misogyny sings of the magistrates' network. The object and objective of this network is war and misogyny. These magistrates are the leaders of the *polis* and have networked in such ways as to create endless war throughout all of Greece. Moreover, their network, as revealed in the myth of Melanion, creates an unwillingness to live in society with women.

The myth of Melanion is taken to excess by the magistrates. Whereas Melanion was seemingly content to secure independence from women by weaving his own nets for hunting and living, the magistrates in their network go far beyond this contentment. The myth of the magistrates, told sometimes by Lysistrata and sometimes by the magistrates themselves, tells of a masculine disconnection from the base material world of weaving to elevate men's network to an ideal form, namely, devising plans for all-encompassing power and self-benefit. Although immaterial in its devising, men's network is material in it effects. As Lysistrata describes, these magistrates network by way of making endless war from which they build their wealth and power. The women and children suffer. In their idealized network, the magistrates take without giving and hence become parasites of the *polis*. The magistrates' network unravels the connection between material and immaterial. The immaterial becomes an idealized *logos* of men's devising dominion. This idealized *logos* carries forward a myth of the pure idea of network, a network elevated from the base material labor of weaving. The myth of the magistrates, as we are told by the magistrates themselves, not just characterized by Lysistrata, announces the objective of men's network: men's independence, that is, securing the means of life and death, namely, network, from women and living their lives loathing women, securing their own benefit through war.

Although networking as devising is idealized for men, it is a vice for and of women. Sources of ancient Greek literature and philosophy, our sources for myths, written by men, typically "express a certain uneasiness towards weaving women."[14] These women come to be called wicked, for their weaving is devising, and works, like men's, to take control and to be free from being controlled. Take, for example, the myth of Clytemnestra, who welcomes her husband, King Agamemnon, home from the Trojan war by throwing the net of fate around him. The net of fate for Agamemnon is not a material net but rather his elaborately devised death at the hands of his wife in return for slaughtering their daughter in sacrifice to war. This net of fate thrown around is not a deed but a representation of network. Immateriality is inescapable, not something a man accomplishes in his loathing of women. Yet, the archaic female network connection to material remains in Clytemnestra's story. Agamemnon does face the material consequences of this net of fate: deathblows. The material scene is featured all the more by the attention paid to his entrance into the space of his death: his adornment with a purple welcome shroud that Clytemnestra ordered her handmaidens to make, woven and unfolded before the King as a cunning welcome device, though signaling in code that only his death will be welcome.[15]

We can make more of network from the myth of Clytemnestra than the story of misogyny. Clytemnestra's story tells of network in terms of materializing fate, and this is just one example of signifying network with fate, and the material outcome of such network as death and madness. When Odysseus casts the net of fate over the suitors, they all die.[16] When Cassandra is caught in the net of fate, she suffers the evils of her death at Clytemnestra's hands.[17] When the Furies seek justice for Clytemnestra, they pursue Orestes's death with their nets of fate, and indeed, Orestes must run to escape.[18] When Zeus throws a net of fate around Io, she goes mad wandering.[19] When Athena throws Ajax into evil nets, he, too, goes mad wandering.[20] When Gorgias argued that Helen may have had the net of fate around her soul, he found a reason to find her blameless against such a force.[21]

Network is associated with death, evil, calamity, treachery, and madness. A suspiciousness of nets is woven into ancient Greek rhetoric. Nets are cunning devices, as they are not noticed, usually, until it is too late, though a trap fashioned well in advance. Nets are *topoi* of entrapment. Their powers to bind liken them to the coils of a snake with the power of death by constriction (*peribolē logou*). The net throws us into ambiguity about immateriality. This immaterial fate is always already (or perhaps never not) material. Death, evil, and all kinds of calamity happen to material bodies, material lives. The materiality is ever-present in immaterial fate.

The material *topoi* of network as calamity originate etymologically and lexically from words that mean "to take," as in to take one's catch from hunting nets, or to take power through military and political domination and war, or to take control of women's bodies and authority (*airein*). Taking by force a city, overpowering a people, killing them, seizing their goods and their women, conquering and enslaving their culture—all are represented by working to "net" a catch. Also represented in working to "net" a catch is, in law, to make a judgment to convict a person, or to prove, as in "*o logos aireei*," reason *proves*. Moreover, choice is represented in the telic taking of network. To choose, to take in preference, to prefer one thing to another, to vote, to elect to office: All show how the ends of judgment are the ends of network. This judgment in turn leads to taking up all in accordance with it, raising up, extolling, even exaggerating all in the name of this judgment. This ancient rhetoric of network celebrates taking. And not just any taking but that characterized by the highest level of taking, a teleology of taking things to the top, a hierarchical march to idealized, immaterial judgments as given truths.

Still, lexical network reveals something other to mythic network, an equal and opposing force to taking: *atelēs*, meaning that which is free from being taken. The *a* marks an alpha privative of *telēs*. *Telēs* is more likely recognizable in English transliteration as *telos*, that end that is taken to be ideal. To be free from *telos* means to be free from an idealized end. To be liberated from *telos* means to be "net," as in "clear of deductions," as in "free from being taken." Archaic myths tell of network as telic and atelic: telic, as in networking to bring the ends of taking, or atelic, networking to avoid the ends of being taken. Still, mostly the atelic myths are another site of gender struggle, wherein masculine ideals lament *atelēs* as networking having come to naught, or idealized ends being left unaccomplished, with loss and defeat and weakness typifying the results. Athena reminds Telemachus that he is the son of a great man who is known for accomplishing all, for being telic. She assures him his journey to find news of his father and bring justice and peace to his home will not be atelic, in vain, but will be accomplished.[22] Homer tells us that the heart within Odysseus ponders thoughts that were not to go unaccomplished.[23] Agamemnon vows after the death of his brother, Menelaus, that there will come a day when Troy shall perish and that this shall not go unaccomplished (*atelesta*).[24]

Atelic rhetorics of network have more to offer than what these constricting, masculinized myths show. The power of atelic rhetorics to be other than issued from *topoi* of loss is exemplified in Penelope weaving.[25] Caught in the net of the suitors, Penelope devises her plan. She weaves. And then she unweaves. She nets and un-nets to be net. She tells her suitors she will choose among them upon completing her womanly work of weaving a death shroud for Laertes. Weaving is telic, as it must be accomplished. All day she weaves. And affirming atelic possibilities of network, all night she unweaves. Her devising and weaving are quite profoundly one and the same. Penelope's story is network's story of the ever-present, originary possibility of being free. Homer tells us that gods have this freedom, hence power, to decide whether to bring something to pass (*teleseien*) or to leave it undone (*ateleseien*).[26] Penelope becomes godlike in her choice to leave weaving undone.

Not only does this opposition within network between *airein* and *atelēs* allow for the possibility of being free in networks designed for taking, but the opposition allows as well for life. Whereas the myth of Melanion is told by the men's chorus to justify the men's networking of civilization built on the hatred of women, Aristophanes tells, on the whole, a different, and better, story with *Lysistrata*. This is a comedy after all, not a tragedy. Lysistrata accomplishes in the end of her women's networking something that is in excess of a constrictive, masculinized network, namely, the gift of the *peplos* to Athena Polias.[27] The gift is said to be appropriate to a free people, to represent their love for each other and their secure and lasting union. This gift, though it seems telic, accomplished in the giving and receiving, can never be accomplished. Obviously, Athena Polias is a statue; she can hardly receive a gift in a material way. Immateriality shows its ever-presence and atelic quality of the gift. Gifts keep on giving; they have what Marcel Mauss describes as a continuing spirit, a life, infused by the giver. As long as the gift is given, the spirit of the gift is a never-ending, eternal, atelic energy.[28] And of course, we know from Mary Douglas's well-known forward of Mauss's classic work, that the theory of the gift offers a theory of human solidarity.[29] The atelic energy of the gift is always already present in network. Still, we must not forget, as Émile Benveniste notes, "to give" in most Indo-European languages has an ambiguity and ambivalence that allows it to mean as well "to take."[30] The atelic energy of network can be part of a poison gift, where that which is given by way of network carries on forever, whether one likes it or not.

Tropical Network

Ancient Greek rhetorics of network in archaic words and myths tell of network as an ambiguous mix of gender and power, materiality and immateriality, and the forces of taking (*airein*), being taken, and being free from being taken (*atelēs*). Far from a teleological genealogy of network's metaphors, from fabric and fabrication to immaterialization, network has always already intermingled the material and immaterial, giving and taking, and paradigms of masculine and feminine power. One is not necessarily a likeness of the other, nor an elevated ideal from the base; rather both sides coexist or comingle as an ambiguous and ambivalent weave in a tropical pattern of amphibolia.

The amphibolia of network can be imagined not only as a trope figuring meaning-making in and of network but also as a trope for imagining how network can be most fittingly thought about, engaged, circumscribed, communicated. The ambiguity of network, as the ancient rhetorics reveal, helps us critique narratives of the digital age that idealize benefits of immaterial network. As described by Stanford intellectual and cultural historian, Ian Beacock, a contemporary communication of digital network is spoken "to the tune of billions of dollars," and the tale is a "Whiggish one" of "the digital ascent of humanity." The story goes, "from our benighted times, we'll emerge into a brighter future, a happier and more open society in which everything has been measured and engineered into a state of perfect efficiency."[31]

Such a story is afflicted with cultural amnesia about ancient rhetorics of network. Ancient rhetorics of network are amphibolic, not teleologic. In other words, in amphibolia, when an idea moves forward, an ambivalent, opposing, contrasting idea arises to throw the matter into ambiguity, to cast the movement into the limen. In amphibolia, teleology gets no traction, unless *teloi* were figured ironically as changes toward the opposite, rather than change for the better. Such irony signals the ancient Greek rhetorical connection between the tropes of *amphibolē* and *cirōncia*.[32] Material life is not teleologic, but rather ironic, full of ambiguity and ambivalence, perhaps its base irony being this: as it lives, it dies. At any moment, any kind of calamity could befall us. Material life does not just get better and better, and material life is inescapable, no matter our desire for or capacity to escape into the ideal immaterial.

This yearning to escape into the ideal immaterial needs turning to its opposite of the base material. As we see in the myth of Melanion, such escape into an ideal, inspired by hatred and division of what is identified as the lowly other base of life, arises at the high and unworthy cost of everything trampled on during the ascent. This ideal is not so ideal for women. Cultural amnesia about ancient rhetorics of network is cultural amnesia of women's lives, work, authority, art, and the very basis of technology.[33] As Virginia Postrel writes, "Textiles are technology—and they have remade our world time and again."[34] Postrel appeals for us to connect to the past, to remember that technology started with textiles and that textiles started with women's network, weaving. If we remember that technology started with textiles, then we remember women in society. Postrel makes an eloquent case for reestablishing a cultural memory of technology and weaving by reminding us that "the ancient Greeks worshipped Athena as the goddess of *technē*, 'the artifice of civilization,'"[35] and of weaving. Our idealization of technology in the form only of digital networks is a problem for material lives, especially for women.

Network's base materialism is women's weaving, and women's weaving is the base materialism of technology. The phrase "base materialism" calls forth Georges Bataille's musings on man's disdain for base materialism, in terms of man's own foot: "But whatever the role played in the erection by his foot, man, who has a light head, in other words a head raised to the heavens and heavenly things, sees it as spit, on the pretext that he has this foot in the mud."[36] Man's ideal denies dependence on and relation to base matter, constructing the base as disgusting, vile, ignominious. Yet, the base cannot be eliminated by the ideal; the ideal depends on the base.[37]

By attending to the ancient rhetorics of network, the words and deeds as offered to us by lexical and mythic sources of network, we see the base and the ideal not in a hierarchy but in an amphibolia, a network of relations that is all-encompassing and characterized by turns of ambiguity, throwing one's arms around in embrace, and throwing ideas and meanings around as to create doubt. Amphibolia is not a simple reversal of the hierarchy of ideal to base, nor a call to return to base sans ideal, nor a dialectical synthesis of the two. Amphibolia acts as a rhetoric of networking relations of, in, and through ambiguity.[38]

Women's lives in society show the stakes of ambiguating rather than idealizing base materialism in relation to the "digital ascent of humanity."[39] Let's consider two images to bring this point into bold relief: Lewis Hine's photograph of a spinner girl (figure 1.1) and a stock image for "network" produced by a Google image search (figure 1.2).

The first image comes from Lewis Hine's body of work documenting the lives of women and children in the textile industry: a little spinner girl, her tiny body dressed in textiles likely woven by others with tiny bodies like hers. She gazes at the world outside, trapped in a spinning industry, standing alone and isolated, offering her work to a world that renders her invisible.[40] The second image comes from a stock photo agency, Shutterstock, and appears in Google searches for images of "network." We see two men, united in a face-to-face handshake, with matching business attire and mirrored stances, in a context of binary code.[41] Each man is seen by the other, hence neither is invisible to the other. Face-to-face, hands clasped, they perform a ritual of recognition and reciprocity, alliance and solidarity. Unlike the spinner girl, whose body and location in the spinners' sweatshop communicates the material conditions that constrain and confine her, the men barely have bodies; they are silhouettes and shadows, suspended in light rays of binary code.

If we were to view the image of digital network teleologically, through a masculinist disconnect of material and immaterial dimensions of network, we would only be able to see ideals communicated, namely those mentioned above of recognition, reciprocity, alliance, and solidarity, and add to these

Figure 1.1. A spinner at the Rhodes Manufacturing Company in Lincolnton, North Carolina, 1908. Lewis Wickes Hine, photographer. Library of Congress Prints and Photographs Division (LC-DIG-nclc-01345).

Figure 1.2. *Digital World Agreement*, found online in a Google image search for "Network." Shutterstock.com.

the shining light of wealth. We would see in this image progress, advancement, and development. The men's silhouettes reflect an idealized achievement of transcendent, immaterial worlds of being. The spinner girl is left behind, rendered again invisible, embodied, entrapped.

When we view this image amphibolically, in response to the tropical call of ancient rhetorics of network, we see ambiguity: these men are and are not embodied. The escape into the immaterial, elite world of pure ideas, of binary code and silhouettes, cannot be a complete escape. The material world lurks; the body cannot be left behind. Moreover, the body, too, appears entrapped, in a net of binary code. The body entrapped in net is emphatic, via the ancient Greek figure of emphasis, meaning something ever-present albeit always in the shadows.[42] The body in network is in the shadows, literally and figuratively, in this stock image of network. Through amphibolia, we turn toward the shadows to see what is haunting us there.

Weaving haunts the digital. Before punchcards were used in computers, they were used to control the patterns on jacquard looms. As Postrel explains, "Weaving is the original binary system, at least 9,000 years old. Warp threads are held in tension, and weft threads go over or under them. Over-under, up-down, on-off, one-zero. Punch cards could control looms because weaving is intrinsically binary."[43] To forget this connection between weaving and coding constrains living well together. We forget the spinner girl. We forget the women of the digital world who are closed out of men's deals in Silicon Valley.[44] We forget about criminal and moral injustice against women working in the digital world, particularly against women who code, who are being sexually harassed and threatened with rape and death.[45] We forget the women and children of developing countries living in what has been described as a "poisonous, post-apocalyptic hell" serving as "the final resting place" for the First World's old televisions, computers, and mobile phones.[46]

How wrong to be so forgetful? Can we live well together as a people with such amnesia? Forgetting is an effect of a teleological rhetoric of the digital network ascent of humanity. Amphibolia, as an ancient rhetorical trope of network, turns us toward network's ambiguity, away from ascent, and calls attention to the immanent material conditions of the immaterial transcendent possibilities. We cannot valorize and idealize an immaterial ascendency of men in solidarity and wealth at the cost of the lives and labors of women, children, and the earth itself.

Rhetorical resources of amphibolia offer more than what haunts us. Amphibolia as network's figure shows forth that which is in excess as well as that which is deficient. We need to see these extremes to bring into focus the standard mean as moderator and measure. The Stranger in Plato's *Statesman* foretells that the weaving art as the political art would be destroyed without excess and deficiency being moderated by the standard mean. The standard mean arises from that which is fitting, kairotic, and opportune.[47] For weaving and political arts care about excess and deficiency in relation to that which is fitting. These arts regard excess and deficiency not as nonexistent but as real difficulties in actual practice.[48] Without such measures in relation to each other and that which is fitting, the woof and warp of the loom cannot achieve and maintain the proper laxity and tension, the fitting balance of separation and unity required for weaving. So, too, in the political arts of living well together. The excesses of Silicon Valley cannot be so valued that we become blind to the deficient conditions of the spinner girl. If we do, no art of living well together can be practiced. Likewise, if the deficient material conditions of the spinner girl were to become the only focus of critique, a stasis at best would be created. Of course, a halt, a moment of suspension, a stasis is a necessary condition for rhetorical change. But we cannot live in stasis, not for long anyway. Rhetoric must generate anew, innovate, give more, create again and otherwise. Amphibolia keeps a needed tension so we can weave our story-cloth of living well together.

In this story-cloth we see digital network entrapment and freedom. We see gender struggles and other dimensions of power struggles. We see the tensions between materialism and immaterialism. We see the gift as it takes and gives. In terms of the contemporary situations of digital networks, the extraordinary encroachments of digital networks in our lives drives discovery of new ways. David Orban proposes that Bitcoin enthusiasts communicate their work as "weaving" rather than "mining." "Mining" focuses on taking, rather than giving, and it enforces allegiance to the same system that honors gold as a monetary base, the very system that Bitcoin culture is trying to leave for all that system's encroachments on privacy. Moreover, Orban asks, "Could women be more easily drawn into the world of Bitcoin if we spoke about weaving the fabric of global financial emancipation? Could this result in a more attractive mental image than the macho world of secretive solitary miners searching for gold?"[49]

I will forgive Orban in this moment his gender essentialism so that I can attend to his question of whether changing the way we talk about network can change the way we live together in network. His question sparks a critique of misogyny in digital network, and calls for weaving freedom. This critique and call offer evidence that something other has gotten through, something forgotten remembered, affirming in turn the atelic possibilities of freedom. And, yes, changing rhetoric can change life.[50]

We can see the possibilities of atelic network in stories beyond Bitcoin. The indigenous Aymara women of Bolivia weave medical devices to help children with heart defects. The designing cardiologist, Franz Freudenthal, told the BBC that "the most important thing is that we try to get really simple solutions for complex problems."[51] Such an approach is figured by an amphibolic turn and relation between the simple and the complex, something a teleological turn just cannot create in its rejection of the simple to idealize and achieve the complex. This simple solution though, of the weaving women of Aymara, is excessively complex, as it exceeds the possibility of mass production. Each device must be woven by hand via the indigenous, archaic weaving and networking of women. These women weaving "parts for hearts" are in turn being networked on social media via *A Mighty Girl*, a digital collection claiming the title of the world's largest of books, toys, and movies for parents, teachers, and others dedicated to raising smart, confident, and courageous girls.[52] We find a tremendous resource for the good life in this amphibolic turn toward the simple weaving of Aymara women to address such complex needs as healing heart disease and parenting mighty girls.

Or consider as another example of the atelic possibilities of network woven in an amphibolic pattern the StEP Initiative in the United Nations: Solving the E-waste Problem. United Nations' web communications describe their role, organization, objectives, work, principles, and education initiatives all through the language of "network."[53] They have developed elaborate digital networks to advance solving this material e-waste crisis. One striking example is the StEP e-waste world map, which provides visual data on the amount of electrical and electronic equipment put on the market and the resulting amount of e-waste generated in most countries around the world. The data on the material quantity and location of e-waste is updated regularly to stay current with changes, and the map provides links to relevant e-waste rules, regulations, and policies. Network got us into this problem of e-waste; network can get us out.

Idealizing digital networks at the expense of forgetting about material origins and effects allows digital networks to proceed as if the ecologically catastrophic horrors of e-material waste do not exist, or perhaps that these horrors must be accepted for achieving the immaterial digital ideals of networked life. The ancient rhetorics of network tell us that such rhetorics of idealization are not fitting for network. Rather, rhetorics of ambiguation, signaled by the words and deeds of amphibolia and *atelēs*, can free us from being so taken by an ideal, can return us to a connection with base materialism that fuels us with resources to create a circuit of solutions all around complex problems. Still, network's amphibolic turns will not allow human beings to rest forever in *atelēs*, let alone experience *atelēs* only as an ideal called "freedom." Network will again take, and yet for being amphibolic, will again offer resources for freeing us from being taken. So is the story-cloth of being networked, of being human in society.

Now, with network's story-cloth in hand, we can imagine the gift of the *peplos*, a robe fitting a free people, in the spirit of the ancient Greek gift of a *peplos* to the goddess Athena at the annual Panathenaic Festival. This gift is given to symbolize a free people. From our genealogy of ancient rhetorics of network, we can see that a free people in network's story-cloth appears as a people capable of both embracing and doubting, as well as recognizing their responsibility to do both, equitably and continually. A free people in network's story-cloth appears as a people capable of and desirous of living in ways other than war. A free people live in the connection of material and immaterial worlds so as to be circumspect always of network's promises and problems. A free people, living in this connection, for example, sees e-waste's ecological catastrophe, and sees the cultural colonization that idealizing the immaterial allows for in gender, race, and class relations. A free people throws down on all sides of such matters to create doubt in the idealized structure of power that networked us into such problems in the first place. This throw down on being taken simultaneously turns a free people to the atelic potential of network to find freedom from such problems. A free people navigate, for example, the domination of masculinist ideals with the ambiguity of shared power within and across genders to address gender-based violence and oppression.

Freedom does not reside in the extremes of excess and deficiency, as the arts of weaving and the arts of living well together show. We see this in network's story-cloth, though ambiguously we see as well that these extremes are ever-present in and for network. May our robe offer safety from extremes.

Notes

1. Gregory Flaxman, "Networked: Politics in the Age of Information," *Media Res*, Oct. 5, 2012. http://mediacommons.futureofthebook.org/imr/2012/10/05/networked-politics-age-information. Accessed October 10, 2014.

2. Ibid.

3. See Armand Mattelart, *Networking the World, 1794–2000*, trans. Liz Carey-Libbrecht and James A. Cohen (Minneapolis: University of Minnesota Press, 2000).

4. The network terms I select, and their summary definitions, are derived from two key sources: S. C. Woodhouse, *English-Greek Dictionary: A Vocabulary of the Attic Language* (London: George Routledge and Sons, 1910); and Henry George Liddell, Robert Scott, Henry Stuart Jones, and Roderick McKenzie, *A Greek English Lexicon* (Oxford: Clarendon Press, 1996).

5. I work with the idea of Homeric catalogs also in *Rhetoric and the Gift: Ancient Rhetorical Theory and Contemporary Communication* (Pittsburgh: Duquesne University Press, 2015), 55–66.

6. *Il.* 2.484–760. Citations for Homer throughout are from the Oxford Classical Texts: Homer, *Homeri Opera: Odysseae*, vols. 3–4 (London: Oxford University Press, 1974); *Homeri Opera: Iliadis*, vols. 1–2 (London: Oxford University Press, 1920). Translations are my own. I consult as well *The Odyssey of Homer*, trans. Richmond Lattimore (New York: Harper Perennial, 1991); and *Odyssey*, 2 vols., trans. A. T. Murray (Cambridge, MA: Harvard University Press, 1984).

7. Benjamin Sammons, *The Art and Rhetoric of the Homeric Catalogue* (Oxford: Oxford University Press, 2010), 9.

8. In the rhetorical tradition, poetry (*poiēsis*) can be distinguished from rhetoric, yet, even in Aristotle's *Rhetoric*, "*poiēsis*" is the term used throughout to signify the "generation" of arguments, or what the Romans would later call "invention." For Aristotle's use of *poiēsis*, see *Rhetoric*, 1362b15, 18; 1367a5; 1379a8; 1381a11, 19; 1381b35, 36; 1382a2; 1388b17. Citations for Aristotle's *Rhetoric* are of the ancient Greek from the Oxford Classical Text: Aristotelis, *Ars Rhetorica* (Oxford: Oxford University Press, 1959). Translations are my own. I consult *Aristotle: The "Art" of Rhetoric*, trans. John Henry Freese (Cambridge, MA: Harvard University Press, 1982); *Aristotle: On Rhetoric*, trans. George Kennedy (Oxford: Oxford University Press, 2007); and *Rhetoric*, trans. Rhys Roberts (New York: Modern Library, 1954).

9. Evy Johnanne Håland, "Athena's Peplos: Weaving as a Core Female Activity in Ancient and Modern Greece," *Cosmos* 20 (2004),170.

10. Ibid.

11. Ibid., 155.

12. *Od.* 8.256–366; See Robert Graves, *The Greek Myths*, vol. 1 (Baltimore: Penguin Books, 1974), 67–71, for his account of the ways in which Aphrodite is taken by the men upon release.

13. Aristophanes, *Lysistrata*, 781–96; citation of the ancient Greek and translation consultation for Aristophanes is from Jeffrey Henderson, ed. and trans., *Aristophanes: Birds, Lysistrata, Women at the Thesmophoria* (Cambridge, MA: Harvard University Press, 2000).

14. Håland, "Athena's Peplos," 165.

15. Aeschylus, *Agamemnon*, 905–10; 980–1013; 1379–83. Citation of the ancient Greek and translation consultation for Aeschylus is from *Aeschylus: Suppliant Maidens, Persians, Prometheus, Seven Against Thebes*, vol. 1, ed. Hugh Lloyd-Jones and trans. Herbert Weir Smyth (Cambridge, MA: Harvard University Press, 2006); and *Agamemnon, Libation Bearers, and Eumenides*, vol. 2, trans. Herbert Weir Smyth (Cambridge, MA: Harvard University Press, 1971). Alexander Galloway begins his entry for "networks" with a reference to *Agamemnon*; see Galloway, "Networks," in *Critical Terms for Media Studies*, ed. W. J. T. Mitchell and Mark B. N. Hansen (Chicago: University of Chicago Press, 2010): 280–82. For an analysis of the agency of "tragic textiles" in ancient Greek tragedy, see Melissa Mueller, *Objects as Actors:*

Props and the Poetics of Performance in Greek Tragedy (Chicago: University of Chicago Press, 2016), 42–68.

16. Homer, *Od.*, 22.33, 41, 43, 65–67, 363, 381–82.

17. Aeschylus, *Agamemnon*, 1035–50.

18. Aeschylus, *Eumenides*, 145, 299–396, 460, 630.

19. Aeschylus, *Prometheus Bound*, 738. For commentary on mad wandering, see Silvia Montiglio, *Wandering in Ancient Greek Culture* (Chicago: University of Chicago Press, 2005), 38.

20. Sophocles, *Ajax*, 59–60. Citation of ancient Greek and translation consultation from *Sophocles: Ajax, Electra, Oedipus*, vol. 1, ed. and trans. Hugh Lloyd-Jones (Cambridge, MA: Harvard University Press, 1994).

21. Gorgias, *Encomium of Helen*, 19, in *The Older Sophists*, ed. and trans. Rosamond Kent Sprague (Columbia: University of South Carolina Press, 1972), 54.

22. *Od.* 2.270–75; 19.137–51.

23. Ibid., 18.344–45.

24. *Il.*, 4.168.

25. *Od.*, 2.94–110.

26. Ibid., 8.570–71.

27. For historical accounts of the ritual of the gift of the *peplos* to Athena, see E. J. W. Barber, *Prehistoric Textiles: The Development of Cloth in the Neolithic and Bronze Ages with Special Reference to the Aegean* (Princeton, NJ: Princeton University Press, 1991). See also, Barber, "The Peplos of Athena," in *Goddess and Polis: The Panathenaic Festival in Ancient Athens*, ed. Jennifer Niels (Princeton, NJ: Princeton University Press, 1992): 103–17; and Håland, "Athena's Peplos," 155–82.

28. Marcel Mauss, *The Gift: The Form and Reason for Exchange in Archaic Societies* (London and New York: Routledge Classics, 2002).

29. Mary Douglas, "No Free Gifts," forward to Mauss, *The Gift*, x.

30. Benveniste, "Gift and Exchange in Indo-European Vocabulary," in *The Logic of the Gift*, ed. Alan Schrift (New York: Routledge, 1997), 34.

31. Ian Beacock, "Humanist among Machines," *Aeon*, June 25, 2015, http://aeon.co/magazine/society/why-we-need-arnold-toynbees-muscular-humanism/.

32. See R. Dean Anderson Jr., *Glossary of Greek Rhetorical Terms Connected to Methods of Argumentation, Figures and Tropes from Anaximenes to Quintilian* (Leuven, Belgium: Peeters, 2000), 17, 39.

33. Virginia Postrel, "Losing the Thread," *Aeon*, June 5, 2015, http://aeon.co/magazine/culture/how-textiles-repeatedly-revolutionised-technology/.

34. Ibid.

35. Ibid.

36. Georges Bataille, *Visions of Excess*, ed. and trans. Allan Stoekl, with Carl R. Lovitt and Donald M. Leslie Jr. (Minneapolis: University of Minnesota Press, 1985), 20.

37. Ibid., 22.

38. Ibid., 4.

39. See Beacock, "Humanist among Machines."

40. Lewis Wickes Hine, photographer, "Rhodes Mfg. Co., Lincolnton, N.C. Spinner. A moments glimpse of the outer world Said she was 10 years old. Been working over a year. Location: Lincolnton, North Carolina," photograph. Lincolnton, NC, 1908. From Library of Congress: *Photographs from the records of the National Child Labor Committee (U.S.)*, accessed June 20, 2015, http://www.loc.gov/item/ncl2004000062/PP/#about-this-item.

41. "Digital World Agreement with Binary Code," Shutterstock, Image ID 61426666, accessed July 11, 2016, http://www.shutterstock.com.

42. See Anderson, *Glossary*, 41–42.

43. Postrel, "Losing the Thread." Postrel uses the phrase "cultural amnesia" to describe our memory loss of the connection between textiles and technology.

44. Two leading digital industries, Facebook and Google, have made public their lack of diversity, in particular gender diversity. Women make up only 15 percent of Facebook's tech workforce. Women at Google fare "better," making up 17 percent of its tech workforce. See Maxine Williams, "Building a More Diverse Facebook," *Facebook Newsroom*, June 25, 2014, http://newsroom.fb.com/news/2014/06/building-a-more-diverse-facebook/; for Google's data, see "We're Working Toward a Web that Includes Everyone," accessed September 20, 2015, http://www.google.com/diversity/index.html. See also Murrey Jacobson, "Google Finally Discloses Its Diversity Record, and It's Not Good," *PBS NewsHour*, May 28, 2014, http://www.pbs.org/newshour/updates/google-discloses-workforce-diversity-data-good/.

45. The Gamergate controversy offers a poignant example: In August 2014, a harassment campaign sought to drive out several women from the video game industry. The campaign of harassment took place in Internet Relay Chat (IRC) channels and online forums such as Reddit, 4chan, and 8chan. The harassment ranged from "doxing" (a digital practice of researching, including hacking, and broadcasting personal information about targeted individuals) to threats of rape, death threats, and a threat of a mass shooting at a university event hosting one of the targeted women, Anita Sarkeesian. She canceled her talk. See Fruzsina Eördögh, "Gamergate and the New Horde of Digital Saboteurs," *Christian Science Monitor*, November 25, 2014, Accessed October 6, 2017, https://www.csmonitor.com/Technology/Tech-Culture/2014/1125/Gamergate-and-the-new-horde-of-digital-saboteurs.

46. Damien Gayle, "Chemical Breakdown: The Toxic Substances Inside Your Mobile Phone," Daily Mail.com, October 5, 2012, http://www.dailymail.co.uk/sciencetech/article-2213366/Chemical-breakdown-What-really-goes-new-iPhone.html. Gayle details the following scene in a suburb of the capital of Ghana: "Piles of cracked computer monitors sit in rancid pools of noxious green slime, while the circuit boards and innards of old PCs leak toxic mercury, beryllium and cadmium, poisoning the earth. Scampering over these mounds of decaying computers are red-eyed gangs of wild, young boys in rags who smash old computer components with rocks to rip out the valuable copperwire inside."

47. Plato, *The Statesman*, trans. Harold North Fowler and W. R. M. Lamb (Cambridge, MA: Harvard University Press, 1925), 284e.

48. Ibid., 284a.

49. David Orban, "Weaving Is a Better Metaphor for Bitcoin, Instead of Mining," *Bitcoin Magazine*, April 25, 2014, https://bitcoinmagazine.com/12311/weaving-better-metaphor-bitcoin-instead-mining/.

50. For essays on rhetoric, materialism, and materiality, see Barbara A. Biesecker and John Louis Lucaites, eds., *Rhetoric, Materiality, and Politics* (New York: Peter Lang, 2009).

51. Ignacio de los Reyes, "The Bolivian Women Who Knit Parts for Hearts," *BBC Health Check*, March 29, 2015, http://www.bbc.com/news/health-32076070.

52. See AMightyGirl.com, accessed September 22, 2015, http://www.amightygirl.com.

53. See StEP-Initiative.org, accessed September 20, 2015, http://www.step-initiative.org. A search of this web page for "network" turned up an array of uses. Here is a sample: members form a global network of more than 470 locations in 55 countries, StEP recognizes the Greener Network Methodology, Step works with the Asian Network for Prevention of Illegal Transboundary Movement, features regional, multisectoral networks for Latin America and the Caribbean, with networks serving as focal points for information exchange, with countries acting as catalysts, networking scientists, policy-makers, educators, as well as citizens on the front lines of the calamity to solve the problem, and on and on the language of "network" is engaged by the United Nations to communicate what is being done, and what is a necessary condition, namely network, required for solving this global calamity.

2

Imagining Confucian Audiences

Tactical Media and the Umbrella Movement

ARABELLA LYON

In *Spectacular Rhetorics*, Wendy Hesford argues that the contemporary human rights spectacle is "a rhetorical phenomenon through which differently empowered social constituencies negotiate the authority of representation."[1] Through visual technologies, activists push the margins of visibility, making the once invisible present to a witnessing world, and thus they gain the authority to represent and shape cultural, political, and legal norms. New media technologies and social networks offer access to hitherto unimagined global audiences and communities. By utilizing visual evidence (seeing-as-believing) and visual construction (seeing-as-power), activists can strategize routes to change through the tactics of engagement. Although generally positive about global activism, Hesford herself resists exaggerating the images' political efficacy, given the passivity of many audiences, the structure of neoliberalism, the ambiguity of images, and the instrumentality of some political discourses.

Images never present a simple truth, but digital images, which can circulate far from their originating contexts, create even more interpretative difficulty. Furthermore, when the invisible is made visible and the inaudible made audible, audiences may not have the experience to interpret wisely. Furthermore, while human rights images are often constructed to shame abusers and to alarm (and titillate) witnesses through visual evidence of abuse, rights images sometimes involve the very different projects of utopic imagining and idealized engagements and recognitions. Visual constructions of a better world require different interpretive strategies for audiences. Add the difficulty of cross-cultural communication to utopic imagining, and global images evade nimble readings. It is little wonder that the study of cultural rhetoric has grown in the digital age.

To demonstrate the complexity of reading cultural images cross-culturally, this essay interprets three memes from the 2014 Hong Kong protests. To accomplish this, however, it must develop a Confucian theory of rhetorical audience, one that differs significantly from the "oratorcentricism" of the Greco-Roman rhetorical traditions.[2] After spending the twentieth century on what Stephen Angles calls "life support," Confucianism is undergoing a resurgence powered by new philosophical and public movements in Asia and the West.[3] New Confucianism has evolved and grown over the last thirty years, first as academic analysis and then as political discourse. Many scholars are arguing that the classical ideals can help explicate contemporary events and provide direction for the development of Chinese democracy and indigenous human rights.[4] There is little controversy in claiming that, twenty-five hundred years after the death of Confucius (551–479 BCE), his thinking can be understood as a critical lens for analyzing contemporary practices. Using the method that Richard Rorty calls "rational reconstruction"—rearticulating the historically limited but durable ideas of a philosopher so that they are relevant to current cultural conditions—this essay reconstructs Confucian ideals of remonstration and audience and uses these ideals as a standard for reading social networks and digital media.[5] For purposes of focus, it examines only one Confucian text, the *Analects*, a set of dialogues and assertions, composed between 479 and 249 BCE. Assuming classical Chinese texts do not describe an impossibly different world and arguing that rereading early Confucian texts is not merely archeological tourism, the essay next demonstrates the currency of the reconstructed theories of remonstration and audience by applying them to digital images created during the 2014 umbrella movement (*yǔsǎn yùndòng*). Ultimately it demonstrates that some of the tactical media deployed during the Hong Kong protests is indebted to the Confucian tradition, augmented by the imaginary of digital images.

Audience Constructed: *Jiàn*, *Shù*, and *Zhong*

Unlike oratorcentric, Western rhetorics ("a good man speaking well"), most classical Chinese theories of political communication assume a more open-ended relationship between the speaker and a powerful audience. Confucian speakers are slow to speak and relenting in attempts to convince another (see *Analects* 4.24, 4.26, 12.2, 12.3, 12.23). In fact, speaking too well is suspect; Confucius warns, "Clever words undermine excellence (*de*)" (15.27). Rather than valuing sophistry or glib persuasion, Confucian speakers ideally remonstrate or demonstrate their ideas within a relationship of trust (19.10). There are weighty differences between the concepts of persuasion (*shùi*) and remonstration (*jiàn*), differences both definitional and temporal.[6] The Latin roots of each concept's English translation clarify the rhetorical implications of the definitional differences. Persuasion's roots, *per* and *suadere*, suggest a result achieved *through sweetness*, while *re* and *monstrare* denote a process of *demonstrating* or *showing again*. A bringing emphasizes the movement of someone; an agent brings an objectified audience to a new place; there is an agent who brings or transforms an objectified audience. Alternatively, remonstration emphasizes the process itself and its reiterative potential. Persuasion may be an effect of remonstration, but it is not the intention. Rather it is a side effect. Clear in their very etymologies, these differences suggest alternative relationships; the audience is imagined as either a moveable object or a witness with judgment.

The differences between remonstration and persuasion can also be seen as marking different temporal processes in acts of communication. Ordinary language philosopher Gilbert Ryle analyzes the difference between process verbs and terminus verbs.[7] Process verbs, like "seek," describe ongoing action, while terminus verbs, like "find," designate an end point of action. "Remonstrate" is a process verb, lacking a *telos*, or end. Process verbs, like "seek" and "remonstrate," describe ongoing action and open possibility. On the other hand, "persuade," and "find," are terminus verbs committed to an end point; for persuasion, the end point is the moment when the audience has been brought along. If the work of political communication is driven by the concept of persuasion, then implicitly Aristotelian deliberations are concerned with ending, the moment the audience is convinced by a strategizing speaker. Usually a transitive verb, the object of persuade tends to be the audience: "I persuaded him to change the rules" or "the politician persuaded us to go to war." Remonstration, rarely a transitive verb, is concerned with ideas and acts; the object of remonstrate is what is shown: "I went to the council to remonstrate against the new rules," "the politicians remonstrated the necessity of retaliation." By emphasizing the act of the speaker, remonstration leaves open the appropriate response of the audience. In short, within the authoritarian state, presenting ideas is safer than pushing a ruler into a wrongheaded action, but even more significantly, a respectful relationship, bounded by mutual duties, precludes a drive to judgment dominated by an orator.

If the objective of remonstration is not moving the audience but representing the ideas and ideals of the speaker, then what relationship is imagined between interlocutors in a Confucian state? Paradoxically, for all of remonstration's emphasis on the speaker's action, the audience does not disappear. Although reciprocal recognition is not its purpose, there is an implied audience for remonstrators, and the audience's separation from the demonstration suggests a mindfulness of their independence, interpretive skills, and powers of resistance, autonomy, and judgment. The lack of an objectified audience in remonstration acknowledges the difference or distance between the speaker and the audience, and it gestures to the next step of forming a relationship mediated by the ideas and acts between them. The act of remonstration proposes that an appropriate deliberative relationship should exist between interlocutors who define a relationship without domination. Furthermore, remonstration intimates acceptance of difference, not a commitment to its resolution. In leaving open the dynamics of communication, remonstration shifts deliberation away from a concern with ends through strategic manipulation and toward acts of recognition and engagement. Acknowledging the audience's ability to decide whether to engage or respond is a significant, foregrounded acknowledgment of difference, especially if one is engaging a potentially dangerous ruler. In effect, there is an assumption that the interlocutors may not share common purposes, though, as I later show, their common humanity is the basis for engaging each other.

Although remonstration implies a relationship with an audience, it can go only so far in teaching us about the proper Confucian regard for, engagement with, and recognition of others.[8] Other concepts, many beyond the scope of this essay, help flesh out the appropriate relationship between interlocutors in a rhetorical situation. Among the many possibilities, harmony, ritual, and filial piety are particularly relevant values for engaging the other. Harmony (*hé*) is a value central to understanding how a Confucian might define a rhetorical situation and justify intervention. Understood as the primary purpose of conventions and relationships, harmony is commonly achieved through ritual (*lǐ*). Following Roger T. Ames and Henry Rosemont, ritual can be defined as the enacted propriety or decency shaped by our roles and relationships. Through ritualized engagements within and beyond their community, people become harmonious as they participate in generative, cultural norms and meet the expectations of others. For a contemporary extension of this idea, consider how the series of actions or rituals inherent in an academic essay, open to some adjudication, are generative of knowledge; when one situates one's work as an extension of earlier work, one creates intellectual roles and relationships. The *Analects* acknowledges that some situations require adjudication or regulation, although the regulations will occur in a continuum with existing cultural rituals: "Achieving harmony (*hé*) is the most valuable function of observing ritual propriety (*lǐ*). . . . But when things are not going well, to realize harmony just for its own sake without regulating the situation through observing ritual propriety will not work" (1.12, see also 1.13). Hence, one should act with a commitment to harmony, the culture, and its values but not follow directives by rote, as ritual propriety is a *creative form*, just not infinitely creative. There is no Confucian revolution but rather an acknowledgment of evolution.[9] This understanding of harmonious change is congruent with remonstration, and it will be evident in my analysis of the representations of relationships between the Hong Kong protestors and their interlocutors.

The concept of the family provides a model for interlocutors, even within politics. The Chinese word for state or nation (*guójiā*) is made of the words "state" and "family," implying that family relationships and extensions of those roles to relationships with friends, neighbors, and teachers are analogous to relationships within the state. Tellingly, for Confucians, relationships among people, especially within the family, are more important than norms, governments, and laws. The normative vocabulary of intimate relationships is considered more compelling than an abstract vocabulary of civic engagement. The fluidity between public and private has not been resolved by more recent Confucians (or for that matter contemporary feminists), in part because Confucians reject a taxonomy that separates ethics from politics.[10]

One might think that the overvaluing of harmony and relationships based on the family metaphor creates an anachronistic basis for modern political and rhetorical relationships. It would seem *that* harmony and family reverence cannot explain contemporary political acts and digital networks. They, however, do provide a framework for the two Confucian concepts most relevant for understanding audience. *Shù* and *zhōng* are the core of Confucianism: "The way of the Master is doing one's utmost (*zhōng*) and putting oneself in the other's place (*shù*), nothing more" (4.15, see also 5.12, 12.2, 15.24). The two concepts highlight the Confucian way.[11] Mitigating Confucian hierarchies and an apolitical emphasis on self-cultivation, the two concepts—*zhōng*, acts of duty to others, and *shù*, personalized understanding of others—together underscore the reciprocal nature of human existence.

Shù is the term with the larger or looser semantic field, and consequently translators have defined it differently. Hall and Ames note that it has been variously translated as "'altruism' (W. T. Chan), 'reciprocity' (Tu Wei-ming), 'consideration' (Waley), 'do not do to others what you do not want them to do to you' (Fingarette), and 'using oneself as a measure in gauging the wishes of others' (Lau)."[12] Although Hall and Ames initially translate it as "deference," emphasizing its bidirectional nature, Ames later uses the clause "putting oneself in the other's place."[13] Hear the Golden Rule. For brevity, I use Hall and Ames's translation, but one should be more analytic in translation. For example, Herbert Fingarette defines *shù* as follows: "To be able from what is close—i.e., oneself—to grasp analogy with the other person, and in that light to treat him as you would be treated."[14] He emphasizes that one cannot put oneself in another's place; apprehending the other is an analogic act. To extend, then, recognition of the other occurs, at least in part, through recognizing oneself. *Shù* challenges individualism, even the Confucian regime of self-cultivation, because it requires a dialectical relationship within the self, one that emphasizes the role of human similarity as the ground for social and institutional responsibility. Furthermore, other scholars, such as Goldin and Nussbaum, emphasize that the social situation, Confucian hierarchies, define the appropriate reciprocity.[15]

Thus, when one's relationship with others is imagined as a continuation of one's self, then the boundaries between individuals are muted: my identity and desire is analogic to your identity and desire. Despite the distance acknowledged in acts of remonstration, the relationship with the other is imagined through identity, not identification. Unlike Burkean identification wherein unity is compensatory to division, or, put differently, unity is polar to division, *shù* acknowledges that recognition occurs through analogical thinking and that the difference between humans is bridgeable through knowing one's own experience.[16] At odds with a powerful, persuading orator manipulating others through their vulnerabilities, the concept of *shù* suggests that communication and engagement is possible through *the recognition of oneself in others*. Common actions arise from the attribution of one's desires to the desires of others. The mystery of consciousness—one's own and others'—is imagined through familial difference, not oppositional difference. *Shù* is not as appropriating as it initially seems, because one's desires are modulated by the cultural norms of harmony and familial relationship. Furthermore—as part of other-regarding—*shù* defines the self in relationship to another, but not in ways that entail symmetrical recognition. One should recognize the other but not require reciprocity (4.14). As with the Golden Rule, the other's lack of recognition does not justify one's lack of recognizing the other. In line with remonstration, an agent has responsibilities that do not require the other's engagement. *Shù* acknowledges human limitations in recognizing and respecting differences, and it allows that interlocutors are limited in their own desires and self-knowledge as well as by their ability to gauge the desires of others. Unlike speakers operating with manipulative understandings of audience, *shù*-driven speakers wish for others the same excellence that they wish for themselves, framed within the respectful values of harmony and family. Speaking freely in an authoritarian state is dangerous, but the ideal centralized state is concerned pragmatically with harmonious relationships between center and margin. *Shù* becomes a means to engage distant, unimagined others even if they do not recognize you.

In the *Analects*, a second word, *zhōng*, also underscores the difficulty of speaking in an authoritarian state. Although sometimes *zhōng* is translated as "loyalty," which is a later meaning, in the time of the *Analects*, it was more closely related to duty and relationship. D. C. Lau interprets *zhōng* as "doing one's best."[17] Implicit in doing one's best is *duty*, giving one's whole heart and mind to an effort with an implication of loyalty to others. In the *Analects*, *zhōng* is used consistently to mark service to others, and service to others involves a self-disciplining commitment to a self-awareness of what is best (1.4, 3.19, 5.19, 12.23, 13.19). Paul R. Goldin argues for an early Confucian meaning of "being honest with oneself in dealing with others" and thus emphasizes the self-awareness necessary for differentiating duty from unreflective self-interest.[18] For example, a leader might dutifully believe that it is in a citizen's best interest to obey the established laws and, through self-interest, forget that those laws may be unfair and outdated. This tension between doing one's best and being honest with oneself in dealing with others will come up again later, but for now, let us return to Lau.

Lau argues for the complementary nature of *zhōng* and *shù*. According to Lau, "It is through *chung* (*zhōng*) that one puts into effect what one has found out by the method of *shù*."[19] To apply this formulation to Confucian communication theory, rhetoric concerns understanding the best possibilities to be achieved by putting oneself in the place of others. Thus, rhetorical method requires consideration of duties to others, but only in terms of self-understanding, perhaps because that is the limit of human vision in the *Analects*. Understanding other minds is a fragile enterprise, and even self-understanding is so limited as to need dutiful reflection. One is expected to understand the other only through personal projections and desires, but one is then to do one's best by the other. The *consequence* of *shù* and *zhōng* is a world in which one does one's best in relationship to the other.

To sum up, in the *Analects*, one can explicate a theory of communication based on remonstration about ideas and rhetor-audience relationships defined through *shù* and *zhōng*. In this frame, the audience is not a stationary target. Even if radical difference is underengaged in the method of *shù* and *zhōng*, the theory of the analogic other is less problematic than many theories of audience that stereotype and objectify. Remember that in book II of the *Rhetoric,* Aristotle characterizes young men as having strong passions that they tend to gratify indiscriminately and old men as cowardly, always anticipating danger. Alternatively, Confucius would have speakers imagine their audiences as themselves, reflect on their own character, and then do their best by the audience. Perhaps *shù* is too personal to be a reliable ethical directive, and the ethical dimensions of *shù* are not clear in that the desires and wishes projected onto the other are not necessarily ethical.[20] Still, as I will demonstrate, it may well be a useful rhetorical method of audience address and political imagining. Supplementing *shù* with the concepts of *zhōng*, ritual, filial piety, harmony, trust, and benevolence modify *shù*'s inherent personal focus. Yes, the Golden Rule does have problems as a political instrument, and it benefits those who ignore the Golden Rule to the detriment of those who follow it. That noted, the critiques of Confucian limitations do not disqualify Confucianism as a communicative theory with descriptive power.

Constructing Confucian Audiences Digitally/Engaging the Emperor Democratically

The relationship imagined between Confucian interlocutors provides a valuable theoretical lens for analyzing contemporary rhetoric. In the time of digitized social networking, a time of distance and difference, it may be possible to understand the other only through personal projections and desires, although it is unclear how one does one's duty for the distant other. The remainder of the chapter tests its theory of Confucian audiences by analyzing the engagements of protestors in the umbrella movement (*yǔsǎn yùndòng*) with a prime audience of Chinese President Xi Jinping. Despite China's history of visual and political satire, this particular test has its limits.[21] It is one small case extending an ancient tradition into a political system heavily inflected with the colonial legacies of Western Marxism and democracy. Furthermore, it ignores other aspects of Confucianism that would disapprove of the protestors' tactics. For example Confucius advises, "Do not plan the policies of an office you do not hold" (8.14, also 1.2, 2.3, 14.26, 16.2), and yet the protestors sought to set policies for the city-state. Obviously the hierarchy inherent in classical Confucianism starkly confronts democratic equalitarianism (3.19, 5.16, 18.7).[22] Even so, Confucian values are visible in the communication practices of some of the protestors. Public acknowledgment of Confucius has returned to Chinese politics. Chinese leaders such as Jiang Zemin (1989–2002) and Hu Jintao (2002–2012) have spoken in ways commensurate with Confucian values, respectively, speaking of the values of "rule by virtue" and "a harmonious society."[23] President Xi even quotes Confucius in his speeches: "When we see men of virtue, we should think of equaling them" (4.17).[24] Confucian sensibilities thus infuse much Chinese political discourse.

The Protest: Beginning in late September and continuing into December 2014, Hong Kong experienced its largest protest since the British handover in 1997.[25] The focal point of the protests was China's shift from long promised universal suffrage for Hong Kong's 2017 elections. In 2007 the National People's Congress Standing Committee (NPCSC) of the PRC had decided that the 2017 chief executive of Hong Kong would be elected by universal suffrage. In August 2014, the NPCSC proposed that a 1200-member nominating committee should elect two or three candidates before presenting them to the Hong Kong voters. Essentially, the Beijing-backed proposal would only allow two or three party loyalists on the ballot, and it thereby destroyed the dream of self-governance. The umbrella movement arose from the modest demand that the city's top leader be chosen by open nomination, with universal suffrage becoming the rallying cry. The protests began with tens of thousands of people, but by December they had dwindled to a few hundred when the police dismantled the few remaining camps.

Due to the conservative nature of Hong Kong's press, social media became a major news source for information about the protests. Jonathan Kaiman of the *Guardian* called the protest "the best-documented social movement in history, with even its quieter moments generating a maelstrom of status updates, shares and likes."[26] Out of all the images circulating as representations of the prodemocracy movement, the yellow umbrella became the most recognizable symbol of Hong Kong's protestors. Initially umbrellas were used simply to block pepper spray and tear gas, but over time, the color yellow, once an imperial color, became known as democracy yellow, and then the yellow umbrella, seemingly playful, became a powerful meme for democratic protest.[27]

The Audience: Arguably the protesters' most important audience was China's Communist leadership, and what follows examines how one leader was addressed through acts of remonstration, the values of harmony and family, and the concepts of *shù* and *zhōng*. Given the distance, physical and political, between the Hong Kong protestors and the Beijing president, digital networks were the only means of address. A sent-down or educated youth of the Cultural Revolution, President Xi is cast as a man of the people. For instance, in October 2014, a photograph of the president won the state-sponsored National News Award, the Chinese Pulitzer. In the official press (*Xinhua*) photograph, Xi is pictured at a construction site, conversing with workers, holding his own umbrella, his pants rolled up. Within China, the photo is significant because it documents him as a man of the people.[28] The international significance of the picture arose when the umbrella was digitally edited to prodemocracy yellow, and the people's President with his now yellow umbrella digitally reappeared in a variety of places, from Lion Rock to the cover of *Time* magazine. These largely anonymous engagements with the president spread rapidly across the internet, appropriating the most respectful image of Xi to reimagine him as a fellow seeker of democracy and justice.

These prodemocracy memes of a sympathetic president do not promote revolutionary events. Adversarial, Western rhetorics might read the images as demands for Xi to join the protestors, but this lens arises from a valuing of confrontation over duty and respect. They are better understood as deliberative considerations of an appropriate relationship between protestors and leaders. Furthermore, they can be productively read as aesthetic statements as much as cyberactivism. That is, I would not go so far as to label these memes "liberation technology" that "can expand political, social and economic freedom," although I do agree with Larry Diamond's claim that "technology can empower those who wish to become political and to challenge authoritarian rule."[29] The challenge here imagines harmonious effort. Although not all the images are the same, the cumulative effect of the many memes of President Xi exhibits an exceptional political practice: one that visually forms facts and relationships in ways that are utopic, engaging, and open to interpretation.

The ultimate desire may be structural change, but the politico-aesthetic memes present democratic practices that can be represented only at a particular moment in history. They are performances in the present, suggesting the future. Following Rita Raley, I argue that "we need not, and indeed should not, think of political engagement strictly in terms of concrete action, organizational movements, or overt commentary."[30] Political and rhetorical engagement also involves aesthetics, utopic imaginings, and idealized engagements and recognitions.

In not limiting rhetorical engagement to those who are present, intimate, and willing to engage, the digitally altered images provide a crowd-pleasing, remonstrative moment when the rhetor presents an idea to an audience who is not necessarily engaged with them. In photoshopping the altered images, rhetors acknowledge the value of the president's engagement and judgment without requiring his presence, engagement, or judgment of the protestors' ideas and desires. By joining the symbol of the prodemocracy umbrella with the image of the approachable president, artists and activists perform acts of Confucian remonstration, understanding Xi as key to the prodemocracy movement as they imply his duty to hear the people's demand for just governance. Within the play of yellow umbrellas, Xi is imagined through the artists' own experiences (*shù*) and so imagined as valuing the prodemocracy movement. That is, through digital means, the protesters visualize a shared lifeworld with the president.

The multiplying political memes remonstrate in multiple ways. First, they do not require an end, or even a response, but simply demonstrate an idea. Although the norms of visual remonstration may differ from the norms of discourse, imaginary openness can characterize each, since both are committed to the idea more than to the dialogue. Second, the memes symbolically align Xi's commitment to the common people with a commitment to political reform without making political demand (*zhōng*). Third, they ad-

Figure 2.1. "Excellent persons (*de*) do not dwell alone; they are sure to have neighbors" (*Analects* 4.25). Image from Facebook page "Yán Channel," (October 23, 2014; image retrieved on January 6, 2015).

dress the necessity of China's good will to Hong Kong's democratic movement without acknowledging the political structure of one country, two systems. In bypassing confrontational demand for politico-aesthetic remonstration, the protestors visualize structural change and directly offer a utopian future to the authoritarian state. An analysis of three photographs, through the Confucian theory of audience inherent in the concepts of *shù* and *zhōng*, suggests the contours of this utopian future.[31]

The first meme shows President Xi front and center in a police confrontation on a nighttime shopping street in Mang Kok (figure 2.1). An endless horizon of brightly lit signs sets the scene as capitalistic, modern, and commercial, interestingly competing with the president and his bright yellow umbrella. As appropriate to his position as an elder statesman and father figure, he stands slightly larger than scale in front of a sea of young protestors. He is the front and center figure, facing three helmeted riot police. Bathed in light, he benevolently faces the police holding only his umbrella.

The anonymous artist presents the president as a prodemocracy leader, standing in harmony with his people. His larger scale suggests his fatherhood, leading children into appropriate acts. In acknowledging the need for the presence of the president, the artist apprehends the other as herself, imagining the president as a protestor (*shù*). That is, through a Confucian lens, one might describe the artist as using her own democratic wishes "as a measure in gauging the wishes of the other" (Lau) and representing the president as sharing the protestors' wish for democracy. As noted earlier, *shù* does not fully coopt another/an other because it is modulated by the values of harmony, family, and propriety. In the imaginary, the students and president are in agreement, he leads them, and they are sufficiently behind him as to represent their will to follow.

Figure 2.2. "When the exemplary person puts a name to something, it can be spoken, and when spoken, it can certainly be acted upon" (*Analects* 13.3). *Time* magazine cover, October 13, 2014.

The artist does not do anything as crude as try to persuade President Xi. Rather the photograph demonstrates an idea of, or desire for, a certain kind of president. It does not matter whether Xi responds or conforms to the demonstration. It does not matter whether the artist is recognized by the other, for her *duty* is to be worthy of recognition (4.14). Following from the requirement of duty and relationship implicit in *zhōng*, to be worthy of recognition, the appropriate relationship must be observed with filial propriety: hence, the president leads the young masses, and they follow him. Even as he is transposed digitally, he is acknowledged as the significant and necessary leader.

The second meme is a digital manipulation of the October 2014 *Time* cover wherein a young protestor emerges from a tear gas cloud with raised umbrellas (figure 2.2). The alternative *Time* cover is digitally edited and envisions the president as the protestor. It works much like the first meme in representing the umbrella president as the wished-for leader, but rather than analyze the image of his bright and leading emergence from a dark cloud of tear gas, I want to consider the how the artist, Sum Andy, gave whole heart and mind through innate loyalty (*zhōng*) to the president.

Implicit in the image is a rectifying of names (*zhèngmíng*). In Confucianism, naming is an ethical act, because using the correct name creates meaning, community, and harmony. Creatively, the artist names the president as the person who should be on the cover of *Time*; after all, he is the leader. In placing Xi amid the tear gas, the artist acknowledges that the president is challenged, even under attack from many sides, but Xi still stands firm and faces the entire world. Ironically, the artist who represents him as a brave leader is among his critics, but he does so in a way that fully respects Xi's centrality to the nation-state. In placing Xi, instead of the protestor, on the cover of *Time*, the artist plays with the tension between the duty to honor his authority and the artist's own self-interested commitment to promoting fair and democratic nomination procedures. He could be considered to be employing *shù*, the method of understanding possibilities by putting one's self in the place of others. Through his own values, he imagines that President Xi shares his democratic wishes; he knows Xi through his own wishes. Against the limits of human understanding, its inherent self-interest, he may employ *zhōng* in that he does his best by the national leader through their relationship. In the imaginary, he demonstrates respect for Xi before the entire (Western) world, thus attending to his duty to do his utmost for the president.

All too human, the artist may be having a problem with being honest about his wishes in relationship with those of others. That is, even as he is doing his Confucian utmost (*zhōng*), he seemingly lacks the required, reflective self-understanding to acknowledge self-interest in democracy. Still, Confucianism requires a remonstration with Xi, a representation of the interlocutor's values for his open regard. Inherent in the act of remonstration is an acknowledgment that interlocutors need not share values. It is the speaker's duty to represent a worldview even as he preserves harmony by not forcing the other to engage. In addition to preserving harmony with the president, the artist seemingly considers secondary witnesses in the English-speaking world who would have seen the original cover; the artist also communicates the centrality of China's leadership. In the meme, the president emerges, looking boldly out at the English-speaking world, and thus, the artist portrays him as the Chinese leader—in all matters—to those outside who might criticize him. In doing so, the artist implicitly respects the president's rank, not only out of respect for Xi and the Communist leadership but also before the world, which might manipulate the protest to the detriment of the ethnicity and nation.

The next meme, the work of Fok Chun Wing (figure 2.3), is a (re)citation of Renè Magritte's *Golconda* (1953). The cultural differences between the cascade of well-dressed men in bowler hats and the casually dressed president, however, are significant. By Magritte's own account of *Golconda*, "The men are all dressed alike, as simply as possible, in order to represent the crowd. . . . The man in the bowler hat simply constitutes the middle class in its anonymity."[32] Inherent in this mid-twentieth-century work is a criticism of anonymity and similitude as a middle-class commonplace. From the perspective of *shù* and *zhōng*, the multiplying leader represents something very different from anonymity: President Xi is known as an excellent individual, as excellent as ourselves. He is not the bourgeois or the too common man but rather the president or sage king. The effect of the picture is not homogeneity but a *copia* of homage. More telling for the democracy movement,

Figure 2.3. "Governing with excellence (*de*) can be compared to being the Northern Star: The Northern Star dwells in its place, and the multitude of stars pay it tribute" (*Analects* 2.1). Image from Facebook page of Fok Chun Wing (October 25, 2014; image retrieved on January 6, 2015).

the picture recognizes the value of shared harmony; hence the possibility that the leader exemplifies the common people, not as their leader or representative but as the repeating of their embodied and harmonious citizenship. The flying objects or presidents do not represent the anonymous crowd but suggest the gathering of citizens or protestors engaged in realizing democratic excellence. The shared requirement of *shù* and *zhōng* is to grasp the other from oneself and dutifully treat him as one would be treated. If we interpret this meme through a Confucian lens, then the analogic nature of relationship flows both ways. As President Xi and his yellow umbrella are multiplied, so too are the protestors in Hong Kong. In effect the artistic act of respecting Confucian hierarchy is twisted into a democratic statement about equality and egalitarianism.

In all three images, the president is represented as central and benevolent. In the democratic lifeworld of the artists, the president is integrated into the protestors' society. He is no longer a figure of alienation but coextensive with the sociopolitical world of the student's Hong Kong. Consequently, he is engaged through the ideal and ethical communication of a Confucian state. That is, he is not manipulated verbally but instead presented through the artists' ideas and ideals of themselves. Granting his centrality to the state, the artists understand him through their self-understanding and their wishes.

They use themselves—their values—as an analog to gauge the value of another. This dynamic, however, can be pushed further in that Confucius said, "Authoritative persons establish others in seeking to establish themselves and promote others in seeking to get there themselves" (6.30). As the artists establish themselves as proponents of democracy, as they self-author, they work to cultivate a democratic President, too.

Imaginary Relationships/Impossible Relationships

Confucian rhetorical theory exerts substantial explanatory power in the digital age, in no small part because it refuses the dichotomy of domination and subjection without denying hierarchy. In working to articulate appropriate relationships between interlocutors who are "differently empowered," Confucianism offers a productive instrument for understanding rhetorical situations and spectacles.[33] Rather than representations that reinforce divisions, these three images, read through Confucianism, develop connections that emphasize duty (responsibility) and relationship, and they do so without denying difference. Rather than manipulate pity, rather than lob shame and blame, these images enact new lifeworlds and reconfigured relationships in the service of ongoing political engagement. Conceiving the other through a lens of obligation and similitude with one's self has disadvantages as well as advantages.[34] Even so, it may be the ideal or the only rhetorical approach to addressing the impossible audience. If the ruler is not available physically, ideologically, and emotionally, and the mass media are not open, then how can one proceed in an urgent situation?

Seemingly, digital art is a viable avenue for remonstration, especially as it presents respectful relationships of mutual responsibility. One purpose of remonstration is to maintain and nurture human relationships, and the images—together envisioned as a spectacle—demonstrate that remonstration can be innovative and even radical in creating and mediating social relationships. Unlike discursive arguments, which unfold through time, the presentational nature of images asks for immediate apprehension. More than simply counterfactuals, contrary to fact, these images present facts not yet in existence and yet visible and apprehendable now. As the images of Xi proliferated on the web, a mass audience sees not just an imaged future but the potential of immediate alliances in Hong Kong. Rather than repudiate a relationship with China, the artists conceive of the existence of a radical present, not just a radically different future.

Through the openness of remonstration, time is grasped in *potential human relationship* and not through a linear unfolding of history or a goal-directed election cycle. In Confucian communication, interlocutors attend to an analytic of relationships rather than goals, vocabulary, grammar, tropes, power acquisition, and targeted action. All these images foreground a respectful relationship with a distant authority. That is, their generative work is to form a new relationship. The artists assume responsibility for creating an ethical relationship, as they assert their rights of democratic freedom, including relational rights such as freedom of speech and association. In acknowledging relationships and the duties they entail, the artists represent a fundamental conviction about the nature of the state and political action. Thus, in emphasizing the shared obligations of a citizenry, Confucian communication theory proposes an approach to contemporary, political thinking that avoids the limitations of liberal individualism, agentless structuralism, and postmodern skepticism, because it founds political engagement in dutiful rather than adversarial relationships. As a corrective to contemporary political and rhetorical theory, Confucianism offers a preliberal citizen who both values society more than self and embraces communitarian ideals. Both these correctives acknowledge that recognizing the other is a basic part of constituting one's subjectivity, and in doing so, they promote a beneficial responsibility.

As the three memes demonstrate, Confucian political communication remains a tool for engaging without enraging. In this particular set of images, the action of recognition is more important than the promotion of universal suffrage, but within the acts of recognition, democratic concerns, vigorous voices, and equality among citizens are foregrounded. Although China remains an authoritarian state and President Xi leads it as one, the response to the Hong Kong movement thus far has been moderate.[35] In November, Xi told the media, "The central government will steadfastly implement One Country, Two Systems and the Basic Law, steadfastly support Hong Kong to advocate the development of democracy based on the law and steadfastly maintain Hong Kong's prosperity and stability."[36]

Notes

1. Wendy S. Hesford, *Spectacular Rhetorics: Human Rights Visions, Recognitions, Feminisms* (Durham, NC: Duke University Press, 2011), 17.

2. I am indebted to Damien Smith Pfister for the coinage "oratorcentricism." It captures the limits of the Greco-Roman tradition handily.

3. Stephen C. Angles, *Contemporary Confucian Political Philosophy* (Boston: Polity, 2012), 2.

4. The literature, even in English, is copious. For an introduction, consider Joseph Chan, *Confucian Perfectionism: A Political Philosophy for Modern Times* (Princeton, NJ: Princeton University Press, 2015); Susan Ogden, *Inklings of Chinese Democracy* (Cambridge, MA: Harvard University Press, 2002); Stephen Angles, *Contemporary*

Confucian Political Philosophy (Malden, MA: Polity, 2012); David L. Hall and Roger T. Ames, *The Democracy of the Dead: Dewey, Confucius, and the Hope for Democracy in China* (Chicago, IL: Open Court, 1999); and Wm. Theodore de Bary and Tu Weiming, *Confucianism and Human Rights* (New York: Columbia University Press, 1999).

5. Richard Rorty, "The Historiography of Philosophy: Four Genres," in *Philosophy in History: Essays in the Historiography of Philosophy*, ed. Richard Rorty, Jerome B. Schneewind, Quentin Skinner (New York: Cambridge University Press, 1984), 57.

6. In *Deliberative Acts: Democracy, Rhetoric, and Rights* (University Park, PA: Pennsylvania State University Press, 2013), I discuss the terministic differences between persuade and remonstrate in more detail (38–43); my primary purpose there is the comparative work of understanding alternatives to persuasion.

7. Gilbert Ryle, *Dilemmas* (New York: Cambridge University Press, 1954), 102–9.

8. Remonstration, despite being the speaker's act, escapes oratorcentricism because it entails a relationship. Sons and ministers are not simply to obey fathers and rulers but are duty-bound to remonstrate in response to immorality or crisis (*Analects* 4.18, 19.10). Julia Tao argues that trust starts as particularized trust or benevolence through the family virtue of love, *ren*, but through *shù* (consideration) and *zhōng* (duty) trust becomes a civic and political virtue, supporting the doctrine of a benevolent government protecting the welfare of the people and developing their moral character (*renzheng*) (1.5, 8.2). She notes the importance of being "able to see oneself as part of a shared humanity, be (ing) able to develop a relational view of the relationship between the self and others, and be (ing) able to extend one's caring from those who are naturally related to oneself to those who are unrelated or are strangers through taking action to support the common good"; see her "Trust Within Democracy: A Reconstructed Confucian Perspective," in *Taking Confucian Ethics Seriously: Contemporary Theories and Applications*, ed. Kam-por Yu, Julia Tao, and Philip J. Ivanhoe (Albany: SUNY Press, 2010), 117. Given the importance of familial, social, and political relationship, the good Confucian political ruler is committed to trusting relationships over military might or food (*Analects* 12.7).

9. As David Hall and Roger Ames urge in *Thinking through Confucius*, "the emergent nature of the cosmic and social order . . . requires active participation in realizing the world" (284).

10. Chinese feminist Chen-chieh Huang discusses the private and the public broadly; "East Asian Conceptions of the Public and Private Realms," *Taking Confucian Ethics Seriously: Contemporary Theories and Applications*, ed. Kam-por Yu, Julia Tao, and Philip J. Ivanhoe (Albany: SUNY Press, 2010): 73–98. More tellingly, despite metaphoric connections between the family and the state, at least one chapter of the *Analects* suggests that preserving intimate and family relationships are more important than the rule of law (13.18).

11. Scholars, such as Lau and See Yee Chan, interpret *shù* and *zhōng* together as complementary components of Confucian relationships.

12. I cannot see *shù* as reciprocity or empathy (Tao). Although he works from

neo-Confucian Dai Zhen, I take to heart Justin Tiwald's sense of *shù* as "sympathetic understanding," a codicil that emphasizes its perspectival nature, diminishes simply projecting one's desires on another, and notes the valuing of the other person; see "Is Sympathy Naive?: Dai Zhen on the Use of Shu to Track Well-Being," in *Taking Confucian Ethics Seriously: Contemporary Theories and Applications*, ed. Kam-por Yu, Julia Tao, and Philip J. Ivanhoe (Albany: SUNY Press, 2010): 145–62. Particularly helpful is the weighing of desires. Still, in the end, "understanding" goes too far in imagining a shared world. One might think that ritual propriety (*li*), the family (*xiao*) metaphor, and harmony (*he*) work toward a shared lifeworld and thus understanding, but it is possible that Confucius saw the need for such intense knitting of relationships because understanding is so difficult.

13. David L. Hall and Roger T. Ames, *Thinking Through Confucius* (Albany: SUNY, 1987), 285.

14. Herbert Fingarette, "Following the 'One Thread' of the *Analects*," *Journal of the American Academy of Religion* 47, no. 35 (1979): 382.

15. Paul R Goldin, "The Theme of the Primacy of the Situation in Classical Chinese Philosophy and Rhetoric," *Asia Major* 18 no. 2 (2005): 1–25; Martha Nussbaum, "Golden Rule Arguments: A Missing Thought?" in *The Moral Circle and the Self: Chinese and Western Approaches*, ed. Kim-chong Chong, Sor-hoon Tan, and C. L. Ten (Chicago: Open Court, 2003), 3–16.

16. See my critique of identification in *Deliberative Acts*, 59–65.

17. *Confucius: The Analects*, trans. D. C. Lau (Hong Kong: Chinese University Press, 1992), 33.

18. Paul R. Goldin, "When Zhōng Does Not Mean 'Loyalty,'" *Dao* 7, no. 2 (2008): 169.

19. Lau, *Confucius*, viii–ix.

20. For examples, see Yee Chan, Fingarette, "Following," and Heiner Roetz, *Confucian Ethics of the Axial Age: A Reconstruction Under the Aspect of the Breakthrough Toward Postconventional Thinking* (Albany: SUNY Press, 1993).

21. John Crespi provides an overview of early-twentieth-century pictorials, political satire, and their transnational origins; "Beyond Satire: The Pictorial Imagination of Zhang Guangyu's 1945 *Journey to the West in Cartoons*," in *Oxford Handbook of Modern Chinese Literatures*, ed. Carlos Rojas and Andrea Bachner (New York: Oxford University Press, 2016). Although I do not read the memes provided as satire, I do acknowledge that another could approach the memes in that light, although s/he would miss the Confucian aspects of recognition.

22. Confucius is not consistently hierarchical. After all, the *Analects* is a compiled text and not without internal tensions. For example, despite concern with family hierarchies, "The Master said, 'In instruction, there is no such thing as social classes'" (15.39).

23. Angle, *Contemporary Confucian Political Philosophy*, 4–5.

24. Jinping Xi, *The Governance of China* (Beijing: Foreign Language Press, 2014), 202–3.

25. Extending beyond this essay's scope, Bo Wang analyzes the Hong Kong protests and their reception in the West in "Plural Local Terms, Dialectical Pro-

cesses, and Co-participants: Re-envisioning Transnational Rhetorical Studies," *JAC* 34 (forthcoming). The documentary *Lessons in Dissent*, directed by Matthew Torne (2014, Hong Kong, EDKO Films, 2014), DVD, provides context for the events leading up to the protests.

26. Jonathan Kaiman, "Hong Kong Protests Bring Crisis of Confidence for Traditional Media," *Guardian*, October 29, 2014, http://www.theguardian.com/world/2014/oct/29/hong- kong-protests-confidence-media.

27. Jennifer Eagleton, "Umbrellas and Yellow Ribbons: The Language of the 2014 Hong Kong Protests," *OUPBlog*, November 4, 2014, http://blog.oup.com/2014/11/language-hong-kong-protests/.

28. Matt Schiavenza, "Why This Seemingly Innocuous Photo of Xi Jinping Is So Important," *Atlantic*, July 7, 2013, http://www.theatlantic.com/china/archive/2013/07/why-this-seemingly-innocuous-photo-of-xi-jinping-is-so-important/278018/.

29. Larry Diamond, "Liberation Technology," *Journal of Democracy* 21, no. 3 (July 2010): 70–1.

30. Rita Raley, *Tactical Media* (Minneapolis: University of Minnesota Press, 2009), 151.

31. Ellie Ng was helpful in locating artists and providing an overview of memes on President Xi. Please see her article "How Chinese President Xi Jing Ping and his Yellow Umbrella Became a Hong Kong Umbrella Revolution Meme." https://globalvoices.org/2014/10/24/how-chinese-president-xi-jinping-and-his-yellow-umbrella-became-a-hong-kong-umbrella-revolution-meme/.

32. Quoted in Siegfried Gohr, *Magritte: Attempting the Impossible* (New York: D.A.P., 2009), 275.

33. Hesford, *Spectacular Rhetorics*, 17.

34. Beyond the scope of this essay are the millennia of critiques of Confucian hierarchy, conservatism, misogyny, idealism, and its obsession with inner nobility, family, and self-cultivation. For the moment, I also sidestep the critiques of the Golden Rule.

35. Xi is committed to a strong Communist state and not to Western civil society. Although he has mitigated the gulags and population control policies, he has also been clear in his opposition to Western-style individual rights and a free press.

36. Eddie Luk and Hilary Wong, "Occupy Central 'is Challenging Beijing,'" *Standard*, November 10, 2014.

3

Big Data and Global Knowledge

A Protagorean Analysis of the United Nations' Global Pulse

E. Johanna Hartelius

The secretary-general of the United Nations (UN) launched in 2009 an initiative called Global Pulse. This program, whose mission is to "accelerate discovery, development and scaled adoption of big data innovation for sustainable development and humanitarian action,"[1] represents the UN's effort to use digital networks for public good. It furnishes UN interventions with "big data," the technoculture term for large aggregates of data and metadata that, because of their complexity and size, require computational management and analysis. Contextualizing its agenda, Global Pulse staff explains,

> A global data revolution is already underway. People—in both developing economies and industrialized ones—are generating a growing ocean of digital data every minute of the day just by going about their daily lives. As people use mobile devices to communicate, buy and sell goods, transfer money, search for information on the Internet, and share their lives publicly on social networks, they leave digital trails that private sector firms are mining to understand the needs of customers, track emerging market trends, and monitor their own operations in real-time. Once anonymized to protect privacy, the same data, tools and methodologies can also reveal insights on changes in human well-being, real-time trends on population behavior or perceptions related to sustainable development issues. Such insights can guide the public sector to respond more effectively to emerging crises and vulnerabilities. Big data represents a new, renewable natural resource with the potential to revolutionize sustainable development and humanitarian practice.[2]

Global Pulse is a massive epistemic undertaking, discerning digital patterns in order to set policy. Its harnessing of big data is intended to measure the "pulse" of the world.

Big data is data about data. It has emerged as a result of our technological ability and cultural need to synthesize and interpret the tiny but copious digital marks generated by individuals, corporations, and institutions. Commonly characterized by the three *V*'s—volume, variety, and velocity—big data is amassed at volumes that are "difficult to process with traditional database and software techniques."[3] Some concrete examples are helpful here: Imagine all transactions ever completed on Amazon.com by its 152 million consumers; imagine all that Amazon "knows" about those consumer profiles. Imagine all the people who use the GPS navigational function on their mobile telephones to get to a destination, correlated with all the other things that they ever do with, and download on, those phones. Imagine all the data points about all the people who have a Facebook page. Now imagine big data in the public sector: all the data collected in hospitals about patients, cross-referenced with the progression of their medical conditions. Think of the possibilities for managing or curing diseases if millions of individual cases and variations were integrated. Or consider all the data collected by election campaigns about registered voters, correlated with their votes.

Of primary concern regarding big data are, in tandem, expertise and motive. Many organizations rely on contracted or in-house data scientists to organize and present raw data as intelligible information. Sound data analysis in the private sector can lead to a corporation's competitive edge in a global market and enormous profit. For nonprofit and nongovernmental organizations, it can lead to coordinated and targeted interventions (e.g., Hewlett Packard's *Earth Insights* has collected data from more than one thousand cameras and climate sensors in South America and made them internationally available to conservation scientists and governments). The technology of big data means that a lot can be known all at once; whoever has access to the measurement wields tremendous influence. Information becomes the foundation for deliberation and judgment, both of which can serve good or evil. Thus big data magnifies the scope of the adage about great power and responsibility. Useful expertise in the context of big data determines success—the definition of success, of course, varies.

The prominent Sophist Protagoras, whose work endures only in fragments and Platonic dialogues but whose ideas have generated conversations about human knowing and being from the fifth century to present day, claimed: "Of all things the measure is man: of the things which are, that they are, and of the things which are not, that they are not."[4] Protagoras's *homo mensura* (human as measurer) thesis is an epistemic and political orientation, a perspective on what exists, what can be known, and what must be done. This chapter turns a Protagorean lens onto new methods for producing and evaluating knowledge with digital measurements. Specifically, I engage three of Protagoras's central themes: *homo mensura*, *dissoi logoi*, and *orthoepeia*. I argue that Global Pulse's public discourse oscillates strategically between a human measure imaginary and a technoscientific agenda. The former consists of references to embodiment and spirit; it evokes traditional diagnostics and measurement as the means to human welfare. The latter presents a geopolitical exigence, revved by big data's promise to make information-as-knowledge integrated, accessible, instant, and useful.

A Protagorean perspective allows me not only to illuminate several of Global Pulse's (hereafter GP) measurement practices but also to question the program's epistemological assumptions and political impact. Working with Protagoras's insights regarding knowledge and civic conduct, I critique GP's elision of difference between data and information, and between information and judgment. In GP's discourse, the act of measurement and the state of comprehension are conflated; to measure (digital activity) is to understand (the human condition and the world). Global Pulse obfuscates its own constructedness as a human measure and belies the consequences and ideological structures of its interventions. The totalizing focus in GP discourse is on sophisticated measurement and the technologies whereby such measurement is afforded. The program is engrossed in its own technological capabilities; its texts seem to say with rather naive exuberance, "Look what our cool machine can do!" Protagoras's perspective on measurement and understanding is strikingly relevant: Humans are compelled to measure, to pursue knowledge. Acting on that compulsion, we not so much discover an external reality as we invent a structure, or a technology, that reflects our own positionality and agenda. That is, we find what we were looking for; not because it was "out there" but because *that* is what we knew to seek.

The quantity and complexity of scholarship on Protagoras and his philosophical legacy mandate that I comment on my chapter's constraints. My intent is not to adjudicate alternative interpretations of Protagoras's philosophies or to assess his influence on classical rhetorical theory. Nor is it to defend the Sophists in general or Protagoras in particular from Plato's scorn. As Poulakos observed, "Hegel, Grote, Nietzsche and others in the nineteenth century did that."[5] Further, Protagoras has been called a phenomenologist, an empiricist, and a relativist, as well as the first humanist, the first political philosopher, and the father of debate. My project here is not to determine whether such ascriptions are merited. And although I am guilty of the kind of hermeneutic appropriation of Protagoras's ideas for the purpose of rhetorical criticism that Schiappa discourages, I take seriously his caution against the modern compartmentalizing of these ideas. As he writes, "Protagoras created his doctrines as parts of a complete and whole way of understanding his world."[6] Insofar as I offer "an integrated account of Protagoras' doctrines," drawing on his own fragments as well as Plato's *Protagoras* and *Theaetetus*, my method resembles that of Zilioli, who concedes the potential perils of relying on Plato to discover Protagoras's ideas but proceeds to do so with the hope that, "on the basis of what Plato himself says, one might identify philosophical consequences, alternative to those drawn by Plato, and less straightforwardly untenable."[7]

Homo Mensura: Epistemology, Ontology, Judgment

The *anthrōpos metron* thesis, more commonly known by its Latin moniker, the *homo mensura* thesis, survives as a fragment and as a central talking point in Plato's *Theaetetus*; this dialogue pursues the question, What is knowledge? During the first of the dialogue's four stages, Socrates's exchange with Theaetetus begins with sense perception: a wind that blows can feel warm to one person and cold to another;[8] what one person would characterize as white, another would describe differently. While this point is fairly obvious to modern readers, it conflicts with Parmenidean bivalent philosophy—predominant at the time and profoundly influential for Plato—which denies that an object can at the same time be both *x* and not-*x* (cold and not-cold). The conversation proceeds from sense perception to thought perception, or perception of cognitive experience.[9] Human experience as a measure of the world is internal as well as external.

When setting up the *homo mensura* thesis as a tool for criticism, a subjectivist stance must be rejected in favor of a relativist stance, as Burnyeat argues. With a relativist orientation, "every judgment is true *for* (in relation to) the person whose judgment it is."[10] It is not that whatever is true for me is true absolutely—Protagoras would reject such a "fundamental distinction between truth and falsehood."[11] Moreover, argues Zilioli, Protagoras's thesis must be considered "an epistemological position with original ontological consequences."[12] He writes, "The key feature of the Measure Doctrine, interpreted as a theory of perceptual truth, is, I believe, ontological relativism, for which the ontological status of the perceived object is relative to the perceiver."[13] Put simply, objects in the world become ontologically determinate when they are perceived and experienced by a human. When they are, the objects are named; in the names inhere judgment (this warm cup of tea is nice; this cold rain is unpleasant). The world exists as a function of being measured and evaluatively "worded." Indeed, as I argue with my case study, digital data about humans in the world exists only when, and because, it is collected *as* data. This is not to say that whatever GP collects is absolute in what it captures but instead that the data is true in relation to the UN's metrics.

Protagoras's relativism, which sustains the mutually implicative ideas of *dissoi logoi* and *homo mensura*, extends to his democratic commitments. Most biographical accounts of Protagoras note with enthusiasm that he was commissioned by Pericles to draft laws for the Panhellenic community of Thurii, a city in southern Italy.[14] As a public figure and educator, his political ethics and pedagogy cohere, grounded in an epistemology with no "external, superhuman authority."[15] In the *polis*, knowledge is relative to human perceptions of what is advantageous to the civic community. What may be understood as just or prudent varies culturally.[16] As Macé explains, "Protagoras' physics in the *Theaetetus* is haunting Protagoras' politics in the *Protagoras*, suggesting that behind the democratic ideals and the claim that anyone can say what he thinks about justice and that each voice is as good as the other, one can hear the *homo mensura*-doctrine."[17] Protagoras demonstrates with his career and in his teaching the implications of the human measure thesis for political judgment. He identifies measurement as a human achievement, a sign of an advanced society, to an audience for whom measurement and its devices were divinely inspired.[18]

The *homo mensura* thesis structures GP in two prevalent ways. First, the unit of measurement is human activity. The basic premise of the initiative is that the "pulse of communities" can be taken by measuring humans' production of digital content and exhaust.[19] This assumption of human activity as a measurement of the world emerges in notably different forms in GP's discourse. In reference to measuring the habits of "industrialized" peoples, GP discusses mobile phone and internet usage, especially as they pertain to consumption, including the consumption of information.[20] By contrast, in references to measuring the activities of peoples in "developing" regions, GP describes tracking movements and behaviors during disasters.[21] For example, a hypothetical household in a developing country illustrates the crisis of "soaring commodity prices, particularly food and fuel."[22] Global Pulse's audience is invited to consider the family's options in coping with sudden hardships (the mother could sell drapes in town twice a week and the father could work longer hours). Describing the catastrophic consequences for the hypothetical family, GP lists a series of data sources that would indicate the crisis: "Health hotlines might see increased volumes of calls reporting symptoms consistent with the health impacts of malnutrition or unsafe water sources."[23] In both the industrialized and the developing case, the small (human) is a measurement geared toward knowledge processed by the big (data, global initiatives, international policy). GP humanizes data collection and global epistemology by making humans the measures of prosperity and calamity.[24]

In this epistemic process, the difference between gathering data and understanding humans dissipates. As the actions of the ordinary person and the habits of everyday life become measures of global events and circumstances, tracking digital activity results in "richer, deeper insights into human experience."[25] The purpose of GP, then, is "analyzing contextual and digital data to gain a real-time understanding of changes in human well-being."[26] Social media content, money transfers, and database searches add up to knowledge not only of producers and agents but also of their worlds. The most cogent engagement with, or recognition of, any distance between people's digital habits and a general experience of life in GP's discourse is the discussion of "measurement challenges."[27] So-called sentiment analysis, "the computational study of opinions, sentiments and emotions expressed in text" is impeded by the challenge of ascertaining rhetorical intent. The recurring solution for GP is technical: increasingly sophisticated algorithms and better experts.[28]

Second, the *homo mensura* thesis structures the GP initiative in the animation of the program itself. GP constructs itself as an organism, a "body" of evidence and research.[29] Through the body, data pulses. The body's organs are the so-called Pulse Labs, dispersed across the globe to develop various projects. Each lab operates somewhat independently, responsive to the cultural particularities of its location.[30] Each moves through a "lifecycle" with every project, that is, executing its purpose for the overall organism.[31] As organs of a body, the Pulse Lab in Jakarta, Indonesia, is separated in space from the Pulse Lab in Kampala, Uganda. They operate differently according to the needs and resources in their respective place; but they are interconnected and coordinated via "*head*quarters" in New York.[32] Their coordinated efforts serve the same purpose, the organism's viability. This animation of GP as itself a human measure appears especially prevalent in discussions of measuring climate conditions, like rainfall,[33] or activities wherein human actors are removed, like timber trade.[34] Absorbing input about these processes, digesting information, the GP body is alive with data. It is a human measure of natural phenomena, comparable to the perceived temperature of wind[35] and the flow of rivers.[36]

As a body, GP possesses sense perception, specifically sight. It "visualizes" data as an epistemic practice of seeing configurations of data points and the significance thereof. Such visualizations include standard graphs and charts[37] but also the popular "word cloud," which displays the most frequently recurring words in a given text by size, font, and contrast.[38] A word cloud represents knowledge of a text by *seeing* its individual part, the words, as data points. Further, GP's sight, evident in its data visualization, has a remarkable aesthetic sophistication. The data visualization of "global legal timber trade," created by a graphic designer, is quite beautiful. GP's vision-as-knowing presents data with an artistic sensitivity that refuses the conventional dullness of technoscientific discourses. As a body that knows the world, GP also knows how to make patterns and proportions pleasing to the eye.

Dissoi Logoi: Binary Code and Global Flux

According to Diogenes Laertius, Protagoras proposed that on every issue (*pragma*), there are two arguments (*logoi*) opposed to each other.[39] There are two sides to every matter; two perspectives compete on any given topic. According to Poulakos's account of the "diplological"[40] thought of *dissoi logoi*, "Protagoras' insight suggests a twofold symbolic world consisting of contrary discourses, and counsels against attempts to fuse them into one unified entity. . . . Everything that gets said can only be understood in terms of something other, something different, and, more specifically, something opposite."[41] *Dissoi logoi* are distinguishable from other conceptual and practical binaries such as dialectic and dialogue insofar as these culminate in the resolution of difference. *Dissoi logoi* resists the "powerful tendency to resolve the tension of contraries."[42] In the theory of anti-*logos*, there is always "another side to the same story waiting to be told."[43] *Dissoi logoi* implies a *logos* that may be infinitely multiple, fragmented, and dispersed.[44] In this section, I discuss *dissoi logoi* in reference first to binary code, second to the notion of flux, and finally to GP's ascription of motives to themselves and other users of big data.

A Protagorean approach to *logos*, which requires attunement to *dissoi logoi*, affords a perspective on big data's binarity. On an elemental level, digital data are constituted by a binary code, 1s and 0s; this code may be thought of as the "is" and "is not" of digital configuration. Digital data encodes *dissoi logoi* into its elemental structure; any digital discourse operates as a mezzo-level signifier, available for measurement and analysis simultaneously as big data and at its most granular base. The "*dissoi*" characterizing the 1s and 0s *logos* constitutes computational code at the highest and lowest levels. For GP, big data is data about data, or a (dissoi) *logos* of *logos*.[45] GP explains that big data is data about both online content ("What people say") and "data exhaust" ("What people do").[46] For GP, online content includes "international and local online news sources, publicly accessible blogs, forum posts, comments and public social media content, online advertising, e-commerce sites and websites created by local retailers that list prices and inventory."[47] Data exhaust, or what people do "as they go about their daily lives,"[48] consists of a "trail of clicks":[49] "passively collected transactional data from the use of digital services such as financial services (including purchases, money transfers, savings and loan repayments), communications services (such as anonymised records of mobile phone usage patterns) or information services (such as anonymised records of search queries)."[50] GP's big data consists of two *logoi*: what people say digitally and what people do digitally. This *dissoi logoi* of saying and doing must be understood in relation to one of the master tropes of big data: that the digital evidence of how people behave reflects reality more objectively than the contents of what they report. As GP notes, people's "perceptions can be inaccurate and thus misleading."[51] Following the "trails" that people leave, however, leads to such valid measurements as can be used reliably to make policy decisions.

Examining the *logos* of GP, it is helpful to trace how Protagoras extended *dissoi logoi* beyond language theory to phenomena in the natural and social worlds. In this move, scholars have noted, Protagoras's Heraclitean influences become evident. Hoffman suggests that Heraclitus understood *logos* as the structural relationship among things in the world *and* among the symbols used to make sense of them.[52] Hoffman writes, "If logos did signify 'ordered composition,' Heraclitus simply took a word that signified the ordered arrangement of language, and expanded its significance to include the ordered arrangement of the world."[53] Schiappa describes this sense of *logos* as the attributes of an object or external situation, which, according to Heraclitus's theory of flux, constantly change.[54] A powerful force dominates a weaker force only to be reversed as circumstances change.[55] To wit, the data of the world—the "given" (from the Latin *dare*, to give)—as well as the data in a data aggregate change in relation to each other. Floods and droughts change in ebb and flow. People living with such changes register their experiences on social media. One cannot step into the same data stream twice. *Dissoi logoi* is a fluctuating discourse representing the flux of the world.[56]

GP's accounts of its programs and their contexts are replete with discussion of fluctuation, volatility,[57] trends,[58] and change[59] in epidemics, crops, news dissemination, weather, and migration. References abound to global interconnectedness and the ways in which change propagates within a system: "Fluctuations in economic conditions—harvests, process, employment, capital flows, etc.—are certainly not new, but it seems that our global system may have become more prone to large and swift swings in the past few years."[60] Constructing fluctuation as the condition for decision-making, GP notes, "Policymakers today are confronting a volatile, hyperconnected landscape in which economic crises, social change and political instability can emerge without warning, affecting vulnerable populations and reverberating around the globe."[61] The structural relations among the component parts of a global economy, refracted through Protagoras's Heraclitean *logos*, carry effects of change from one part of the system to another. As these changes occur, "a massive amount of data is regularly being generated and *flowing*" into GP's analysis.[62]

An illustration of GP's use of flux as a condition of its big data *logos* is the notion of "real time." With an emphasis on the urgent present and immanent future, the concept implies that time passed is un-real, no longer present or accessible for intervention. Real time is defined by GP as "information which is produced and made available in a relatively short and relevant period of time, and information which is made available within a timeframe that allows action to be taken in response."[63] That GP highlights its capacity for real-time intervention is hardly surprising given that it distinguishes Big Data for Development (or BD4D) from traditional development data (surveys and census materials), which is found inevitably lagging.[64] With real-time data, direct relief is possible, for example, in a food shortage situation, preventing a potential danger from turning into a severe famine. Of interest for my purposes is how the notion of real time assumes a constant flux of time that turns real time into no-longer-real time. Urgency is presented as a rationale for GP's program, and speed one of its benefits.[65] With big data, GP can manage global flux. Important to note is that, if *dissoi logoi* is a fluctuating discourse signifying the flux of the world, it makes a difference that the discourse of big data fluctuates differently than the discourses of Protagoras's day. The natural and social world may or may not be in flux differently now than then—it is likely that some things have changed—but what GP emphasizes is the change in the measuring and/as symbolizing instrument. The technological *dissoi logoi* of GP (big data) capture flux more adeptly than do lower-tech discourses.

The *dissoi logoi* of big data are discernible in GP not only at the level of programming (i.e., the 1s and 0s of information exchange code) but also in the story that GP tells about motives and ethics. In accounts of its own inception, chronicling the emergence of big-data science, GP describes how the private sector has been profiting from big-data analysis for years; now it is "time for the development community and policymakers around the world to recognize and seize this historical opportunity to address twenty-first century challenges, including the effects of global volatility, climate change, and demographic shifts, with twenty-first century tools."[66] In short, what has long been used for profit ought now to be used for "the common good."[67] To affirm this distinction, GP uses the phrase "Big Data for Development,"[68] whose objective is to "help policymakers and development practitioners gain richer and timelier insights on the experiences of vulnerable communities and implement better-informed and more agile interventions."[69] In private and corporate usage, big data is an "economic asset, like currency or gold."[70] By contrast, data in GP's narrative of its own epistemology is "responsible."[71] Mobilizing big data for "development" is a given good as a function of its contrast with the dubious motives of the private sector. Any data point of interest may be the same for a consulting firm as for GP—a Facebook update about a noncommunicable disease like asthma or diabetes or a Google search for local job listings—but GP presents opposing rationales (*dissoi logoi*) for the data science of private agents and of BG4D. GP brings "development" to targeted areas with digital precision, using digital precision for the greater good.

Orthoepeia: Precision and Coherence

In what may seem like a departure from relativism, Protagoras valued "rightness" or "correctness" of language, *orthoepeia*. He is credited by Diogenes Laertius as the first to distinguish verb tenses and parts of speech;[72] Aristotle notes that Protagoras organized nouns by gender as masculine, feminine, and neuter;[73] and Plato describes in the *Cratylus* that Callias learned the correctness of names from Protagoras.[74] *Orthoepeia*, or *orthos logos*, is a standard of accuracy and correctness of "diction."[75] The original denotation of *orthos*—"right," as in a ninety-degree angle, the slightest diversion from which would be wrong, or *not* right—translates readily into English, bridging appropriateness and measured precision. Inseparable from Protagoras's relativism, it is a contingent notion of rightness, oriented in relation to particularities; *orthoepeia* is a human measure of what is right contextually. Inseparable also from *dissoi logoi*, it discerns rightness against dual tensions.

Contemporary scholarship on *orthoepeia*, however scant, authorizes approaching the concept in terms of precision and coherence, rather than simple rightness. This orientation to the concept reveals why Protagoras's emphasis on rightness of language may be understood as perceptiveness to cultural particularity, rather than as a break with his relativism. Protagoras taught precision and coherence as standards of rhetorical and civic practice. For him, being right, or being precise, is not an objective status like a ninety-degree orientation but a relationship of precisely right sensibility in a particular context. His students analyzed poetry and literature in order to hone the ability to identify failures of internal consistency.[76] Analyzing Homeric poetry and assessing whether the poet's expression cohered structurally and aesthetically, they learned to "detect inconsistencies in the seemingly self-evident views of an opponent."[77] In the Protagorean pedagogy, a student who could successfully scrutinize poetry and find flaws and irregularities would be able to do the same with political arguments and actions. This directedness toward argument, especially argument in the context of politics, indicates how closely Protagoras theorized rightness with dualism and with the unpredictable currents (flux) of political life. *Orthoepeia* is instructively thought of not simply as tyrannical grammar but as coherence in reasoning directed toward civic goals.[78]

In data analysis, precision, that is, attention to structural coherence, and detection of internal anomalies are imperative. Demonstrating these capacities, most of GP's discourse details the program's epistemic precision. According to GP, "precise" describes its big data, information-gathering procedures, coordinated efforts, targeted strategies, and the real-time intervention discussed above. GP's technologies and practices are sophisticated, advanced, and ideally suited to certain ends; they enable "enhanced early warning," "real-time awareness," and "rapid impact evaluation."[79] GP is precise, using data just so, training experts in exact methods, such as tracking the coordinates of people's migration during a natural catastrophe.[80] GP explains, "Big Data analytics refers to tools and methodologies that aim to transform massive quantities of raw data into 'data about the data'—for analytic purposes. They typically rely on powerful algorithms that are able to detect *patterns*, trends, and *correlations* over various time horizons in the data, but also on advanced visualization techniques as 'sense making tools.' Once trained (which involves having training data), algorithms can help make predictions that can be used to detect *anomalies* in the form of large *deviations* from the expected trends of relations in the data."[81] The tools of GP enable *orthos logos*, data that is precisely right. GP pursues *orthoepeia* in order to provide a "precise tool of measurement for the sake of more productive control and more effective display."[82] The objective is to "get the picture right."[83]

GP's discussions of data management return continually to the detection of anomalies. The program tracks anomalies in how communities access vital services in crises[84] and how populations use digital devices.[85] The gauging of "socioeconomic anomalies" is an "overarching challenge" in the study of "human ecosystems."[86] If assessing an opponent's argument according to the Protagorean standard of *orthoepeia* means identifying inconsistency, assessing a data stream with the same method might mean identifying glitches, indications that something is amiss. The stream metaphor is instructive again insofar as anomalies may be either radically foreign (a twig swept by the current) or internal to the substance of the stream itself (a sudden whirlpool of bubbles) but nevertheless a deviant phenomenon. The analysis in either case focuses on "the various elements of a stretch of discourse itself."[87] Just as the strength of an argument for Protagoras's students could be determined by its internal consistency, the value of epistemic data, GP suggests, may be assessed less in reference to an external absolute than to its own internal coherence. Subsequent to this assessment of *orthoepeia* for both is the move toward prudent action.

Conclusion: *Euboulia*, Data, and Understanding

Within a relativist epistemology, expertise is a dicey proposition, particularly as a foundation for teaching. The challenge Protagoras faces from Plato is this: If knowledge is a function of perspective, and everyone's truth is contingent, then what can a teacher teach a student? Can there ever be such a thing as wisdom? Just as dicey are the methods by which one arrives at a value judgment. Are teaching truth and teaching goodness implicative by necessity? What good is Protagoras the tutor in the world of his own theorizing? Protagoras's response, mediated by Socrates in the *Theaetetus*, connects the *homo mensura* principle to civic action. As students are instruments of measurement, gauging the world of physical and cultural experience, so is the role of the teacher to sharpen the instrument. Protagoras understands the teacher as "one who effects improvement in our perceptions, feelings, or thoughts."[88] His pedagogy shifts students' dispositions; it moves their souls from a "pernicious" to a "sound state,"[89] and their tastes from bitter to "the opposite."[90] To teach, in other words, is to use *logos* to facilitate students' capacities as measurers.

Protagoras's pedagogy is designed not to make his students' knowledge truer, as Socrates' dialectic purports to do, but to make their active judgments *better*. Early in Plato's *Protagoras*, Protagoras describes such notions as *euboulia*, the ability to conduct one's private affairs and public charges with prudence. *Euboulia* is the fruit of Protagoras's students' labor; it is what they are capable of when they, as human instruments of measurement, have been fine-tuned. It entails the ability to consider *dissoi logoi*, assessing the duality of situations and of reasoning. This ability, in turn, is buttressed by a sense of *orthoepeia*; students who have a cultivated skill with *dissoi logoi* recognize that a rejection of the monistic does not mean disregard for coherence. On the contrary, *dissoi logoi* requires attunement to coherence in such a way that, relying on the ideal of *orthoepeia*, an opponent's failure to reason coherently may be critiqued.[91] *Euboulia* is thus the culmination of Protagoras's pedagogy and the synthesis of his politics and philosophy.

To review, my analysis of GP, which ties three Protagorean themes to the development of *euboulia*, and connects *euboulia* to the active judgments made at the intersection of science and politics, suggests that GP oscillates between a human-measure imaginary and a technoscientific conviction. The imaginary, a rhetoric of humanness and embodiment, moderates the conviction, the familiar teleological belief that science will eventually achieve total knowledge, that total knowledge is a sign of human supremacy and achievement, and that technologies are invented to facilitate science. The imaginary, GP's humanizing discourses, ostensibly tempers the allegiance to technoscience; constructing humans as the units of measurement, and the measuring mechanism itself as animate, softens the entire initiative's structural reliance on the multiple technologies of advanced science. These technologies, the opposite pole of the oscillation, "anonymize" data by removing their traceability to human sources;[92] the protection of individual privacy that converts human measurements into dehumanized information is constructed as technology's service to humanity.

Two things are particularly significant about this oscillation: first, it facilitates the elision of distance or difference between data, information, and judgment. Measurement itself—and action and an instrument—is constructed as knowledge. Indeed, one of the most salient patterns in the GP discourse is the conflation of data and understanding. That "researchers are now able to understand human behavior at the population level rather than the individual level" is attributed to technoscientific sophistication, specifically the development of "algorithms driving computer applications."[93] Even with recurring commentary on the intricate challenges of data analysis, GP ultimately leaves the impression that to have data is to understand. Big data "allow decision makers to track development progress, improve social protection, and understand where existing policies and programmes require adjustment."[94] Data about such phenomena as rainfall and drought, epidemics, migrations, and internet access amount to understanding humans. To measure these things is to understand the human experience. Data points are easily translated in GP's discourse into information; information is translated into useful knowledge; knowledge into readiness to judge and act.[95]

That this conflation requires a radical reduction of epistemic processes is difficult to detect in GP's language, partly because the conflation (or transition from one to the other) appears consistent with scientific procedure, requiring no explicit commentary; partly it passes unremarked because translating data into knowledge and action is what the public implicitly asks of GP. Configuring data into information about a phenomenon and then contextualizing that information to generate knowledge is GP's mandate; here, the technoscientific end of the aforementioned oscillation again becomes evident. GP's big data technology produces knowledge of the world that is not only superior to the data gathered by smaller and/or slower methods but also more reliable than the data gathered by attending to human descriptions—what people do is "truer" than what they say. Data trails can be followed with precision, whereas "sentiment analysis" is necessarily muddled by human ambiguity. What must be explicated in GP's discourse is the astounding ease and speed, not just technically but rhetorically, with which data becomes understanding. The distance between the act of measurement and the state of comprehension is traversed in a way that mimics rhetorically the technological facility of global science. No epistemic or rhetorical gap is detectable in GP's texts between data and understanding. To measure is to know.

Second, the oscillation between the human and the technological is significant because it produces GP's entry into geopolitical exigencies. *Euboulia*, prudent and concrete judgment, encompasses both the assessments made by Pulse Lab researchers in the generation of data and the extension of those assessments into the sphere of political decisions. GP emphasizes its advisory function; it is a "catalyst" for those who make decisions "in order to better protect populations from hunger, poverty, and disease."[96] It "builds public understanding" in order to, for example, "manage climate risks."[97] It "guid[es] the development of regulatory frameworks" in order to "solve problems."[98] It ensures that "access to better information sooner makes it possible to keep international development on track . . . and strengthen resilience to global shocks."[99] At once enacting and facilitating *euboulia*, GP enables strategic judgment, assisting those who make decisions regarding families' private affairs and communities' public business.

Yet while GP goes to great lengths to describe its powerful technologies for gathering data and the profound difference that its methods will make for strategic interventions, neither the technologies nor the interventions are adequately contextualized. Beyond the suggestion that for-profit organizations have long used big data for greedy ends, and now the time has come from nonprofit organizations to use big data for good, GP does not situate its project in relation to big data in the private or public sector. Nor does it provide an informed discussion of the reality that the public/private distinction has over the past decade become blurred precisely by transnational technocultural movements. And while the idea of allowing big data to serve the greater good on a global scale is compelling—many organizations already do laudable work in this vein—GP offers virtually no commentary on how the "common good" will be determined, or by whom. There is almost no discussion of how deliberations within and beyond the United Nations will assess the prudence not only of particular interventions but also of the measurements themselves. Thus, a striking discrepancy emerges between the sophistication of GP's big-data technology and the naïveté of its remarks on serving the "greater good," particularly insofar as they do not acknowledge ideological assumptions or implications. Among the giant agents of big data (Hadoop, IBM, Cloudera, Hortonworks) GP seems comparatively unreflective about the political significance of integrated measurements and knowledge on an unprecedented scale.

In the context of GP's measurements and strategies for corrective action, it is instructive to think capaciously about the Greek "metron," implied in Protagoras's *homo mensura* thesis and notoriously difficult to translate. Especially intriguing is the idea that the thesis bespeaks human tendencies and predispositions. Humans are measurers. As Payne notes, the thesis "implies a definition of human existence as an activity of 'measuring' the 'things' of reality."[100] Thus, among the many activities that occupy humans, measuring things is a defining characteristic. Payne suggests further that knowledge by way of measuring and evaluating is ineluctably directed toward perfecting. Humans measure things because that is what we are wont to do; we judge things because we measure them; and we endeavor to improve, indeed perfect, the things that we judge. Our measurements become part the world—join the world—as we experience it. We produce machines that measure things in the world, and the machines produce measurements. According to Payne, "human *techne*, or the artistic impulse, is driven by the perfected image of the world, and reality is transformed by the things people create and the actions they take."[101] Protagoras's perspective on being and knowing reveals that human *technē* is employed not only for intervention but also for perfection. Thus, as I agree with Van Ophuijsen that "we should not be too squeamish, inhibited, or apologetic about resorting to a form of *eikos* argumentation in our own reconstruction or reinvention of a philosophically interesting Protagoras."[102] And I submit that we ought to exercise caution when tempted to merge the impulse to perfect with a Sophistic "suggestion of the possible" for the world.[103]

Notes

1. "About," *United Nations Global Pulse*, accessed February 27, 2015, http://www.unglobalpulse.org/about-new.

2. Ibid.

3. "Big Data for Development: Challenges and Opportunities," *United Nations Global Pulse*, May 2012, http://www.unglobalpulse.org/sites/default/files/BigDataforDevelopment-UNGlobalPulseJune2012.pdf, 13. In a discussion of the capacity of big-data analytics, GP references "data about the data," 17.

4. Plato, *Theaetetus*, 152a. The gendered translation is used in this chapter interchangeably with references to the *human* measure.

5. John Poulakos, *Sophistical Rhetoric in Classical Greece* (Columbia: University of South Carolina Press, 1995), 2.

6. Edward Schiappa, *Protagoras and Logos*, 2nd ed. (Columbia: University of South Carolina Press, 2003), 23.

7. Ugo Zilioli, *Protagoras and the Challenge of Relativism: Plato's Subtlest Enemy* (Hampshire, England: Ashgate, 2007), 13, 7. Kastely critiques the traditional assumption that Plato misrepresents the Sophists in his dialogues as a reflection of his antipathy. He characterizes the *Protagoras* as "not an attack on the sophists but an entangled and comic exploration of sophistic and Socratic difference"; in "Plato's

Protagoras: Revisionary History as Sophisticated Comedy," *Rhetoric Review* 15, no. 1 (1996): 31. Similarly, Poulakos cites W. K. C. Guthrie's remarks on the exaggeration of Plato's animosity toward Protagoras: "I do not understand how anyone can read the brilliant and sympathetic speech of Protagoras in the *Protagoras* from 323c to 328c and still hold that Plato in his representations of the best of the Sophists was setting out to blacken their memory," *Sophistical Rhetoric*, 107, endnote 8.

8. Plato, *Theaetetus*, 152a.

9. Michael Mendelson, *Many Sides: A Protagorean Approach to the Theory, Practice, and Pedagogy of Argument* (Dordrecht, the Netherlands: Kluwer Academic Publishers, 2002), 8.

10. Myles Burnyeat, "Protagoras and Self-Refutation in Plato's *Theaetetus*," *Philosophical Review* 85, no. 2 (1976): 172.

11. Noburu Notomi, "A Protagonist of the Sophistic Movement? Protagoras in Historiography," in *Protagoras of Abdera: The Man, His Measure*, ed. Johannes M. van Ophuijsen, Marlein Van Raalte, and Peter Stork Leiden (The Netherlands: Brill, 2013), 34. The difference, Burnyeat claims, is "crucial" and resolves the self-refutation controversy plaguing the *homo mensura* thesis in the *Theaetetus* as well as in its reception in philosophy; see 174 and 186–88.

12. Zilioli, *Protagoras and the Challenge*, 15. See also Ugo Zilioli, "Protagoras Through Plato and Aristotle: A Case for the Philosophical Significance of Ancient Relativism," in *Protagoras of Abdera: The Man, His Measure*, ed. Johannes M. van Ophuijsen, Marlein Van Raalte, and Peter Stork Leiden (The Netherlands: Brill, 2013), 237–39.

13. Zilioli, *Protagoras and the Challenge*, 35.

14. Rosamond Kent Sprague dates this to 444 BCE and notes that Protagoras and Pericles's "acquaintance was marked by mutual esteem"; in *The Older Sophists* (Columbia: University of South Carolina Press, 1972), 3.

15. Adriaan Rademaker, "The Most Correct Account: Protagoras on Language," in *Protagoras of Abdera: The Man, His Measure*, ed. Johannes M. van Ophuijsen, Marlein Van Raalte, and Peter Stork Leiden (The Netherlands: Brill, 2013), 106.

16. As most commentators on Protagoras's legacy note, relativism has been critiqued by classical and modern philosophers on two main points: as an incoherent and indefensible tenet and as ethically suspect.

17. Arnaud Macé, "Privatising Perception: Plato's Protagoreanism (*Theaetetus* 154B–157C)," in *Protagoras of Abdera: The Man, His Measure*, ed. Johannes M. van Ophuijsen, Marlein Van Raalte, and Peter Stork Leiden (The Netherlands: Brill, 2013), 213.

18. Van Berkel contextualizes the *homo mensura* thesis in relation to fifth-century foundation myths about heroes who introduce measures "crucial to the survival and functioning of communities," Tazuko A. van Berkel, "Made to Measure: Protagoras' *Metron*," in *Protagoras of Abdera: The Man, His Measure*, ed. Johannes M. van Ophuijsen, Marlein Van Raalte, and Peter Stork Leiden (The Netherlands: Brill, 2013), 55. Van Berkel argues that measurements would have been considered emblematic of

accountability and cultured civilization, instruments in a society that values duty (e.g., in the case of measuring and repaying debt).

19. "Big Data for Development," 9.

20. Ibid., 16.

21. A foundational premise of GP's project writ large is that "the data revolution is not restricted to the industrialised world; it is also happening in developing countries—and increasingly so," "Big Data for Development," 9. GP uses the terms "industrialised" and "developing" without commenting on the ideological assumptions inherent therein.

22. Ibid., 13.

23. Ibid., 14.

24. GP's discussion of the humanity of what it measures is permeated by a kind of "everyman" idealism. By emphasizing the significance and value of every human's words and actions, GP constructs every human as a human measure. This resonates with Protagoras's ethics, particularly his insistence that, as a function of their innate moral precepts, all persons are suited for civic participation and deliberation about moral dilemmas. Beresford characterizes this democratic emphasis on the ordinary, that which is human on a small scale; Adam Beresford, "Fangs, Feathers, and Fairness: Protagoras on the Origins of Right and Wrong," in *Protagoras of Abdera: The Man, His Measure*, ed. Johannes M. van Ophuijsen, Marlein Van Raalte, and Peter Stork Leiden (The Netherlands: Brill, 2013), 153. Further, this ordinary humanness in political moments is what Garver captures with "civic knowledge," attempting to reconcile experts' and nonexpert citizens' roles in the *polis*; Eugene Garver, "Can Virtue Be Bought?" *Philosophy & Rhetoric* 37, no. 4 (2004): 353–82. He argues, "Civic knowledge will be a tertium quid between the expertise of the alien, Protagoras, who teaches but does not practice civic virtue, and the forms of civic excellence not based on knowledge, possessed by statesmen whose sons turn out badly," 357. The connection is here evident again between Protagoras's pedagogy and his political and ethical egalitarianism. Benitez examines it as manifested in the *Protagoras*, wherein Socrates's and Protagoras's pedagogies, indeed worldviews, are at stake in the debate; see Eugenio Benitez, "Argument, Rhetoric, and Philosophic Method: Plato's *Protagoras*," *Philosophy & Rhetoric* 25, no. 3 (1992): 240.

25. "Big Data for Development," 12.

26. "Research," *United Nations Global Pulse*, accessed March 5, 2015, http://www.unglobalpulse.org/research/projects.

27. "Big Data for Development," 28.

28. Ibid., 28, 33. See also "Data, Data Everywhere," *Economist*, February 25, 2010, http://www.economist.com/node/15557443.

29. "Big Data for Development," 20–23.

30. "Pulse Lab Diaries," *United Nations Global Pulse*, accessed March 5, 2015, http://www.unglobalpulse.org/lab-diaries.

31. "Research," *United Nations*.

32. "Pulse Lab Diaries," *United Nations*.

33. "Big Data for Development," 16.
34. Ibid., 19.
35. Plato, *Theaetetus*, 152a.
36. See note 56.
37. "Big Data for Development," 10, 21, 38.
38. Ibid., 19.
39. Diogenes Laertius frag. 68.B6a Diels-Krantz.
40. Poulakos, *Sophistical Rhetoric*, 35.
41. Ibid., 58.
42. Mendelson, *Many Sides*, 50.
43. Ibid.
44. *Dissoi Logoi* is the title of a Sophistic text of unknown authorship. It discusses contrasting definitions of what is good and bad, seemly and disgraceful, just and unjust, truth and falsehood. For example, death is good for gravediggers; illness is good for doctors; a corroded tool is bad for its user but good for the blacksmith. Offering her own translation, Sprague dates the text subsequent to the Peloponnesian War; in *The Older Sophists*, 279.
45. Schiappa, *Protagoras and Logos*, 162.
46. "Information Sheet," *United Nations Global Pulse*, September 2013, http://www.un.org/millenniumgoals/pdf/GP%20Backgrounder-General2013_Sept2013.pdf, 1.
47. Ibid.
48. Robert Kirkpatrick, "A Conversation with Robert Kirkpatrick, Director of United Nations Global Pulse," *SAIS Review* 34, no. 1 (2014): 4.
49. "Data, Data Everywhere." The connotative clash between the metaphor of a trail, a potentially intriguing path left behind with signs of activity, and the metaphor of exhaust, unwanted residue that is potentially toxic, goes unremarked in GP's text.
50. "Information Sheet," *United Nations Global Pulse*, 1.
51. "Big Data for Development," 27.
52. David C. Hoffman, "Structural *Logos* in Heraclitus and the Sophists," *Advances in the History of Rhetoric* 9, no. 1 (2006): 1–32. Hoffman cites G. S. Kirk to define Heraclitean *logos* as structured or ordered composition involving interchangeable elements.
53. Hoffman, "Structural *Logos*," 15. Hoffman uses Heraclitean *logos* to account for the unity of Protagoras's multiple areas of interests and inquiry: "If, like Heraclitus, the sophists understood *logoi* as structures not only in language, but in the psyche, in human societies, and in the physical world, then it would make perfect sense for them to pursue cosmology and political science right along side with the criticism of poetry and the composition of paradoxical speeches," 22.
54. Schiappa, *Protagoras and Logos*, 113.
55. What is commonly referred to as Protagoras's "stronger/weaker" fragment is present in this chapter as an inherent assumption of the *dissoi logoi* thesis. As Schiappa explains, "Of the two *logoi* in opposition concerning any given experience, one is—at any given time—dominant or stronger, while the other is submissive or weaker"; *Protagoras and Logos*, 107. The stronger/weaker fragment has often been

read as evidence of Protagoras's lack of moral compass; indeed it represents the notion that rhetoric itself lets bad arguments prevail over good ones. An alternate interpretation relying on Heraclitean flux redeems the stronger/weaker idea by focusing on binary fluctuations, wherein what is relatively strong and weak changes. It should be noted that there is some debate regarding the extent to which Protagoras's ontological indeterminacy necessitates a Heraclitean worldview. Zilioli argues, put simply, "Stability is allowed for in the world of the relativist"; in *Protagoras and the Challenge*, 52. That a thing does not in itself have "one unique, determinate essence (since their indeterminate essence is variably determined by different perceivers in the perceptual act)" does not necessarily mean it is changing; *Protagoras and the Challenge*, 40. For an argument against Protagoras's Heraclitean influence, see Notomi, "A Protagonist," 11–36.

56. On data streams, see, for example, "Big Data for Development," 22. Heraclitus's theory of flux endures primarily in his aphorism that one cannot step into the same river twice; change is continuous. See Mendelson, *Many Sides*, 69–70.

57. "Big Data for Development," 6, 11, 13.

58. Ibid., 10, 20, 37.

59. Ibid., 6.

60. Ibid., 11.

61. "Information Sheet," *United Nations Global Pulse*, 1.

62. "Big Data for Development," 8, emphasis added.

63. Ibid., 15.

64. Explaining the shortcomings of traditional monitoring systems, GP notes, "By the time hard evidence finds its way to the front pages and the desks of decision makers, it is often too late or extremely expensive to respond. The main triggers will often be known—a drought, rising temperatures, floods, a global oil or financial shock, armed conflict—but even with sufficient macro-level contextual information it is hard to distinguish which groups are affected, where, when, and how badly," "Big Data for Development," 11.

65. Drawing on Paul Virilio, Roger Stahl argues thoughtfully that real-time technology places crises in an eternally urgent present (in his case the ticking clock of the infinite war); local time is superseded by the technologies of global screen time. This acceleration and global spectacle of threat may foreclose important reflective and deliberative moments. Roger Stahl, "A Clockwork War: Rhetorics of Time in a Time of Terror," *Quarterly Journal of Speech* 94, no. 1 (2008): 73–99.

66. "Big Data for Development," 6.

67. Ibid., 13; see also 42.

68. "About," *United Nations Global Pulse*; Kirkpatrick, "A Conversation," 5.

69. "Big Data for Development," 41.

70. Steve Lohr, "The Age of Big Data," *New York Times*, February 11, 2012, http://www.nytimes.com/2012/02/12/sunday-review/big-datas-impact-in-the-world.html?pagewanted=all&_r=0.

71. Kirkpatrick, "A Conversation," 5.

72. Sprague, *The Older Sophists*, 4–5.

73. Aristotle, *Rhetoric*, 3.5.1407b6.

74. Plato, *Cratylus*, 391c.3–5. See Christopher Lyle Johnstone, "Sophistical Wisdom: *Politikē Aretē* and 'Logosophia,'" *Philosophy & Rhetoric* 39, no. 4 (2006): 282; Sprague, *The Older Sophists*, 15.

75. Poulakos, *Sophistical Rhetoric*, 87.

76. Plato, *Protagoras*, 339a. See Andrea Capra, "Notes and Discussions: Protagoras' Achilles: Homeric Allusion as a Satirical Weapon," *Classical Philology* 100, no. 3 (2005): 276; Sprague, *The Older Sophists*, 16.

77. Rademaker, "The Most Correct Account," 88.

78. I return in the conclusion to the civic significance of *orthoepeia*, specifically in relation to Protagoras's teaching of *euboulia*, or sound judgment and action. To foreshadow, Protagoras's instruction was grounded in the assumption that *orthos logos* as a rhetorical habit of the *polis* would render proper decision-making and conduct.

79. "Information Sheet," *United Nations Global Pulse*, 1.

80. Kirkpatrick, "A Conversation," 4.

81. "Big Data for Development," 17, emphasis added.

82. Nathan Crick, "The Sophistical Attitude and the Invention of Rhetoric," *Quarterly Journal of Speech* 96, no. 1 (2010): 34. Crick explores the inventive potential of Sophistical *logos*, particularly the imagination of new possibilities for action; precision in the use of *logos* productively reveals distinctions, and therein opportunities.

83. "Big Data for Development," 27.

84. Ibid., 6.

85. Ibid., 39.

86. Ibid., 33.

87. Rademaker, "The Most Correct Account," 102.

88. Burnyeat, *Theaetetus*, 23. See also Zilioli, *Protagoras and the Challenge*, 68.

89. Plato, *Theaetetus*, 167b.

90. Ibid., 166e.

91. Schiappa, *Protagoras and Logos*, 185–87.

92. "Big Data for Development," 24–25.

93. "Data, Data Everywhere."

94. "Big Data for Development," 4.

95. I have argued elsewhere that the conflation of measurement and evaluation lead to argumentative fallacies and a culture that relishes metrics but neglects standards. E. Johanna Hartelius and Gordon R. Mitchell, "Big Data and New Metrics of Scholarly Expertise," *Review of Communication* 14, no. 3–4 (2014): 288–313.

96. Kirkpatrick, "A Conversation," 3.

97. "Big Data Climate Challenge," *United Nations Global Pulse*, accessed March 6, 2015, http://www.unglobalpulse.org/big-data-climate.

98. Kirkpatrick, "A Conversation," 5, 7.

99. "Big Data for Development," 3.

100. David Payne, "Rhetoric, Reality, and Knowledge: A Re-Examination of Protagoras' Concept of Rhetoric," *Rhetoric Society Quarterly* 16, no. 3 (1986): 189. See also van Berkel, "Made to Measure," 51. The notion of "man" as the measurer of all things gets a brief mention in George Kennedy, *The Art of Persuasion in Ancient Greece* (Princeton, NJ: Princeton University Press, 1963), 13.

101. Payne, "Rhetoric, Reality, and Knowledge," 194.

102. Johannes van Ophuijsen, "Introduction: Protagoras of Abdera: Amicus Homo Magis Amica Veritas?" in *Protagoras of Abdera: The Man, His Measure*, ed. Johannes M. van Ophuijsen, Marlein Van Raalte, and Peter Stork Leiden (The Netherlands: Brill, 2013), 8.

103. John Poulakos, "Toward a Sophistic Definition of Rhetoric," *Philosophy & Rhetoric* 16, no. 1 (1983): 42.

4

On Fear and Longing

Gorgias and the Phobos *and* Erōs *of Visual Rhetoric*

Nathan Crick

> Thus it is natural for the sight to grieve for some things and to long for others, and much love and desire for many objects and figures is engraved in many men.
>
> —Gorgias, *Encomium of Helen*

The September 1, 2015, edition of the *New York Times* features a story titled "For Thousands of Exhausted Migrants, a Journey's End in Munich." The writer describes the scene outside the Munich train station where hundreds of migrants, many of whom had traveled more than two thousand miles through war-torn areas, were being welcomed by volunteer corps of German citizens handing out "drinks, diapers and mobile routers so the new arrivals could connect with friends and send word of their arrival."[1] But most striking is the photograph that accompanies it, showing an attractive, smiling young German woman extending her hand out to an equally happy young boy. In her hand is a teddy bear. The article explains that "for many of the migrants, Germany was their dream destination, a safe haven and an opportunity to begin new lives after enduring war and strife in the Middle East or Northern Africa."[2] Drawn by its strong economy and humane asylum policy, these migrants hoped for a better life away from hunger, poverty, and war. Only a few of the expected eight hundred thousand migrants will achieve their dreams, the article warns, but the photographic evidence promises otherwise to eyes desperate for hope.

Two days later, in the September 3, 2015, edition of the *New York Times*, an article appeared titled "Image of Drowned Syrian, Aylan Kurdi, 3, Brings Migrant Crisis into Focus."[3] Unlike the previous article, in which the photograph was just an accompaniment to the story, the story here is about the photograph itself. In it, we see three-year-old Aylan. The boy's hair is recently cut. He is wearing a short-sleeved red shirt, blue shorts, and small blue running shoes. He is on the beach, where one would expect to see him playing with family and friends. Only he is alone, "his round cheek pressed to the sand as if he were sleeping, except for the waves lapping his face."[4] The story is about the hundreds of migrants who have drowned trying to escape their homelands. Aylan, his five-year-old brother, and his mother drowned when the flimsy fifteen-foot inflatable raft they were given to cross the Mediterranean by smugglers flipped over and sent them into the sea. Only the father survived, after spending three hours attempting to keep his family alive. But the death of his son had an impact. The photo, writes the journalist, "appears to have galvanized public attention to a crisis that has been building for years. Once again, it is not the sheer size of the catastrophe—millions upon millions forced by war and desperation to leave their homes—but a single tragedy that has clarified the moment."[5] The power of the visual image of the dead body of a single boy roused in the nations of the world an affective response that thousands of previous deaths had not.

The clear rhetorical power of these types of viral images, which are picked up and shared widely and immediately across multiple networks, is unmistakable in our digital environment. That the visual image has a unique impact, different than that of writing or speech, has long been recognized. Marshall McLuhan was the most famous to announce this transition in the twentieth century, arguing that "in the age of the photograph, language takes on a graphic or iconic character, whose 'meaning' belongs very little to the semantic universe, and not at all to the republic of letters."[6] For McLuhan, photography stood in the most direct contrast to what we might call the republic of *logos*, primarily because it produced in its viewers a psychic "gestalt" whose meaning was revealed in a single moment rather than cumulatively over time.[7] But other media we might normally consider equally visual had different effects. Print advertisements, for McLuhan, gained their power from their iconic quality that focused "a large region of experience in tiny compass."[8] Television was a high participation medium in which viewers felt a "total involvement in all-inclusive *newness*."[9] And movies were closest in form to the novel by offering a "private fantasy experience" unrolled on the big screen in sequence.[10] Whatever the specific impact of different media, however, McLuhan assured us that with the new electronic (and now digital) technology, we have taken the decisive step from "the age of Typographic man to the age of Graphic man."[11] The diffusion of digital media throughout advanced capitalist societies is thus an intensification and acceleration of tendencies already at work at the end of the nineteenth century and through the twentieth century, greatly magnifying the scale of impact as well as speeding up the pace by which the affective force of any image is registered across multiple audiences. Contemporary rhetorical theory has only recently come around to the study of images. In seeking to develop a theoretical and critical vocabulary to study images, rhetorical theorists started with the premise that, in Gronbeck's words, "visuality always has been integral to rhetorical consciousness."[12] Consequently, rhetorical critics commonly approached visual artifacts as they would textual artifacts, by extending the traditional rhetorical project of understanding the relationship between text and context, specifically the relationships between the said and the unsaid, the explicit and the taken for granted. As Olson, Finnegan, and Hope explain, this "does not mean to study images or artifacts in isolation from larger textual performative context in which an audience might encounter them, but rather in precise relation to those contexts that give them shape and meaning."[13] Rhetoricians would immediately recognize this interpretive strategy as connected with the enthymeme. Hariman and Lucaites, for instance, argue in the case of their study of iconic photographs, "in terms of classical rhetoric, photojournalism operates as an enthymematic process by which the audience applies social knowledge and shared values to complete a speaker's argument."[14] Finnegan, in her study of image vernaculars, defines them as "enthymematic modes of reasoning employed by audiences in the context of specific practices of reading and viewing visual cultures."[15] In other words, although we encounter artifacts of visual rhetoric through the eye, we interpret them through the mind in ways similar to how we interpret any type of text. After all, the power of images like the teddy bear or the drowned boy have rhetorical power only insofar as they are understood within the narrative context of the 2015 European migrant crisis.

Yet there are significant limitations to privileging the enthymematic interpretation of images as if they could be treated as simply another kind of "word." If we reach back further into rhetorical history, for instance, specifically in the extant fragments of the fifth-century BCE Sophist Gorgias, we find a competing strategy of interpretation whereby the power of sight (*opsis*) is explicitly *contrasted* with the power of words (*logos*) to understand this unique impact. This distinction occurs within his *Encomium of Helen*, a fictional defense of Helen of Troy that attempts to absolve her of guilt by attributing her actions to external causes that made her act against her will. Two of these possible causes are identified as words (*logos*) and love (*erōs*), although as it soon becomes clear, the latter really refers more narrowly "only to external beauty as the object of desire."[16] That is to say, whereas *logos* refers to the forms of speech like poetry, witchcraft, philosophy, and argumentation that alter our cognitive judgments through language, *erōs* refers to the power of sight (*opsis*)—specifically the sight of such things as military arms, bodies, statues, and paintings—to arouse what for Gorgias are largely noncognitive, affective reactions of love (*erōs*) or fear (*phobos*). In other words, he discusses *opsis* in terms of affect, or what Teresa Brennan defines as any "physiological shift accompanying a judgment."[17] As Shaffer sums up his "novel strategy, Gorgias invokes the persuasive power of speech and the compelling force of visual impressions to remove the calumny from Helen's name."[18] For our purposes, however, what is important is that Gorgias bases this defense on the unique quality of the visual image to compete with and even overwhelm the more rational sensibility he associates with words and reasoning.

Gorgias was not unaware that these two causal mechanisms can influence the other, of course. As Michele Kennerly observes, "Although *opsis* technically pertains to what the bodily eyes behold and not the mind's eye," Gorgias effectively "extends its meaning from present sights to prospective sights, which are *phantasia's* domain."[19] That is to say, the effects of *opsis* are not limited to the physical body in the present but endure in the memory and are transformed through interaction with language within the realm of our imaginative, emotional visualization (*phantasia*) as well as our future judgments. As she concludes, "impressions goad judgment, and their lingering nature works to sustain the sense that the judgment was justified."[20] Nonetheless, what is important is that Gorgias suggests that we first must distinguish the unique, distinct, and immediate impact of the image *on* the soul (*psuchē*) before we are able to understand the interaction between *opsis* and *logos* and judgment (*krisis*). The differences between *opsis* and *logos* must be attended to before we can understand their points of contact and cooperation. What I wish to argue is that as rhetorical theory seeks to understand the persuasive power of visual images encountered in digital networks that are often accompanied with little to no textual explanation or context—that is to say, in which the image itself must "strike" the viewer immediately—we must balance enthymematic readings with affective interpretations.

As indicated by Gorgias's emphasis on the immediate force of seeing a beautiful object, to study affect is not the same as the traditional study of emotion or Aristotelian *pathos*. Whereas *pathos* is evoked by a conscious construction of words that orient ourselves in certain ways to objects, events, people, or ideas that we may or may not have directly encountered, affect is something that immediately strikes because of the nature of a particular encounter with a sensible object. According to Erin Rand, when we talk about the "affective turn" in rhetorical theory, we refer to "the body's capacity to affect and be affected by it, and the vital forces and intensities that exceed linguistic capture."[21] Drawing from the work of Eric Shouse, for instance, Damien Smith Pfister distinguishes affects from feelings or emotions. Affects, he writes, are "non-conscious bodily reactions to being affected by one's environment" (such as the immediate reaction to the pain of a needle).[22] A feeling "is what happens when the affect begins to be cognized" (as when a needle pain becomes distinguished from other pains designated with a word, like "sharp").[23] Lastly, an emotion "is the rhetorical expression of that feeling, based on socially constructed labels that attempt to reflect recurring feeling-states" (as when a patient grits her teeth in preparation for a doctor's shot).[24] To attend to affect in rhetoric is thus to investigate the potential persuasive effects of precognized bodily reactions that come in response to direct stimuli, not just for individuals but for groups. As David Gruber argues, incorporating affect within rhetorical theory means "continuing the intellectual move beyond discursivity and trying to account for how the material world of biological bodies and environments constitute national or political subjects."[25] As shown graphically by the photograph of Aylan Kurdi, the power of affect may influence the fate of whole populations or nations once it enters the *psuchē* of the public and begins to interact in a sphere of conscious judgment.

However, in contradistinction to the claim that interest in affect is an early-twentieth-century phenomenon, I argue that Gorgias made the "affective turn" in rhetoric long before the phrase was made popular in social theory.[26] Moreover, I believe that analysis of digital images can benefit from a turn back to Gorgias, a turn already begun by Michele Kennerly and Debra Hawhee, both of whom have recognized the importance of Gorgias's conception of *opsis* within the larger movement to integrate conceptions of affect into contemporary rhetorical theory.[27] But I wish to stress, more than they have, how visual affect, understood as a product of *opsis*, necessitates a more autonomous interpretation of affective impact separate from, and even at times in tension with, readings that rely on more traditional text-in-context approaches. This in no way denies the relevance of more conventional enthymematic interpretation of images. As the two cases of the migrant photographs demonstrate, the political significance of an image of a teddy bear or a drowned boy can be understood only by tracing a history that goes back not only months but decades (or, indeed, centuries). Yet turning back to Gorgias emphasizes two critical aspects of interpretation. First, his theory of *opsis* emphasizes that what makes photographs like that of Aylan stand out among other photographs is their unique affective power, which comes from the direct interaction of the image on the eye and then the eye on the body. No symbolic interpretation of these photographs does justice to this immediate, affective force. Second, his theory of *logos* reminds us that even after accounting for the force of affect, the lasting political and rhetorical power of the image is nonetheless contingent on the capacity for *logos* to transform affect into emotion and to situate sense perception within a larger sphere of cognitive meaning and judgment. Thus, while I wish to emphasize how our affective responses to images are often autonomous from and often exist before our more reflective, rational, and linguistic interpretations, affect remains impotent to bring about any enduring change in our habits or reforms in our political institutions without those latter interpretations.

The Force of *Opsis*

Because of the strong connection between the Sophists and rhetoric, the attention that Gorgias paid to the power of *opsis* has largely been overshadowed by the bold claims that he makes for the power of *logos* in his *Encomium of Helen*. Gorgias announces: "Logos is a powerful master and achieves the most divine feats with the smallest and least evident body."[28] The phrase typically rendered as "powerful master" is *dunastēs megas*, also interpreted as "mighty lord" or "great ruler." For Poulakos, this pronouncement established the structure of political power of Athens as a "logocracy" in which the "command of logos is the means par excellence to personal and political power."[29] As a Sophist and teacher of *logos*, Gorgias's articulation of this power structure served his own interests. Guthrie used the metaphor of a genie to describe the power that *logos* gave to its user, claiming that Gorgias believed that "'the word' was a despot who could do anything, but like a slave of the lamp it would be at the service of those who took his courses."[30] With claims such as these, it is not surprising that readers would focus almost exclusively on what Gorgias says about the power of language while giving his description of the power of image only a passing glance.

However, a closer reading of the *Helen* shows that *logos* is *not* a powerful lord over *all* things. Gorgias forcefully argues that visual impressions, whether caused by natural objects or by works of art, have a unique force unto themselves, thus putting the affective power of *opsis* on par with the cognitive and emotional power of *logos*. Sometimes, these forces stand in opposition. As Shaffer explains, Gorgias in the *Helen* "illustrates the agonistic clash between the rational appeal of *logos* and the compelling force of visual stimuli," namely, "because intensely felt visual impressions ignite desire and desire overwhelms the holder, the power of sight is immense, even blinding."[31] On the other hand, these two powers can work together to produce momentous events such as the war with Troy that was later coordinated through the speeches of Agamemnon but first inspired by Helen's beauty. But even when they work together, they operate through different paths in the *psuchē*. Before we can understand how *logos* and *opsis* interact, therefore, we must first understand how they operate separately.

That Gorgias presents a theory of visual persuasion in a defense of Helen is no accident. Helen, of course, was legendary for possessing a rare beauty that had extraordinary power over the desires of men. As Gorgias put it, "For one body was the cause of bringing together many bodies of men thinking great thoughts for great goals."[32] In this respect, the effect of her beauty was very closely aligned with that of the Greek goddess of persuasion, Peithō. As Gordon reminds us, "In the Greek context, the reference to *peitho* has an erotic tradition behind it, and it cannot be separated from seduction."[33] Yet in characteristic sophistical fashion, Gorgias playfully reverses the traditional narrative of seduction. Instead of the Trojan prince Alexander (Paris) being overtaken with desire for Helen's beauty, it is Helen who becomes infatuated with the body of Paris. Gorgias defends her hypothetical reaction this way: "If, therefore, the eye of Helen, pleased by the figure of Alexander, presented to her soul eager desire and contest of love, what wonder?"[34]

The answer to that very question stands at the heart of our understanding of the unique power of visual rhetoric. For what is striking about how Gorgias frames that question is the order of causation that it implies. Notably, *erōs* here is not itself a direct cause of Helen's action. Her "eager desire" is itself an effect of *opsis*. Indeed, Gorgias attributes agency not to Helen's *psuchē*, or soul, but to her *eyes*, which are the first to receive visual appearances. Helen's eyes thus operate as an independent agent with their own pleasures and pains that in turn arouses some desire or fear in the soul—all apparently without the mediating influence of language or thought. As Shaffer reads this argument, "the sight of Paris creates an impression on Helen's psyche so powerful that it produces *eros*, which results in her unexpected or non-rational physical reaction."[35] It is precisely Gorgias's suggestion that visual images might have a persuasive force distinct from and even antithetical to *logos* that makes his theory so provocative.

Particularly given the general reputation of the Sophists as relativists and humanists, the fact that Gorgias explicitly advances a realist understanding of ontology and a materialist understanding of human perception may come as a surprise. However, Gorgias is quite clear in his commitments. In accounting for the power of *opsis* to influence the *psuchē*, he asserts, "The things we see do not have the nature which we wish them to have, but the nature which each actually has."[36] For him, what constitutes the "appearance" of *opsis* is the way in which a thing's actual nature directly impacts our sensory apparatus and triggers certain affective and cognitive reactions. Gorgias accounted for this fact through a theory of sense perception derived from Empedocles, who posited that we have a "specific affective apparatus with which we are able to apprehend the empirical world, namely the *pores* or apertures of our senses."[37] From this perspective, each particular affective apparatus is designed to receive the material input from only one specific kind of stimulus, which is why, in *On Not Being*, he denied the ability for the content of one sense to be translated into another without distortion. As he explains, "each object ought to be judged by its own sense, but not by another."[38] In short, Gorgias suggests that our bodies are really plural entities, with our sight and our hearing and our touch operating semi-autonomously to receive specific material sense perceptions from a real world outside of us.

To describe the causal effects of real things on the *opsis* and in turn on the *psuchē*, Gorgias provides three examples: the sight of enemies in armor, the sight of Paris's body, and the sight of works of art. In each case, one finds an example of the mind directly stamped by the character of the object and causing an immediate physiological response of either longing or fear, thus showing how "the sight engraves upon the mind images of things which have been seen."[39] Despite being a treatise on *erōs*, Gorgias actually begins with an example of *phobos*. He explains how "when belligerents in war buckle on their warlike accouterments of bronze and steel, some designed for defense, others for offense, if the sight sees this, immediately it is alarmed and it alarms the soul, so that men often flee, panic-stricken, from future danger [as though it were] present."[40] Notably, Gorgias does not say that the armor of bronze and steel represents anything, as if the soldier who witnesses this is able to size up the strength of the enemy and calculate the nature of the threat. He speaks of the act of arming the body as an event, as something that accompanies raucous sounds and the clanging of metal and glinting of bronze and iron in the sun. He goes on to explain: "It has happened that people, after having seen frightening sights, have also lost presence of mind for the present moment; in this way fear extinguishes and excludes thought."[41] By "fear" he means the degree to which we are impacted by the force of the raw image itself, and not necessarily by its cognitive meaning.

But our encounters with beautiful things are no less compelling, for it is in the nature (*phusis*) of the things themselves to impact a sensitive soul. We have already seen the impact that Alexander's body had on Helen's eye, creating in her soul an eagerness and a striving for love. But it is also the case that works of art produce such an affect through conscious design. Thus, as Gorgias notes, "whenever pictures perfectly create a single figure and form from many colors and figures, they delight the sight, while the creation of statues and the production of works of art furnish a pleasant sight to the eyes."[42] Once again, it is the *eye*, not the soul, that stimulates the appetites of the human body through the sheer compulsion of beautiful forms. That Gorgias would compare equally the affective force of encountering a real human being, like Paris or Helen, and that of encountering an artistic creation, like the statues of Phidias, simply proves that our longing is not merely a cognitive act. It is primarily an affective reaction of the body to beautiful forms that please the sight. This longing then arouses a more complex emotional state in the *psuchē*, where it interacts with other parts of the soul.

Importantly, however, Gorgias does not argue that sense perceptions (even if they are caused by real objects) have the same meaning or affective impact for all individuals. He argues rather that "the sight engraves upon the mind images of things which have been seen."[43] What he emphasizes here is the directness of the affective impact, not the meaning of that affect. The *meaning* of those sense perceptions is rather constructed in part through one's culture, language, and habit, just as the positive or negative valence of that affect is determined by one's situated position. For instance, the impact of seeing belligerents arm themselves for war, some for defense, others for offense, will depend on the viewer's recognition not only that these are preparations for war but also that these preparations are real and not dramatic (as in a tragedy) and that one is positioned as an enemy rather than an ally. In sum, when Gorgias speaks about affect, he refers to the immediate response of a sense-perception whose meaning does not require any higher cognition but is immediately recognized *as* something.

To summarize Gorgias on the force of the image: There are real objects in the world that have a direct and immediate impact on our different sense perceptions, our sensory apparatus reacts to these stimuli partially independent of any conscious thought or cultivated attitudes, our affective reactions are rooted in our biological nature as human beings, affects tend to be localized in the present and take the form of either intense longing or overwhelming fear, and these affects entangle with *logos* to produce rational or irrational responses. That Gorgias places the compulsion of *opsis* in opposition to *logos* becomes clear at the end of his speech, when he remarks, "If it is a disease of human origin and a fault of the soul, it should not be blamed as a sin, but regarded as an affliction. For she came, as she did come, caught in the net of Fate, not by the plans of the mind, and by the constraints of love, not by the devices of art (*technē*)."[44] In this remarkable argument, he interprets the effects of sight (in this case only of longing) as a disease and an affliction caused by the susceptibility of the body to affect. But how does this compulsion interact with the *technē* of *logos*?

The Power of *Logos*

What is particularly striking after looking closely at his theory of *opsis* is how closely it seems to mirror Gorgias's claims about the power of *logos*. First, like *opsis*, *logos* seems to operate akin to material interaction, as when Gorgias says that "persuasion has the same form of necessity, but it does not have the same power" or that "the effect of speech upon the condition of the soul is comparable to the power of drugs over the nature of bodies."[45] Similar to *opsis*, then, the causal chain of *logos* seems to take a peripheral route around conscious thought, namely, through the affected physical body. Second, *logos* attains (also like *opsis*) its power by arousing emotional reactions. Gorgias argues, "Speech is a powerful lord, which by means of the finest and most invisible body effects of the divinest works: it can stop fear and banish grief and create joy and nurture pity."[46] And in elaborating his analogy between *logos* and drugs, he says that "just as different drugs dispel different secretions from the body, and some bring an end to disease and others to life, so also in the case of speeches, some distress, others delight, some cause fear, others make the hearers bold, and some drug and bewitch the soul with a kind of evil persuasion."[47] At first glance of these passages alone, *opsis* and *logos* appear to work in very similar registers.

Yet the key to understanding the difference between *opsis* and *logos* is found within these passages themselves, specifically the lists of the particular effects of *logos*. *Opsis* was restricted to two specific affects that stood opposite to one another— *erōs* and *phobos*, which can be summarized simply as the desire to grasp or possess something and the urgency to repel or flee from something. This can be understood in terms of the two basic animal impulses of sex, on the one hand, and survival, on the other. But the emotions aroused by *logos* are much different. While fear and pain are present in the description of *logos*, they are not only *created* but also *stopped* and *relieved* by *logos*. *Logos* also creates joy, which is much different than desire. Joy is a state of happiness with one's current state of affairs and does not have the pain that accompanies the feeling of absence brought about by desire. Boldness, too, is different than desire, for it creates courage in the face of a threat and in this way overwhelms fear. Lastly, and most importantly, *logos* has the ability to increase pity, which does not exist in the nonhuman animal world. Pity is defined in terms of the capacity to sympathize with the condition of another being through the imaginative capacity to put oneself in the other's position. Far from either desiring or fleeing from an object, pity simply asks one to *understand* the other by getting outside of oneself, much in the way a tragedy arouses pity and fear in an audience within an act of catharsis. Of poetry, for instance, Gorgias observes that "fearful shuddering and tearful pity and grievous longing come upon its hearers, and at the actions and physical sufferings of others in good fortune and evil fortunes, through the agency of words, the soul is wont to experience the suffering of its own."[48] Gorgias attributes great power to *logos*, but this power is of a vastly different character than the affective force of *opsis*, because of its reliance on *phantasia* for its actualization.

One way to get a sense of this different relationship Gorgias imagines between *logos* and emotions like pity and boldness is to turn to one of his more cryptic fragments in which he speaks about the nature of tragedy, which is both a visual spectacle and a linguistic performance. He is quoted by Plutarch as saying, "Tragedy produces a deception in which the one who deceives is more just than the one who does not, and the one who is deceived is wiser than the one who is not."[49] In this case, to "deceive" or "be deceived" refers to the elaborate production and willing consumption of works of dramatic art in which an audience emotionally participates in the lives of actual characters on stage. As Consigny argues, Gorgias praised deception (at least that which was not *concealed* as deception) for its ability to encourage people to suspend their disbelief for a moment and get outside themselves so that "they may be able to see the world from the perspective of the dramatist and in so doing recognize that their own views are themselves limited and biased."[50] In other words, whereas affects are better described as striking or impacting the sensory apparatus of an individual, emotional responses to tragedy are better understood as more complex and self-conscious actions on the part of an audience to participate in the lives of others through empathetic responses.

Another way to understand this distinction is to consider how affects and emotions are seen to function within the literature of the rhetoric of social movements. James Jasper, for instance, makes a very clear distinction between the two types of mental and physical states. On the one hand, "affects are simply positive or negative evaluations of a person, group, place, activity, object, or idea."[51] They represent the way things strike us. Consequently, activists often rely on what he calls "patterns of affect" to arouse feelings of shock—the case often being, for instance, that "positive feelings toward one's home and surroundings, coupled with strong negative affect toward a proposal that seems to threaten these, are common raw materials for protest."[52] One might imagine, for instance, a Greek social movement arising at the prospect of sending all of the greatest Greek heroes to fight in a far-off land of Troy. On the other hand, emotions, while always built on the ground of affect, are "constituted by shared social meanings, not automatic physiological states."[53] Emotions thus involve higher-level cognitive appraisal and "are conditioned by our expectations, which in turn are derived from knowledge about conditions in the world."[54] Compassion, for instance, is a highly complex emotion that involves higher-level rationality and imaginative reconstruction. Lawrence Blum writes that "compassion is not a simple feeling state but a complex emotional attitude toward another, characteristically involving imaginative dwelling on the condition of the other person, an active regard for his good, a view of him as a fellow human being, and emotional responses of a certain degree of intensity."[55] Consequently, arousing these emotions in the rhetoric of social movements requires active narrative reconstruction of the situations in which another human being is suffering, not simply a visual representation that produces shock. Indeed, the entire *Encomium of Helen* itself is intended to arouse these very emotions and is consistent with the rhetorical strategies of social movements that rely on testimonial, description, and quotation.

That the arousal of complex emotions is in part reliant on situational understanding and imaginative reconstruction thus resolves the apparent paradox that comes from the fact that Gorgias also attributes to *logos* an explicitly rational function, "rational" here referring to the use of reason, not to the reliance upon truth. Indeed, Gorgias rejects truth as a criterion for judging persuasion, arguing that "all men who have and do persuade people of things do so by molding a false argument."[56] Yet this fact is not due to the ethical failure of the speakers but because of the innate incapacities of human beings—our inability to have "memory of things past and [awareness] of things present and foreknowledge of the future."[57] Yet he nonetheless celebrates the capacity of *logos* to give reasons, encouraging his readers to study the argument of astronomers, compulsory debates in the assembly or the law courts, and the contests of philosophical arguments to see how one who possesses the power of language makes "belief in an opinion subject to easy change."[58] What is important with respect to understanding the relationship between *logos* and *erōs/phobos* is that Gorgias is not making a sharp distinction between the capacity of *logos* to change cognitive beliefs and its power to arouse emotions; rather, for him, complex emotional responses are themselves partly derivative of our rational capacities to imagine the world differently.

This relationship between *logos* and *phantasia* has been noticed by Kennerly and Hawhee, both of whom emphasize how persuasion often occurs in response to emotions that react not to immediate stimuli but to imagined environments that are produced by *logoi*. Kennerly, using Gorgias's *Helen*, proposes a concept of "rhetorical transport" that occurs when "images of absent stimuli are generated and made present through civic *phantasia*, [and] we find ourselves transported, and the judgments of the here, the now, and the self altered."[59] Similarly, Hawhee proposes a theory of "rhetorical vision" drawn largely from Aristotle, arguing that words facilitate vision because "they have the capacity, in bringing energetic images before the eyes, to compete with, or perhaps even overtake, what is already before the eyes of the audience, to lift them out of the present life by flooding their eyes with active images from the past or projected into the future."[60] The net effect of both rhetorical transport and rhetorical vision, then, is what Hawhee describes as "an active, vivid engagement with the rhetor's words, such that hearers begin to engage in deliberative *phantasia*, leading to belief formation and decision-making."[61] The point to emphasize with respect to Gorgias is that the route between imagination and judgment always passes through emotion.

The emotions that *logos* arouses, then, are far more complex than the affects of *opsis*. In sight, our organs of sense perception react instantaneously to a particular object that stimulates some basic biological impulse of longing or fear. In contradistinction, to construct a *logos* is to create a more complex account of the world in *phantasia* to which we then respond dramatically. That is to say, emotions aroused by *logos* are always mediated by some symbolic representation that exceeds the provincial boundaries of any particular phenomena. In the case of Paris, for example, he could not have persuaded Helen to leave by the power of a single word or command. In fact, the reason Gorgias recommends looking to astronomers, law courts, and philosophical arguments is to show the type of persuasion that Paris must have engaged in to bend Helen's will to his own. One can imagine him crafting an illusion of her as a caged animal under the thumb of her Greek overlord, Menelaus, all the while deluding her judgment that her flight to Troy would guarantee her happiness and her safety behind the high walls of his city. He may even have inspired a feeling of pity in her for Paris's own suffering as an unrequited lover. Gorgias says that this persuasion has the same power as compulsion, but in a different form. But it is that form that makes all the difference, for it is a form that requires the voluntary participation by the imagination of Helen and the construction, through symbols, of a wider environment in which she feels compelled to play a specific role in attaining the objects of her desire. This is as different from fleeing from loud noises or being drawn toward beautiful bodies as night is from day.

The Boy on the Beach

On September 3, 2015, Charles Homans, the deputy digital editor of the *New York Times Magazine* wrote an editorial titled "The Boy on the Beach." It asked a single question: "Why this boy?"[62] The question addressed the photo of Aylan Kurdi. If this photo had stood out because it captured, in graphic realism, the horror of war and forced migration, then the question would not need to have been asked. But it did nothing of the kind. As Homans explained, Aylan was just one of twelve thousand children killed as the result of the Syrian Civil War, with two million others living as refugees: "The Internet is full of images of dead Syrian children, and it's hard to imagine that the people who were transfixed by Aylan on Facebook yesterday had not seen at least some of them."[63] But whereas the hundreds of other images of dead children did not evoke a mass response, after the initial release of this photograph, "images that have since appeared on the front pages of the major American and European newspapers and flooded Twitter with video montages and sorrowful memes, the social-media equivalents of the stuffed animals and bouquets that pile up at the sites where children have died in car accidents or shootings."[64] Whereas thousands of other children die or suffer in obscurity, Aylan became the subject of a global funeral on social media, thus demonstrating the capacity for digital networks, through both voluntary sharing and mass distribution, to create virtually instantaneous responses to digital images that have a uniquely striking affective force.

In the case of Aylan, Homans identifies a single characteristic of the image that gave it such affective force: Aylan's shoes. Certainly, Homans admits, there is the impact of the way he lays on the beach, appearing to be asleep. Homans writes, "I have a young son, a couple of years younger than Aylan but close enough to him in size that every detail of the photo—down to the angle of repose that, as more than one artist noticed, so precisely echoes that of an exhausted child asleep in his crib—was terribly familiar."[65] But it was Aylan's small blue shoes, with one Velcro strap loose, that really cemented its affective force. The shoes were like any other little boy's shoes that any American child could be wearing. They are the same type of shoes that millions of parents have put on their children's feet and, perhaps, become frustrated with when they are so easily kicked off during play. They may even have been the same brand of shoes. There is thus affective power residing in those shoes that renders them resources for affective shock, particularly when those shoes are now placed in a scene of tragedy. Consequently, from "the vantage of the still-new parent, who has recently learned to view the entire world through the lens of the myriad risks it poses to a child—all that really separates you are a few strokes of misfortune. On Wednesday morning, someone like you was helping Aylan put on his shoes."[66] The image of the boy thus becomes the image of one's own son, grandson, or nephew. It provokes a horror that the same parents that carefully dressed that boy in the morning also sent him to his death.

Yet as forceful as the affect of this photo is, Homans also takes care to point out what is ultimately the conservatism of its character—indeed, the conservatism that is inherent in all affect. Since affect's power to shock relies on already established patterns of response that over time come to exist at a precognitive level, affect almost always serves to reaffirm some level of embedded cultural or personal bias. Aylan is different. That is to say, precisely because he doesn't fit the stereotypical mold of the dirty, emaciated, and alien other. One does not have to go far to find entire galleries of such suffering. But "then a boy in nice, clean clothes washes up on the beach near a resort town."[67] This photo stands apart from all the others precisely because it breaks down the tribal divisions that separate our experience from that of others who live differently from us and, for a moment, allows us to empathize with the suffering of others—but only because that other finally looks like us.

As Gorgias would readily point out, then, the capacity for *opsis* to produce fear or longing—in this case the longing for the child to show us that he is not in fact dead—is insufficient to produce the kinds of judgment that are required to bring reason to the debate. As important as affect is to shock the senses and rouse the mind, affect alone usually leaves us right where we started, unless those images, through the rhetorical transformations of *logos*, can then be transformed into richer emotions and made accessible to judgment. Homans is correct to conclude that Aylan, through the image of his dead body, "is at once an emissary from a distant war of unfathomable, baroque atrocity and a figure of awful closeness. You look at the pictures, and you want, more than just about anything, for him to get up. You find yourself searching for more pictures, pictures of the living boy."[68] But as long as our consciousness remains focused only on the boy, the image closes off the more complex emotions that allow us to make sense of the complex network of forces that brought about his death and impedes the rational thought necessary to produce a political solution. It becomes just another suffering body you "scroll by in the Twitter feed, interspersed with GIFs from old sitcoms and news about Tom Brady," making you feel "dimly aware of amassing one more incremental unit of guilt for your acquiescence to the world as it is."[69] Without the transformative effects of *logos*, affects become just one more thing to consume without seriously altering our emotions or judgments, on par with the seemingly infinite objects of *erōs* that are now tailored to our tastes.

To be sure, digital media have increased the potential scope and scale of influence that any image or video might come to possess once picked up by and distributed through conventional news outlets and social media. But this influence, more often than not, simply serves to reinscribe the same patterns of affect already present in an audience. In this regard, the digital image functions no differently than any photographic image in the public sphere. For instance, in describing the rhetorical impact of the graphic image of Emmett Till, the fourteen-year-old African American boy murdered in 1955, whose mutilated face was featured in *Jet* magazine and other periodicals, Harold and DeLuca identify the affective power of the image insofar as "it provokes a physical response that temporary proceeds and exceeds 'sense'—a reflexive shudder, an involuntary retching."[70] Yet this reaction, they acknowledge, was not universal. They also acknowledge that because of the prevalence of racist attitudes under Jim Crow, images of the tortured black body for many racist whites were sought after as keepsakes and sources of longing and that "lynched and mutilated black bodies, far from inspiring sympathy, primarily inspired revelry and celebration among participating whites."[71] Although this reaction is extreme, it nonetheless points to the conservative nature of affect that influences the reactions even among sympathetic audiences who still feel the gap between themselves and the other. And this conservative nature is magnified to the degree that digital networks become ever more (to borrow a McLuhan term) "tribalized," meaning that affective responses are themselves disciplined and determined by the group with which one identifies.

Despite crediting *opsis* with an extraordinary capacity to directly influence at affective responses in the present by stamping images upon the mind, then, Gorgias ultimately suggests to us the necessity of speech for producing more lasting, long-term ethical and political consequences. It is, after all, through his own encomium that he "tried to end the injustice of blame and the ignorance of opinion."[72] No image alone could have done that. I believe this statement remains as true as ever after the rise of digital media—indeed, even more so. With the increasing emphasis on the use of digital image and video to propel social movements, the result has been an increase in the reliance on affect to arouse shock and direct political response. But as Homans indicates, this has also led to a constant stream of affective images all competing for our attention. Yet until these affective responses become translated through *logos* and integrated into a unified *phantasia* that can make sense of the complex environment and direct public judgment over the long term, they remain affects only. This means that social change, in any meaningful sense, thus relies more than it ever has on the power of the word to make "what is incredible and unclear seem true to the eyes of opinion."[73] For *opsis* connects us with the world as it is, or at the very least with what it appears to be; but only *logos* reveals to us the world as it is not yet but might be. We therefore place false hope in the emancipatory potential of the image, even when its autonomous affective force is taken into consideration. At a time when digital networks are creating tightly closed social systems that different affects find difficult to penetrate, the importance of using *logos* to generate the more complex states of sympathy and understanding between different networks becomes the rhetorical challenge of this century.

Notes

Epigraph. Gorgias, "Encomium of Helen," in *The Older Sophists*, ed. Rosamond Kent Sprague, trans. George Kennedy (Indianapolis, IN: Hackett, 1972), section 18.

1. Emmarie Huetteman and Katarina Johannsen, "For Thousands of Exhausted Migrants, a Journey's End in Munich," *New York Times*, September 1, 2015, http://www.nytimes.com/2015/09/02/world/europe/for-thousands-of-exhausted-migrants-a-journeys-end-in-munich.html.

2. Ibid.

3. Anne Barnard and Karam Shoumali, "Image of Drowned Syrian, Aylan Kurdi, 3, Brings Migrant Crisis into Focus," *New York Times*, September 3, 2015, http://www.nytimes.com/2015/09/04/world/europe/syria-boy-drowning.html?hp&action=click&pgtype=Homepage&module=first-column-region&region=top-news&WT.nav=top-news.

4. Ibid.

5. Ibid.

6. Marshall McLuhan, *Understanding Media: The Extensions of Man* (Cambridge, MA: MIT Press, 1994), 196.

7. Ibid., 201.

8. Ibid., 226.

9. Ibid., 335, emphasis in original.

10. Ibid., 286.

11. Ibid., 190.

12. Bruce Gronbeck, "Foreword: Visual Rhetorical Studies: Traces of Power Through Time and Space," in *Visual Rhetoric: A Reader in Communication and American Culture*, ed. Lester Olson, Cara A. Finnegan, and Diane S. Hope (New York: Sage, 2008), xxiv.

13. Lester C. Olson, Cara A. Finnegan, and Diane S. Hope, "Visual Rhetoric in Communication: Continuing Questions and Contemporary Issues," in *Visual Rhetoric: A Reader in Communication and American Culture*, ed. Lester Olson, Cara A. Finnegan, and Diane S. Hope (New York: Sage, 2008), 2.

14. Robert Hariman and John Louis Lucaites, *No Caption Needed: Iconic Photographs, Public Culture, and Liberal Democracy* (Chicago: University of Chicago Press, 2007), 298.

15. Cara A. Finnegan, "Recognizing Lincoln: Image Vernaculars in Nineteenth-Century Visual Culture," in *Visual Rhetoric: A Reader in Communication and American Culture*, ed. Lester Olson, Cara A. Finnegan, and Diane S. Hope (New York: Sage, 2008), 62–63.

16. Bruce McComiskey, *Gorgias and the New Sophistic Rhetoric* (Carbondale: Southern Illinois University Press, 2002), 140.

17. Teresa Brennan, *The Transmission of Affect* (Ithaca, NY: Cornell University Press, 2004), 5.

18. Diana Shaffer, "The Shadow of Helen: The Status of the Visual Image in Gorgias's *Encomium Of Helen*," *Rhetorica* 16, no. 3 (1998): 251.

19. Michele Kennerly, "Getting Carried Away: How Rhetorical Transport Gets Judgment Going," *Rhetoric Society Quarterly*, 40, no. 3 (2010): 275.

20. Ibid., 276.

21. Erin J. Rand, "Bad Feelings in Public: Rhetoric, Affect, and Emotion," *Rhetoric & Public Affairs* 18, no. 1 (2015): 161.

22. Damien Smith Pfister, *Networked Media, Networked Rhetorics: Attention and Deliberation in the Early Blogosphere* (University Park: Pennsylvania State University Press, 2014), 125. See also Eric Shouse, "Feeling, Emotion, Affect," *Media/Culture* 8, no. 6 (2005): http://journal.media-culture.org.au/0512/03-shouse.php.

23. Pfister, *Networked Media, Networked Rhetorics*, 125.

24. Ibid., 125–26.

25. David R. Gruber, "The (Digital) Majesty of All under Heaven: Affective Constitutive Rhetoric at the Hong Kong Museum of History's Multi-Media Exhibition of Terracotta Warriors," *Rhetoric Society Quarterly* 44, no. 2 (2014), 152.

26. See Patricia Ticineto Clough, ed., *The Affective Turn* (Durham, NC: Duke University Press, 2007).

27. See Kennerly, “Getting Carried Away,” and Debra Hawhee, “Looking into Aristotle’s Eyes: Toward a Theory of Rhetorical Vision,” *Advances in the History of Rhetoric* 14, no. 2 (2011): 139–65.

28. Gorgias, “Helen,” section 8.

29. John Poulakos, *Sophistical Rhetoric in Classical Greece* (Columbia: University of South Carolina Press, 1995), 275.

30. W. K. C. Guthrie, *The Sophists* (Cambridge, UK: Cambridge University Press, 1971), 25.

31. Shaffer, “The Shadow of Helen,” 255–56.

32. Gorgias, “Helen,” section 4.

33. Jill Gordon, *Plato’s Erotic World: From Cosmic Origins to Human Death* (Cambridge, UK: Cambridge University Press, 2012), 34.

34. Gorgias, “Helen,” section 19.

35. Shaffer, “The Shadow of Helen,” 253.

36. Gorgias, “Helen,” section 15.

37. Scott Consigny, *Gorgias: Sophists and Artist* (Columbia: University of South Carolina Press, 2001), 54.

38. Gorgias, “On Not Being,” in *Early Greek Political Thought from Homer to the Sophists*, ed. and trans. Michael Gagarin and Paul Woodruff (Cambridge, UK: Cambridge University Press, 1995), section 81.

39. Gorgias, “Helen,” section 17.

40. Ibid., section 16.

41. Ibid., section 17.

42. Ibid., section 18.

43. Ibid., section 17.

44. Ibid., section 19.

45. Ibid., section 12, section 14.

46. Ibid., section 8.

47. Ibid., section 14.

48. Ibid., section 9.

49. Gorgias, in *The Older Sophists*, DK10.

50. Consigny, *Gorgias*, 191.

51. James Jasper, *The Art of Moral Protest: Culture, Biography, and Creativity in Social Movements* (Chicago: University of Chicago Press, 1997), 410n21.

52. Jasper, *The Art of Moral Protest*, 107.

53. Ibid., 109.

54. Ibid., 110.

55. Lawrence Blum, “Compassion,” in *Explaining Emotions*, ed. Amélie Oksenberg Rorty (Berkeley: University of California Press, 1980), 509.

56. Gorgias, “Helen,” section 11.

57. Ibid., section 11.

58. Ibid., section 14.

59. Kennerly, “Getting Carried Away,” 270.

60. Hawhee, “Looking into Aristotle’s Eyes,” 159.

61. Ibid., 159.

62. Charles Homans, "The Boy on the Beach," *New York Times Magazine*, September 3, 2015. http://www.nytimes.com/2015/09/03/magazine/the-boy-on-the-beach.html.

63. Ibid.

64. Ibid.

65. Ibid.

66. Ibid.

67. Ibid.

68. Ibid.

69. Ibid.

70. Christine Harold and Kevin Michael DeLuca, "Behold the Corpse: Violent Images and the Case of Emmett Till," in *Visual Rhetoric: A Reader in Communication and American Culture*, ed. Lester Olson, Cara A. Finnegan, and Diane S. Hope (New York: Sage, 2008), 264.

71. Ibid., 260.

72. Gorgias, "Helen," section 21.

73. Ibid., 13.

5

Impure Imaginations

The Rhetorical Humors of Digital Virology

Christopher J. Gilbert

> Illnesses have always been used as metaphors to enliven charges that a society was corrupt or unjust.
>
> —Susan Sontag, *Illness as Metaphor* (1978)

> We are healthy only to the extent that our ideas are humane.
>
> —Kurt Vonnegut, *Breakfast of Champions* (1973)

The communication phenomenon of "going viral" is a cultural obsession. Indeed, "spreadable media" thrive in ecologics of digitally networked communication.[1] Whether because of cunning acts of persuasion, the ennui of mass audiences, or even plain delight (or disgust), certain media viruses become communicable—that is, relatively popular and readily transmissible—and so circulate widely "through the networks of the mediascape."[2] Furthermore, virology now means that sociopolitical, cultural, and/or economic value is ascribed to (if not realized out of) dispositions or depictions that assume a catchable quality. Examples abound, from footage of late-night talk show host Jimmy Fallon's "epic" lip sync battles and material of cats in various nonsensical situations to ads like IKEA's spoof of Apple with the promotion of its 2015 "bookbook." Still, among the most popular media viruses is a pair of animated shorts produced in 2011 and 2013 by Chipotle Mexican Grill, which touted sustainable farming practices along with images of debauched modes of food production across the Web, in movie theaters, on television, and in an iOS video game. The videos are exercises in juxtaposition and exemplars of integrated marketing content, with one ("Back to the Start") depicting an idyllic family farm corrupted by corporate agriculture only to be salvaged by the probity and pluck of a farmer, and another ("The Scarecrow") portraying the evil machinations of Crow Foods Incorporated—an epitome of agribusiness—in conflict with an innocent, even savior-like, Scarecrow. In "going viral," the videos reflect common attributes of moralistic advertisement in addition to media contaminants: being branded, buzz-worthy—and blighted.

To take Chipotle's viral videos seriously is to revamp common ways of seeing mediated sickness. Viral media content follows a "model of communication" that sees contamination as information spread.[3] But this model betrays a problem with viral metaphors that rely on Henry Jenkins's notion "If it doesn't spread, it's dead." Such an ethos of scatter to survive too often serves as a hermeneutic for critiquing audience engagements with digital media messages.[4] The metaphors thereby overlook the homeopathic, even pathological, considerations embedded in the very *cultural relations* that provoke communicability. Possibilities for shares, views, and mash-ups are then played up while a more robust study of how viral content disposes us to think about publics and publicity gets downplayed. What is worse, "spreadability" glosses over a view of media viruses as rhetorical expressions of collective investments and shared feelings about matters of concern.

This essay approaches the rhetorical humors of media viruses as signs of cultural anxiety. Ancient physicians like Hippocrates (father of epidemiology) and later Galen, philosophers such as Plato and Aristotle, and even the historian Plutarch fixated on maladies. Along with "sickness and death in reality," ancient cities also equated "disorder in the state with a sickness of its inhabitants."[5] Classical rhetoricians in particular used grammars of "physical contamination" to portray communication pathologies.[6] With this in mind, I develop a "festering" metaphor that emerges out of *stasis*—or the idea that contaminated media content captures strife—in order to highlight digital rhetorics of "(im)purification." I also consider how something like *katharsis*, or the purgation/purification of "bad" feelings, makes viral media never merely personal but in fact driven by a *principium collectivus* that manages collective orientations.

The apparent ephemerality of Chipotle's videos, not to mention other popular media viruses, epitomizes the communicability of cultural dis-ease. To begin with, Chipotle's digital rhetoric is among the most popular (and controversial) in the viral mediascape. "Back to the Start" went online in August 2011 before appearing as a "pseudo-PSA" in movie theaters across the United States and airing on CBS during the 2012 Grammy Awards.[7] It also won the Grand Prix for Branded Content and Film at the Cannes Film Festival. James Poniewozik of *Time* lauded it for its "images both wholesome and unsettling."[8] The film garnered nearly six million views on YouTube within a week of its publication, and two years later *Forbes* was still keeping track of its views.[9] Additionally, "Back to the Start" laid the groundwork for "The Scarecrow," which generated even more ire and as much acclaim. Many people praised the second film "as an innovative piece of marketing and beautiful work of art."[10] Others were less enamored with its corporate activism, especially insofar as it "elides Chipotle's real-life meat sourcing with the aesthetic of a vegetable harvest."[11] Yet as one commentator put it, though expressly unfair and distortive in its portrayal of conventional farming, the video is meant to stick with you, "whether you love it or hate it."[12] The same was true of Chipotle's four-episode comedy series on Hulu, *Farmed and Dangerous* (2014), which critiques industrial agriculture. Put simply, if Chipotle's marketing objective is sustainable agronomy, its digital virology is "sustainable advertising" that stays lodged in ecologies of public moods and cultural memories.[13]

This rootedness is partly why what Ekaterina Haskins calls the "ephemeral visibility" of digital rhetoric is not necessarily fleeting at all, for the attitudes and affects it provokes are difficult to discard.[14] The "stickiness" of Chipotle's impressions, along with the tension that is sustained in different (if not contradictory) feelings about them, is even more pronounced when virtuality and materiality intersect.[15] Consider the magnitude of Chipotle's messages of "food with integrity" in the context of its link to a serious E. coli outbreak in late 2015 and early 2016. Viral videos like "Back to the Start" and "The Scarecrow" were consequential because they exploited the fact that US food culture is actually haunted by the imperfections of industrial agriculture.[16] This is no doubt why, in the wake of the E. coli outbreaks, comedian and host of *Last Week Tonight*, John Oliver, dubbed Chipotle America's "emotionally abusive boyfriend." The ephemerality of Chipotle's virology comes back to haunt not only the corporation but also the customers when people literally grow ill. Viral rhetoric festers: It is both generative of and conducive to the decay of images and ideas of, say, "good" food. How telling, then, that the outbreak of disease left Chipotle "stuck in marketing limbo" and that crisis managers were quick to advocate for the production of a YouTube video aligning a farm-to-table ethic with food safety practices.[17] Viral media campaigning such as this is inextricably tied to a reorganization of the very culture of public health.

In approaching public health for its *nomos* (cultural organization) as well as its *phusis* ("natural" state), I examine digital virology as "a rhetorical disorder" that transmits (im)purities in order to (im)purify errors of body and mind.[18] Chipotle, for instance, typifies the fact that corporate entities are increasingly expected to cure the very ills that they, or at least their industries, help to cause. (There is no lack of irony in the contemporary phenomenon of "cause marketing.") To make my case, I engage rich accounts in classical humorism that align disease with cultural dis-ease.[19] Media virality represents what Ted Striphas might call a "language shift" in discourses of media circulation,[20] not just "a dominant digitoral art form" that "enframes, contextualizes, bounds, manages, regulates, and controls" how certain media contents or communication events catch on.[21] Ancient humoral theories countervail dominant theories of viral rhetoric insofar as they show this shift for an imagined balance of the cardinal humors (sanguinity, choler, melancholia, and phlegm), which once formed a metric for judging virtues and vices not unlike Aristotle's tenet of a "golden mean." If there was balance, one was in good humor; if there was *im*balance, one was ill, or at least out of humor (as in Ben Jonson's 1599 text, *Every Man Out of His Humour*). Moreover, equilibrium arose when the "bad" was purified by the "good," whether through bloodletting, diet, induced fevers, or other methods of dampening afflictions. Important here is the idea that rhetoric drives the circulation of perverted appetites, corrupted reason, and sickened spirits, never mind the dispensation of *pharmaka*.[22] Viral marketing campaigns therefore evoke an older sense of communication as "a homeopathic remedy" wherein "the disease and the cure are in cahoots."[23] So it is fitting to imagine not only dispositions of bodies (i.e., "healthy" participants in public culture) and bodies politic (i.e., disease-ridden collectivities) via rhetorics rooted in a *politeia*, a way of life, but also a "constitution" that emphasizes the humoral composition of communities.

Following Hippocrates, bodily humors conditioned mental states, or frames of mind. Humor was also recognized as "an excited state of *public* feeling,"[24] or what David Hume called the "humours of the people,"[25] and so offers a critical nexus for examining the relationship between the tempers of a single person and the general mood of a public. A revaluation of media viruses as particularly rhetorical contagions demands a feeling for viral rhetoric that works through digital expressions of cultural anxiety by imagining how and why communication pathologies seem to exhibit "good" citizenship, "indifferent" viewership, "melancholic" social consciences, and so on.[26] Perhaps most importantly, it makes digital virology less about spreadability than about the magnification of cultural disturbances that get contained in the diffusion of "single cultural units" like hashtags, photographs, and videos.[27] Such containments fester within communities and media such that they, like bad blood or pesky phlegm, require remedies or releases to restore balance in human relations.

Chipotle's ads participate in this viral imagination, evocatively picturing dire fallout from subtherapeutic antibiotics, animal maltreatment, and genetic modification in food production, while situating the company as the source of large-scale change despite its own "industrial" dependence. They also debase American food culture by linking corporeal conditions to ethical choices, animating so-called food with integrity to purify principles of both production and consumption. This alliance of somatic experience and sense-making is why it is so intriguing that Jenkins, with his creditable concern for human agency and participatory cultures, remains so committed to rational choice that he misses not only how people act "irrationally" and still voluntarily transmit images and ideas but also how viral media "operates beyond traditional paradigms of reason and authority."[28] Affective investments can be as persuasive as, if not more compelling than, reasoned arguments, as can the draw to "feel good foods" and the repulsion of foodborne illnesses. Besides, is a pathogen not a producer of experience, especially if media content and not just people can be pathogenic? To some extent, all persuasive media is "viral" since it "pollutes" human capacities for judgment even as it influences the rhetorical and cultural matters that impact collective life.[29] We are therefore wise to dwell on those pathogens that seem to infect our public culture.

Ultimately, imagining "viral" videos as intensifying rhetorical humors reveals a pathogenic view of cultural relations that thrive on (im)purification. Chipotle's viral campaign against "Big Food" exemplifies how viral media provide moments of pause for pondering the "contamination of logic" in cultural exchanges.[30] It is so disturbing in part because it relies on the curative power of a "pure imagination" in rhetorical expressions of ill-humored corporate and consumer practices. The campaign also exaggerates the fact that civic harmony and strife rely heavily on what we *actually* consume, from aesthetic entities to tangible foodstuffs to big mistakes in commercial activities. Plus, Chipotle's own (im)pure products give reason to reconsider the ancient wisdom that viruses constitute cultural values and the materiality of bodies politic, making us communicable as much as communicative beings.

From Ancient Antecedents to the Humors of Digital Rhetoric

Real and imaginary diseases haunted the ancient Greek (and Roman) imagination.[31] Dis-ease in general was central in disordering, as well as restoring order to, society. For instance, in Greek mythology, the *nosoi* were those diseases (and their cures) that were sent by Zeus as penalties for earthly improprieties. In Homer's *Iliad*, a cloudburst of pestilential arrows represents an epidemic brought on by Apollo after his priest is dishonored. In fact, arrows were regularly harbingers of disease in classical lore, and they almost preordain the "vectors" of transmission that digital diseases tend to follow. Consider the goddess Artemis and her use of "disease-carrying arrows" to protect nature, animals, and agricultural lands.[32] The Roman festival Robigalia aimed to ward off agricultural blights. Diseases also indicated a tension between unseen, supernatural powers and the phenomenal space of civilization. Simply, diseases could "masquerade as concrete objects of perception" and instigate corrupted images of public health.[33]

Importantly, these conceptions picture diseases as not only physiological conditions but also responses to, or symptoms of, shared dispositional disorders. For some, like the pre-Socratic polymath Empedocles, rhetorical purification precipitated public healing. Hippocrates expanded Empedoclean theories in his orientation to sickness as indicative of tensions between oppositional forces within the body, not just among natural elements in the cosmos, in earthly germs, or in moments of divine punishment.[34] Aristotle, too, channeled homeopathic understandings and allopathic measures to construct his rhetoric of *katharsis*, which identified in the virtues of public speech and theatrical performance the amplification of communal tensions, or vices, for the purpose of purging them. The management of proportions, mixings, excesses, and deficiencies was also promoted for its ability to prevent dis-ease by balancing "viral" transmissions with ethical conduct. What is more, purification was never simply personal since audiences to an oration or spectators of a play observed each other, and so gleaned cues about how to address tragedies or manage inappropriate behaviors. This is no doubt why Plutarch identified "disease, treatment, and health" as the "three conditions of the body politic."[35] This association of infirmity with cultural strife warrants renewed attention to viral media, first, because it displays public screens as powerful communicative spaces that mediate relations among rhetorical substances and lived realities[36] and, second, because it reveals the powerful epidemio-logics behind cybernetic productions and perceptions of public health. More specifically, a rhetorical perspective on communicability stresses the cultural cum pathogenic qualities of digital virology.

Hippocrates defined disease as "a *systemic* problem" of humors in the body.[37] For instance, *blood* was related to the liver and to air (heat and moisture). If one was excessively *bloody*, s/he was considered "sanguine" (courageous, optimistic—or, in ancient times, bloodthirsty). *Phlegm* was associated with the lungs and the brain, along with water (cold and moisture). If one was overly *phlegmatic*, s/he was deemed apathetic or unexcitable. *Choler*, or yellow bile, was linked to the spleen and fire (heat and dryness). Those markedly *choleric* were bad-tempered or irascible. Lastly, *melancholia*, or black bile, was connected to the gall bladder and earth, or the ground (cold and dryness). A *melancholic* person was miserable, discontented, or sad. Each of these conditions was also variable. An individual might be prone to melancholia, but such a disposition could fluctuate in accordance with emotional, mental, and physical states. Consequently, one could be habitually melancholic but, for a moment, sanguine. Of course, while Hippocratic humorism honed in on the etiology of diseased bodies (and souls), it also harnessed the body politic itself as a balance of humors. This, in turn, linked disease to bad civic health and to the corruptions and corrections of rhetoric.[38] A body politic that was too sorrowful, for example, could be compelled through a public speech or a tragic play to feel more sanguine. The point is that rhetorical humors made impressions of and on cultural anxieties.

It is easy to see humoral theory's impact on ideas about *katharsis*. After all, Hippocrates held that the so-called four temperaments moved from bodily activities to collective dispositions. The restorative force of *katharsis* was gleaned from artful reanimations of sociopolitical ills in dramatic portrayals of cultural problems. Put simply, an art form was cathartic if it diffused turmoil. This is why Aristotle described *katharsis* as a doctrine of purgation *and* purification: It did away with contaminating agents that disturbed communal values and norms. Kenneth Burke examined catharsis for its organization of civic rituals of victimage, which operate on guilt, purification, and redemption.[39] In a Burkean schema, Chipotle's videos do the dramatic work of constructing victims (consumers) and a villain ("Big Food") in order to resolve the apathy and unexcitability in consumer orientations toward unethical food production with the courage and optimism embodied in Chipotle's "pure" products. What is more, prudence and reflectiveness become the rhetorical humors for cleansing contaminated feelings about food cultures that "seem to surround us, like a thickness in the air."[40] I say more about the humoral situation in my analysis of Chipotle's videos below, but for now it is important to acknowledge that *katharsis* draws attention to specific aspects of Chipotle's (im)pure imaginations: One entity must be mortified to "purify" others; the communicative agent of dis-ease is as (if not more) influential as a communicable disease (i.e., toxic food); and for a body politic to relieve or release some tension it must be made to feel it first, which is to say that such a tension—or, such an imbalance of rhetorical humors—might not exist in any articulable sense until it is created or incited, in this case by viral media content.

Crucial here is a turn to balance. That is, humoral theory inspires consideration of moods like anger, sorrow, disinterest, or optimism, as they structure public feelings while goading an inquiry into just how much purifications restore a *stasis*, or status quo. In Burke's philosophy, a scapegoat is integral to the restoration of order. With a marketing campaign like Chipotle's, consumers are not really expected to change their consumption habits but instead are urged to blame Big Food for altering their choices of where and what to buy. The point, though, is still to patronize large, corporate eateries. A we-are-what-we-eat ethos thus translates a public matter into a personal decision. Nevertheless, via humorism, the emphasis is much less on a quasi-religious deliverance from secular sin than on a patent recomposition of a body politic and therefore a movement from what is sinful to what is insalubrious for collective life.

This is the reason Galen—a Greek physician for the Roman empire who carried on the Hippocratic Corpus—conceptualized disease as an "impairment of bodily activities,"[41] holding that physicians "should restore the balance" of dis-ease in individuals and bodies politic.[42] Galen's position makes sense: He survived the Antonine Plague and witnessed the "institutional" response ordered by Marcus Aurelius, which relied on symbolic explanations of the ills within the res publica. It also hastened the sorts of "purificatory procedures" (bloodletting, the measurement of bodily outputs according to dietary intake, auscultation, etc.) that followed from Hippocrates's idea that diseases thrive in the bad atmospheres humans create for themselves.[43] Today, Chipotle's videos are contact points for examining the bad blood that sustains consumer relations, environmental degradation as a byproduct of food production, and shared feelings that weigh into the anger (or apathy) that follows from factory farming.

Dis-ease, then, is an "imbalance or disorder of the beliefs, knowledge, feelings, and appetites that give the *psyche* its content and structure."[44] Mary Douglas calls this "peculiar behavior," and it highlights the theoretical stakes in imbrications of *nomos* and *phusis* as catalysts for rhetorical ecologies. From Gorgias through Aristotle, dis-ease was deemed a byproduct of "the ethical accords of community."[45] Plato used this outlook to degrade rhetoric as a perversion of natural (i.e., good) appetites.[46] Hippocrates saw "the medical art" as a study of "general modes of existence," not just singular "conditions of ill health,"[47] aligning natural law with the second nature of cultural habits. "Digital contagions" are thus "cultural forms of contagion" that shape principles, perspectives, and conduct.[48]

Much viral media is largely built on promises of recuperation. When Chipotle infected public imaginations with an image of "some kind of sinister Mass Food-Producing Corporation That Does Cruel and Hideous Things to Its Cows and Chickens,"[49] the company exposed misconducts endemic to Big Food and also offered its own Food with Integrity™ as a curative measure. The emphasis of virology on "corrupted humors" appears as what Burke might call a "disease of communion."[50] Hence my turn to communicative pathogens that seem, in a "digital economy," to magnify the deep, perhaps primal, viral qualities of rhetoric.[51] Many rhetorical theorists follow their ancient forbears in using medical metaphors to identify rhetoric's materiality, to rationalize homeopathy, to evaluate tendencies to scapegoat, and more.[52] From the ancients forward, rhetoric has been recognized as vital to the constitution of bodies politic. Insofar as rhetoricians strive to shape public encounters, much as physicians dispense physics to alter physiologies, there is an ethic to rhetoric as an art capable of producing "'contagion'-like 'transmissions.'"[53] For some older Sophists like Gorgias, this meant not only that rhetorical afflictions were not inherently bad but also that they could persuade unwitting sick persons to take medicine. They could also draw forth "dreaded outcomes" to treat illnesses like sepsis or purify temperaments.[54] One outcome: Rhetoric embodies pathways to Good or Bad citizenship.[55]

This embodiment urges a "struggle for rhetorical ownership of illness: how it is possessed, assimilated in argument and in cliché,"[56] not to mention how it bears on the burdens of civic responsibility and the quest for healthy public discourse. Stephen Olbrys Gencarella follows Empedocles to advocate for a "purifying rhetoric" that harnesses healing arts.[57] Others, too, promote "a politics that begins modestly from where we are (sick) instead of aiming at where we are not (health)."[58] Impurifications in computer systems take shape as malware, which damage or disrupt the "behaviors" of programs, applications, and even hardware.[59] In a similar manner, viral marketing takes advantage of the cybernetics of media viruses that infect networks of sociopolitical interaction. Accordingly, "viral aesthetics" fold into "operational aesthetics," turning the rhetorical force of "sharing" media into an engine of cultural exchange.[60] Such an irruption of branded communication events expresses the viability, and virality, of consumer choice in making corrections to corrupted ways of seeing or doing things.[61] One important consideration here is how the rhetoric of humorism ultimately sizes up "the ways cultures interact."[62] Media virality is a means of judging what matters based on so many admixtures of cultural goods (and bads).[63] In today's networked society, a "digital culture of bodies and viruses" constitutes a framework for understanding "embodied media dispositions," on the one hand,[64] and, on the other, a view of what helps or harms public health. Most accounts, though, are overdetermined by the "spreads-like-a-disease" metaphor or the ill-conceived definition of "health as a deficiency of disease."[65] Skeptics of "manufactured virality" as the rhetorical pathogen for "viral information spread" are right to challenge the opinion that viruses are to digital media as diseases are to human populations.[66] What is most needed is not a doubling down on conceptions of spreadability; instead, we need a revival of virology as a way of identifying collective cultural investments and modes of enframing communication.

One way to revive virology in this way is to see media viruses as festering, not spreading. Perceptions of virality emerge as videos are viewed, stories or sentiments are shared, and so on, approaching what Malcolm Gladwell might deem the "tipping point" of cultural epidemics.[67] But this does more to intimate "an affective, circulating, and evolving series of encounters" than it does to expose a contamination, so to speak.[68] Hippocrates was concerned with bodies as contact points unto themselves, not just with the infectious potential of bodies in contact. Magnitude, in this sense, is about contagiousness in numbers and about the *status* of a media virus that contains a corrupting (or correcting) influence on the balance of rhetorical humors. So media viruses become outgrowths of "viral atmospheres" and "viral encounters."[69] Chipotle's viral marketing campaign evinces media virality as *stasis*, not spreadability. In common parlance, *stasis* suggests *in*activity, even immutability. Yet in both modern medical terms and the classical tradition, *stasis* signals a "stoppage" in the order of humors. Moreover, it signifies a "state" in which order "is disrupted and its normal functions undermined."[70] Most rhetoricians dub *stasis* an inventional resource derived from points of contention, or "issues." As early as Aristotle, though, *stasis* was seen as a sickness unto itself, the "root image" being "*nosos*, or disease,"[71] making "good" rhetoric about the alleviation of public malaise. Consequently, "disorder in the state (*stasis*)" signaled an ailing body politic.[72] Viral media provide a way to imagine how images and ideas contaminate us with the charismata of rhetorical contagions.[73]

Chipotle's videos associate with public dis-ease by aligning the company's food, corporate values, approach to consumer-as-citizen orientations, and ethics of caveat emptor with the failures and breakdowns in commitments to corporate structures of human flourishing. They therefore caricature what contaminates and what purifies, animating the power of "viral dynamics" to express what matters through a "characterization of normative cultural values."[74] The result is not a simple exploitation of common sense but rather the use of cultural networks to influence politics, profit, and publicity around public health to foster virtues and vices as the shared outcomes of being what we eat—or the media we engage and exchange.

Coming Down with Chipotle

Chipotle's attempts to purify public sentiments of food production came at a time when growing interests in animal rights (weighed against the demand for cheap meat) met questions about the ethical implications of factory farming. Still, even as the early-twenty-first-century food movement established a public discourse "ubiquitously re-iterating the message that industrial agriculture wreaks ecological havoc, endangers human health, and exploits workers in order to produce food that's overly processed, overly cheap, and overly globalized,"[75] it never "cured" a culture of apathy that encircles the slaughter of animals for food and the widespread use of antibiotics in intensive agriculture. In other words, the zeitgeist orientations that contribute to irascibility around Big Food are offset by a rather mainstream unexcitability around holding food producers accountable—books like *Fast Food Nation* and documentary films like *Food, Inc.* notwithstanding.[76] This is not to discredit the growth of "animal rights extremism," so-called humanewashing, or even the sort of agricultural elitism that is frequently paired with farm-to-table initiatives. However, it is to say that Big Food remains fairly protected from public scrutiny, particularly as a result of robust ag-gag laws. It also sustains consumers as newfangled colonialists that ignore the undemocratic nature of corporate farming and its Upton Sinclair-esque industry brutalities.[77] Chipotle's campaign makes what Carol Adams calls the "absent referents" of food production not only present in images of how we get food products but also a cultural object of dis-ease.[78]

Chipotle has long tried to rebalance the rhetorical humors around food production and consumption, namely, with billboards, posters, and radio spots situating the company as a wholesome alternative in the "fast-foodiverse."[79] The admixture of its advertising efforts exemplifies how challenging it is to track the "networks of circulation and forms of communication" that constitute viral media.[80] To say that media content goes viral, then, is to label transmissions that are clearly influential or consequential but in ways that capture the (im)purification of cultural values more than they indicate particular effects on any one individual or community. This is why so many critics suggest that Chipotle's campaign put "debilitating figments of the imagination" and "virus-like time bombs" into cultural ecosystems.[81] It is difficult to imagine a farm that builds itself, or a pigpen that grows out of the ground. It is also difficult to imagine a sentient scarecrow subjugated by a crow. Yet Chipotle's pair of video shorts made such fantasies imaginable with their respective mockeries of the human ingenuity behind factory farming. The first, titled "Back to the Start," tells "the story of a farmer who, after cramming his livestock into pens and pumping them full of hormones, suffers a crisis of conscience and returns to his roots as a sustainable farmer" (figure 5.1).[82] The distinctively adenoidal voice of country music icon Willie Nelson covering Coldplay's "The Scientist" plays in the background like a musical jeremiad against industrial agriculture: "Questions of science / Science and progress / Don't speak as loud as my heart. . . . I'm going back to the start." So it goes from perversity in food production to cartoonish pastoralism.

A sequel of sorts appeared two years later. "Scarecrow" portrays a "dystopian fantasy world" wherein "all food production is controlled by fictional industrial giant Crow Foods,"[83] and scarecrows no longer protect crops but rather labor for a cadre of malevolent, mechanical crows. Like the farmer, an anthropomorphic scarecrow suffers a crisis of conscience after seeing chickens being injected with growth hormones, cows being confined in tiny enclosures, and environmental degradation being perpetrated through corporate farming practices (figure 5.2). "Scarecrow" has a soundtrack as well: singer-songwriter Fiona Apple's "Pure Imagination," originally sung by Gene

Figure 5.1. Screenshot of Chipotle's "Back to the Start" (2011). A farmer looks on as his formerly free-range pigs are enclosed in pens.

Wilder in Mel Stuart's creepy 1971 film version of Roald Dahl's book *Charlie and the Chocolate Factory* (1964). Both videos are centerpieces of marketing efforts for Chipotle's Cultivate Foundation, the philanthropic arm of the larger corporation, to which proceeds from the two songs are donated when they are downloaded from iTunes. Both are also part of a "Cultivate a Better World" initiative. And both went viral, making them part of a more pervasive pattern of marketing by managing the dis-ease of public moods.

After "Back to the Start" first appeared, the allegory—as Chipotle called it—was affirmed by *Advertising Age* and others as A-list branding. TED named it one of its many "ads worth spreading."[84] Within a year it became a wrench in the spokes of Big Ag and a blemish on the face of fair food movements. In part, the video went viral for the alleged purity in its portrayal, encapsulating "things like animal welfare, monoculture, locally sourced produce, water pollution," and more.[85] It also typifies the sort of unthinking consumer habits that seem to be represented through the emplacement of farm animals in pens and factories that construct themselves, and in a return to "old" ways with a farmer who breaks down corporate machinations and literally cranks his farm back into form (predictably, before loading his newly fresh farm food into a truck emblazoned with the phrase "Cultivate a Better World"). However, a difference between this video and its sequel lies in how collective apathy trumps shared despair when it comes factory farming. Or, as one commentator put it, "Scarecrow" converted "apathetic customers into

Figure 5.2. Screenshot of packaged flesh on conveyor belt from Chipotle's "Scarecrow" (2013). The scarecrow rides a conveyor belt as machines transfer genetically modified cow meat from production lines into inorganic meals.

brand ambassadors,"[86] tipping the balance from apparent unconcern to excitability around the status quo of industrial agriculture.

From the first, "Scarecrow" inspired discomfort. One reason is that it "doesn't look like a commercial or promo at all" but rather "like a music video or short film."[87] Along with its biting message about factory versus family farms is a caricature of contaminated food systems. The Scarecrow, with his straw hat, stick limbs, and orange coveralls, toils among many other laborers under the watchful, red eyes of "drone crows" on the Crow Foods Incorporated factory floors.[88] On the "floors," webs of conveyer belts transport artificial meals behind the scenes, out of sight and thus out of the public mind. The Scarecrow is visibly tormented by this situation and is even pecked by a crow when he looks with downtrodden eyes at consumers of the unnatural products he helps to produce. Clearly, Crow Foods signifies a paradise lost. Most of the Scarecrow's work entails patching up its surface features: he realigns a broken plank on a billboard containing a picture of poultry and an advertisement for natural food (which conceals a chicken being injected with antibiotics) (figure 5.3); he repairs a massive metal statue of a cow (which covers up the abuse of livestock within the structure itself); and then he travels from the city to the country by bus through sprawling fields of faux farm-fresh produce, which are sprayed with chemicals by colossal robotic crows. In the end, after discovering a red pepper reminiscent

Figure 5.3. Screenshot of a chicken injected by a robot arm from Chipotle's "Scarecrow" (2013). An anthropomorphic, robotic scarecrow clutches a chicken that is being injected with growth-promoting antibiotics.

of Chipotle's logogram, the Scarecrow plucks vegetables from his home garden, cooks them, and serves them in burritos from a makeshift urban market stand like a good neoliberal hero.

All of this is set to Apple's remake of "Pure Imagination," which intones: "Come with me / And you'll be / In a world of pure imagination / Take a look / And you'll see / Into your imagination." The song formerly celebrated the unbridled inventiveness of Willy Wonka, Roald Dahl's infamous chocolatier who muddles ingredients to create newfangled confections and then invites a lucky few to come see and taste his concoctions. Apple's version mocks a view of paradise by showing the Crow Foods factory for what people *should* see when they imagine food that comes from "other" fast-food chains. It also asks those people to share in the blame while urging them to feel that, via imaginations, worlds can be remade. This notion that shared fantasies of media ecologies and everyday pollutants bring together "the indexical and the artificial" makes it even more appropriate that "Scarecrow" yielded a viral parody (figure 5.4).[89] Dubbed the "honest version," the parody channels quasi-gallows humor to restore Chipotle's video to its proper proportions. If Chipotle's video encourages a courageous-optimistic view, the parody extracts a measure of irascibility-prudence, exploiting a balance of rhetorical humors that enframes the dystopian tale. It also positions parody as a Galen-like "purificatory procedure."

In the parody, the Scarecrow is "us," which is to say Chipotle. Crow Foods is "not us." And, following its revision of "Pure Imagination," to go (viral) with Chipotle is to see a world of pure *manipulation*. "This is just a way to advertise," goes the new song. On the one hand, the parody renders Chipotle delusional in its own pretensions of purity. On the other, it calls out a corporate logic that repeats Platonic relations of rhetoric to cookery. Chipotle has since been exposed for its own impurities, which the company characterizes as exceptions (even amid multiple outbreaks of E. coli, Salmonella, and Norovirus, the latter of which the company blamed on an employee who worked while sick).[90] Many critics in the food industry have even noted that the videos do little more than distort "the reality of conventional farming in

Figure 5.4. Screenshot of "Pure Manipulation" parody. The Scarecrow's experiences are captioned by an omniscient narrator that sees through Chipotle's diagnosis of industrial agriculture.

the U.S."[91] The fact is that some of Chipotle's products *do* contain GMOs. Not all of its meat sources (i.e., Niman Ranch) keep animals outdoors, nor do they allow them to roam freely in open fields. When there are supply shortages, Chipotle, to maintain its menu offerings, readily accepts meat from livestock that have been given antibiotics. The company also uses a mixture of local and nonlocal ingredients, again depending on availability. Moreover, there are more than 1,500 Chipotle locations across the globe, which in epidemic terms implies that the company could bear an immense viral load. The viral videos therefore embody networks of food production and consumption that connect people like zoonoses and food-borne pathogens, and influence those consumer outlooks that contribute to cultural ethics. The company also relies on an industrial system of food transportation and distribution and so, despite its bucolic imagery, sustains rather than supersedes the status quo. That Chipotle, a for-profit food chain, would want to make its customers "feel all the feelings" for its brand is anything but remarkable,[92] except that its intensification of public moods is not persuasion per se but rather an operation and organization of affective associations with diet and lifestyle. In this sense, Chipotle emerges as, at best, the lesser of many evils, even among corporate contaminants.

Nonetheless, Chipotle's "values integration" offers a powerful image of media virality.[93] From its "cute-but-revolting factory farm imagery and heart-tugging animal cruelty" in its viral videos to the "exploding cows, eight-legged chickens, amusing dialogue, and cartoon-like bad guy executives" in its TV satire,[94] Chipotle typifies "going viral" as a process of making contact across digital and other networks. It is thus a locus in which cultural values collapse into one another and reveal particular temperaments that connect people, policies, politics, and products: for instance, the producers of branded content, the consumers of "responsibly raised" food, the industry of agribusiness, the merits of concentrated feeding operations, federal farm bills, labeling practices, and more. The humor in Chipotle's videos is not comical but rather built on a framework that makes disregard for animal welfare and food production practices a symptom of civic ennui. Yet it can be taken as its own kind of *im*balancing act insofar as the courage and optimism that emerges out of its own corporate duplicity conceals more prudential responses to factory farming. However, insofar as Chipotle's campaign advertises media virality, it also amplifies the rhetorical conditions of public health.

When Susan Sontag wrote about the despair that comes with disease, she lamented how the hope of consumerism comes to infect approaches to a public and its problems. Even more, she suggested that viral logics possess communities that are repressed.[95] This idea assumes a much more potent sense of rhetorical impurity when viewed in terms of media impressions, which combine metaphors of audience magnitudes, collective feelings, and perceptual marks that are left on bodies politic. Viral videos impress by constituting "virtual communities" through networks of collective feelings that permeate the surfaces—that is, the media—of public life.[96] And they repress when they contain alternative viewpoints in dialectics of poisons and remedies, pollutions and impurifications, virtues and vices. Today, cultural ecologies fold into digital networks, making the communicability of dis-ease even more about the very legible, and very open, scatter of images and ideas about the public good.[97] Chipotle's campaign should haunt us, most pointedly because of the neoliberal fantasies that linger in its advocacy of corporate physicians that treat the very problems they take part in creating. The irony is that, if the singular scarecrow is actually *us*, the public, then our task is to re-produce the very food we consume—and then to share it with others. Talk about an ethic of being what you eat.

Conclusion: Eating Crow?

At one point in "Scarecrow," an overhead shot of the city in which Crow Foods thrives reveals a billboard marked: "EAT CROW." This could be a nod to Chick-fil-A's imploration to "Eat Mor Chikin." But to "eat crow" is also to accept the humiliation that comes from disclosing lapses or faults, particularly after previously standing firm on specious convictions. The prima facie righteousness in Chipotle's viral video campaign is counteracted by the ridiculousness of the company's very pronouncements of its own purity. What is more, the communicability of Chipotle's expressions of cultural dis-ease strike an imbalance in rhetorical humors precisely because they do less to spread a message about the need for change than leave a collection of affective and logical discrepancies to fester in the body politic.

Nowhere is the rhetorical force of Chipotle's festers more pronounced than in the fallout from its problems with food safety within a couple years of its viral marketing campaign. After multiple outbreaks, Susan Berfield of *Bloomberg* called Chipotle's advertising "the source of much schadenfreude," or gratification derived from the suffering of others.[98] The outbreak, in other words, transmogrified a happy fantasy of fresh food into a haunting image of rhetorical as well as literal ingredients gone bad.[99] This is in large part why Chipotle is a test case in the contamination of viral media that, when called out for the corruptions within its "good" impurities, actually disrupts while dwelling in and on conventional ways of seeing. Additionally, because branded content maintains such "prominence within the affairs of everyday life" by over and again exploiting the "rhetorical force and moral authority" of cultural values,[100] there is a sense in which viral media is a convention of controlling dis-ease unto itself. If a particular image or idea catches on, it often does so as a result of tried-and-true "patterns of preferences for consumer goods" as popular vehicles for judging what matters, even as those patterns exist within "networks of security, instrumentality, and normative control."[101] If rhetoric exists as a key constituent of public health—which is to say the sociopolitical well-being of a body politic, from its discourses through its food production and consumption habits—then media viruses can embody the *wrongs* of a society. They can also trouble the very fitness of particular rhetorics as remedies for a stated dis-ease. From a critical-vegetarian perspective, it is easy to see Chipotle's videos and still find reprehensible the inordinate amount of energy that goes into the production of almost all meat in the United States (and some vegetables, for that matter). Moreover, advocacy for animal welfarism oftentimes still ends with the conversion of those animals into products for human consumption. But even a viral caricature of actual production practices can give pause to the relationships between consumer values and civic commitments.

One of Hippocrates's tenets is therefore an apt sentiment with which to close: A curative is compromised when the patient's interests are in conflict with the public good. There is hardly a better reason to recover the notion of humorism as a rhetorical means of influencing collective temperaments. Nor is there a better rationale for reconsidering media viruses as points of stoppage that contain errors of the body and mind with underlying communicative and ideational pathogens. This is all especially the case when virology is juxtaposed to laments about how digital technologies are frequently regarded as diseases that do damage to the body politic. It is also the case when communicability in digital networks is studied for the ways in which certain viral media seems to eat away at public cultures, and not simply for the appearances of its ready transmissibility among audiences. As author E. A. Bucchianeri suggests in *Brushstrokes of a Gadfly* (2011): "If you are what you eat, you are what you see and hear." Or, following Hippocrates, the symptoms that we can see, hear, and feel are the byproducts of what we "consume." Viral media content is therefore a symptom as much as it is a rhetorical expression of dis-ease in public communication.

Notes

1. Henry Jenkins, "If It Doesn't Spread, It's Dead (Part One): Media Viruses and Memes," Confessions of an Aca-Fan, February 11, 2009, http://henryjenkins.org/2009/02/if_it_doesnt_spread_its_dead_p.html.

2. Douglas Rushkoff, *Media Virus! Hidden Agendas in Popular Culture* (New York: Ballantine Books, 1996), 9.

3. James W. Carey, *Communication as Culture: Essays on Media and Society* (New York: Routledge, 1992), 34.

4. What Richard Dawkins offhandedly dubbed "virus-makers" have, in the digital age, collapsed distinctions between the production, consumption, and circulation of content across online networks. This has led scholars like Jenkins to conclude that viral metaphors conceal the role of human actors in the transmission of ideas, emphasizing disproportionate power relations between producers and users, or even eliding the extent to which commercial motives drive access and distribution. See Marita Gronnvoll and Jamie Landau, "From Viruses to Russian Roulette to Dance: A Rhetorical Critique and Creation of Genetic Metaphors," *Rhetoric Society Quarterly* 40, no. 1 (2010): 46–70; and Barbara Warnick, *Rhetoric Online: Persuasion and Politics on the World Wide Web* (New York: Peter Lang, 2007), 8. Note as well that media virality is not simply about prevention and treatment. Instead, it is about exploiting passageways for the movement of contaminated content, which inspires a potent sensibility when one considers the perceived prevalence of *actual* diseases and the panic that tends to follow their news coverage (i.e., the 2014 Ebola outbreak). It also bespeaks a peculiar combination of anxieties and desires that draws the rhetorical "force of contagion" from overwrought portrayals of extreme outcomes. See Robert Payne, *The Promiscuity of Network Culture: Queer Theory and Digital Media* (New York: Routledge, 2015), 8; see also Jonah Berger and Katherine L. Milkman, "What Makes Online Content Viral?" *Journal of Marketing Research* 49, no. 2 (2012): 193.

5. Roger Brock, *Greek Political Imagery: From Homer to Aristotle* (New York: Bloomsbury, 2013), 69.

6. Victoria Wohl, "Rhetoric of the Athenian Citizen," in *The Cambridge Companion to Ancient Rhetoric*, ed. Erik Gunderson (New York: Cambridge University Press, 2009), 175.

7. Gabriel Beltrone, "Willie Nelson Covers Coldplay's 'The Scientist' for Chipotle Ad," *Adweek*, August 30, 2011, http://www.adweek.com/adfreak/willie-nelson-covers-coldplays-scientist-chipotle-ad-134529.

8. James Poniewozik, "And the Grammy for Best Animated Pig in a Burrito Ad Goes To . . . ," *TIME*, February 13, 2012, http://entertainment.time.com/2012/02/13/and-the-grammy-for-best-animated-pig-in-a-burrito-ad-goes-to/.

9. Maggie McGrath, "The Advertising Game: How Brands Like Chipotle, Google And Gap Rise Above Competitors," *Forbes*, October 21, 2014, http://www.forbes.com/sites/maggiemcgrath/2014/10/21/the-advertising-game-how-brands-like-chipotle-google-and-gap-rise-above-competitors/#3dba746b3689.

10. Elizabeth Weiss, "What Does 'The Scarecrow' Tell Us About Chipotle?" *New Yorker*, September 23, 2013, http://www.newyorker.com/business/currency/what-does-the-scarecrow-tell-us-about-chipotle. See also Tim Nudd, "Ad of the Day: Chipotle Makes Magic Yet Again With Fiona Apple and a Dark Animated Film," *AdWeek*, September 12, 2013, http://www.adweek.com/news/advertising-branding/ad-day-chipotle-makes-magic-again-fiona-apple-and-dark-animated-film-152380.

11. Weiss, "What Does 'The Scarecrow' Tell Us."

12. David Vinjamuri, "Chipotle Scarecrow Makes Enemies to Win Customers," *Forbes*, September 13, 2013, http://www.forbes.com/sites/davidvinjamuri/2013/09/13/chipotle-scarecrow-makes-enemies-to-win-customers/#5a11fd4e5d01.

13. Rich McEachran, "Is Chipotle's Scarecrow the future of sustainable advertising?" *Guardian*, Wednesday, October 2, 2013, http://www.theguardian.com/sustainable-business/chipotle-scarecrow-sustainable-advertising.

14. Ekaterina Haskins, "Ephemeral Visibility and the Art of Mourning: Eyes Wide Open Traveling Exhibit." In *Rhetoric, Remembrance and Visual Form: Sighting Memory*, ed. Anne T. Demo and Bradford Vivian (London: Routledge, 2012), 89–112.

15. I borrow this metaphor of "stickiness" from Sara Ahmed. See Sara Ahmed, *The Cultural Politics of Emotion* (New York: Routledge, 2004).

16. Karin Klein, "Chipotle's 'Scarecrow' video: Lighten up, folks, it's just an ad," *L.A. Times*, September 26, 2013, http://articles.latimes.com/2013/sep/26/news/la-ol-chipotle-scarecrow-video-20130926. See also Twilight Greenaway, "What's Eating Us About That 'Hauntingly Beautiful' Chipotle Ad," *Smithsonian Magazine*, September 20, 2013, http://www.smithsonianmag.com/arts-culture/whats-eating-us-about-that-hauntingly-beautiful-chipotle-ad-10363916/?no-ist.

17. Julie Jargon, "Why You Won't Soon See Chipotle Ads," *Wall Street Journal*, December 25, 2015, http://www.wsj.com/articles/why-you-wont-soon-see-chipotle-ads-1451083903. See also James B. Stewart, "Chipotle's New Mantra: Safe Food, Not Just Fresh," *New York Times*, January 14, 2016, http://www.nytimes.com/2016/01/15/business/chipotles-new-mantra-safe-food-not-just-fresh.html.

18. Judy Z. Segal, *Health and the Rhetoric of Medicine* (Carbondale: Southern Illinois University Press, 2005), 75. See also Peta Mitchell, *Contagious Metaphor* (New York: Bloomsbury, 2012), 139.

19. See Ruth Padel, "External and Internal Forces of Disease," in *In and Out of*

the Mind: Greek Images of the Tragic Self (Princeton, NJ: Princeton University Press, 1992), 49–59.

20. Ted Striphas, "The Internet of Words," *Chronicle Review*, August 11, 2014, http://chronicle.com/article/The-Internet-of-Words/148179/.

21. Jonah Sachs, *Winning the Story Wars* (Cambridge, MA: Harvard Business School Publishing, 2012), 187; Alexander R. Galloway and Eugene Thacker, *The Exploit: A Theory of Networks* (Minneapolis: University of Minnesota Press, 2007), 108.

22. See Jacques Derrida, "Plato's Pharmacy," in *Dissemination* (London: Continuum, 1981): 67–186. See also Adam David Roth, "Reciprocal Influences between Rhetoric and Medicine in Ancient Greece" (PhD diss., University of Iowa, 2008), http://ir.uiowa.edu/etd/3.

23. John Durham Peters, *Speaking into the Air: A History of the Idea of Communication* (Chicago: University of Chicago Press, 1999), 6.

24. "humour | humor, n." *OED Online* (Oxford: Oxford University Press, 2014), accessed April 1, 2015, http://www.oed.com/view/Entry/89416?rskey=7VCK5I&result=1&isAdvanced=false, emphasis added.

25. David Hume, *The History of England, from the Invasion of Julius Caesar to the Revolution in 1688* (London: W. Bulmer and Co. Cleveland Row, 1803), 490.

26. James E. Combs and Dan Nimmo, *The Comedy of Democracy* (Westport: Praeger, 1996), 47.

27. Limor Shifman, *Memes in Digital Culture* (Cambridge, MA: MIT Press, 2014), 56. Viral media also seems to cure itself insofar as its popularity spikes before it spirals out of the public imagination to make way for the next digital virus.

28. Elizabeth Winkler, "The Viral Imagination," *Los Angeles Review of Books*, November 17, 2014, http://lareviewofbooks.org/essay/viral-imagination.

29. See Stephen Olbrys Gencarella, "Purifying Rhetoric: Empedocles and the Myth of Rhetorical Theory," *Quarterly Journal of Speech* 96 (2010): 231–56. See also Laurie E. Gries, *Still Life with Rhetoric: A New Materialist Approach for Visual Rhetorics* (Boulder: University Press of Colorado, 2015).

30. Derrida, "Plato's Pharmacy," 149.

31. G. E. R. Lloyd, *In the Grip of Disease: Studies in the Greek Imagination* (New York: Oxford University Press, 2003), 202.

32. Alexander Rubel, *Fear and Loathing in Ancient Athens: Religion and Politics During the Peloponnesian War* (New York: Routledge, 2014), 128.

33. Brooke Holmes, *The Symptom and the Subject: The Emergence of the Physical Body in Ancient Greece* (Princeton, NJ: Princeton University Press, 2010), 134.

34. See J. N. Hays, *The Burdens of Disease: Epidemics and Human Response in Western History* (New Brunswick: Rutgers University Press, 1998), 294. Also note that Empedocles opposed the human consumption of meat, or "the din of slaughter." Pythagoras and later Plutarch likened such a din to the human penchant for violence. Hippocrates, too, advocated a plant-based diet. What is more, the Hippocratic oath relies on an appeal to earthly (i.e., ethical) medical practice even as it is driven by appeals to Apollo (often depicted as the ultimate physician), Asclepius (the god of healing arts), and Hygeia (the god of health and humors).

35. James F. McGlew, *Tyranny and Political Culture in Ancient Greece* (Ithaca, NY: Cornell University Press, 1993), 73.

36. Kevin Michael DeLuca and Jennifer Peeples, "From Public Sphere to Public Screen: Democracy, Activism, and the 'Violence' of Seattle," *Critical Studies in Media Communication* 19, no. 2 (2002): 125–51. It is worth noting that "to screen" is also to test for a disease, which makes the public screen a site for trying out images and ideas to see if they go viral—and a site for critics to engage moments of cultural anxiety.

37. Hays, *Burdens of Disease*, 12. Note that some classical conceptions aligned temperaments with seasons and even elements in the universe (i.e., earth, water, fire, and air) or framed them in terms of hot and cold conditions of both environments and the animate bodies therein.

38. See Brock, *Greek Political Imagery*, 69. See also Kostas Kalimtzis, *Aristotle on Political Enmity and Disease: An Inquiry into* Stasis (Albany: SUNY Press, 2000), 5; and Donna J. Haraway, *Simians, Cyborgs, and Women: The Reinvention of Nature* (New York: Routledge, 1991), 7.

39. See Kenneth Burke, "On Catharsis, or Resolution," *Kenyon Review* 21, no. 3 (1959): 337–75; and Kenneth Burke, "Catharsis—Second View," *Centennial Review* 5, no. 2 (1961): 107–32. See also Debra Hawhee, "Welcome to the Beauty Clinic," in *Moving Bodies: Kenneth Burke at the Edges of Language* (Columbia: University of South Carolina Press, 2009), 125–55.

40. Ahmed, *Cultural Politics of Emotion*, 10.

41. I. S. Yeo, "The Concept of Disease in Galen," *Uisahak* 12, no. 1 (2003): 54–65.

42. Hays, *Burdens of Disease*, 13.

43. Jacques Jouanna, *Greek Medicine from Hippocrates to Galen: Selected Papers* (The Netherlands: Brill, 2012), 127.

44. Pedro Laín Entralgo, *The Therapy of the Word in Classical Antiquity*, ed. and trans. L. J. Rather and John M. Sharp (New Haven, CT: Yale University Press, 1970), 131–32. See also George A. Kennedy, *Classical Rhetoric and Its Christian and Secular Tradition From Ancient to Modern Times*, 2nd ed. (Chapel Hill: University of North Carolina Press, 1999), 30.

45. James Crosswhite, *Deep Rhetoric: Philosophy, Reason, Violence, Justice, Wisdom* (Chicago: University of Chicago Press, 2013), 125.

46. Plato approached the body as a site of unpredictability, confusion, and chaos, and so promoted "humorless" reason.

47. Debra Hawhee, *Bodily Arts: Rhetoric and Athletics in Ancient Greece* (Austin: University of Texas Press, 2004), 88.

48. Mitchell, *Contagious*, 130, 5.

49. Alexandra Petri, "The Problem with Chipotle's Scarecrow Commercial—and the Funny or Die Parody," *Washington Post*, September 20, 2013, http://www.washingtonpost.com/blogs/compost/wp/2013/09/20/the-problem-with-chipotles-scarecrow-commercial-and-the-funny-or-die-parody/. See also Carey Polis, "Chipotle 'Scarecrow' Ad Parody Is Right on Point," *Huffington Post*, September 19, 2013, http://www.huffingtonpost.com/2013/09/19/chipotle-scarecrow-parody_n_3953692.html.

50. Holmes, *Symptom*, 14; Kenneth Burke, *A Rhetoric of Motives* (Berkeley: University of California Press, 1969), 22.

51. Gries, *Still Life with Rhetoric*, xx.

52. See N. Katherine Hayles, *How We Became Posthuman: Virtual Bodies in Cybernetics, Literature, and Informatics* (Chicago: University of Chicago Press, 1999), 213.

53. Debra Hawhee, "Kairotic Encounters," in *Perspectives on Rhetorical Invention*, Janet M. Atwill and Janice M. Lauer, ed. (Knoxville: University of Tennessee Press, 2002), 30, 32. See also Janet M. Atwill, *Rhetoric Reclaimed: Aristotle and the Liberal Arts Tradition* (Ithaca, NY: Cornell University Press, 1998), 74.

54. Combs and Nimmo, *Comedy of Democracy*, 46.

55. See Henry Jenkins, Sam Ford, and Joshua Green, *Spreadable Media: Creating Value and Meaning in a Networked Culture* (New York: New York University Press, 2013), 22.

56. Susan Sontag, *Illness as Metaphor and AIDS and Its Metaphors* (New York: Anchor Books, 1988), 181–82, 3.

57. Gencarella, "Purifying Rhetoric," 238, 252. See also Stephen Olbrys Gencarella, "The Myth of Rhetoric: Korax and the Art of Pollution," *Rhetoric Society Quarterly* 37, no. 3 (2007): 251–73.

58. C. T. Ricciardone, "'We are the Disease': Truth, Health, and Politics from Plato's *Gorgias* to Foucault," *Epoché* 18, no. 2 (2014): 301.

59. Consider the role of viruses as coded computer contagions that upset the integrity of a digital system, corrupting files, breaching data stores, crashing hard drives, and subjecting operating systems to veiled forms of control and criminality (even as such contagions are controlled in turn by those individuals and entities that practice cyber security). See Cynthia J. Davis, "Contagion as Metaphor," *American Literary History* 14, no. 4 (2002): 828–36; Seth Godin, *Unleashing the Idea Virus* (New York: Do You Zoom, 2001); Karine Nahon and Jeff Hemsley, *Going Viral* (Malden: Polity Press, 2013); Jussi Parikka, *Digital Contagions: A Media Archaeology of Computer Viruses* (New York: Peter Lang, 2007); Priscilla Wald, *Contagious: Cultures, Carriers, and the Outbreak Narratives* (Durham, NC: Duke University Press, 2008).

60. See Alan Liu, *The Laws of Cool: Knowledge Work and the Culture of Information* (Chicago: University of Chicago Press, 2004); and Angela Ndalianis, *The Horror Sensorium: Media and the Senses* (Jefferson, NC: McFarland, 2012).

61. Robert E. Moore, "From Genericide to Viral Marketing: On 'Brand,'" *Language and Communication* 23, no. 3 (2003): 352.

62. Parikka, *Digital Contagions*, 2.

63. See Thomas B. Farrell, "Sizing Things Up: Colloquial Reflection as Practical Wisdom," *Argumentation* 12, no. 1 (1998): 1–14.

64. Parikka, *Digital Contagions*, 204; Payne, *Promiscuity of Network Culture*, 135.

65. Sampson, *Virality*, 2; George E. Yoos, *Politics and Rhetoric: Coming to Terms with Terms* (New York: Palgrave MacMillan, 2009), 130. See also Gries, *Still Life with Rhetoric*, 27–29; W. J. T. Mitchell, *Cloning Terror: The War of Images, 9/11 to the Present* (Chicago: University of Chicago Press, 2011), xi; and Sontag, *Illness as Metaphor*, 156.

66. Payne, *Promiscuity of Network Culture*, 68; Joseph Turrow, *Playing Doctor: Television, Storytelling, and Medical Power* (Ann Arbor: University of Michigan Press, 2010), 228.

67. Malcolm Gladwell, *The Tipping Point: How Little Things Can Make a Big Difference* (Boston: Back Bay Press, 2002).

68. Thomas Rickert, *Ambient Rhetoric: The Attunements of Rhetorical Being* (University of Pittsburgh Press, 2013), 44.

69. Sampson, *Virality*, 58, 79.

70. Kalimtzis, *Aristotle*, 16.

71. Ibid., 18.

72. Brock, *Greek*, 69.

73. Sampson, *Virality*, 28.

74. Payne, *Promiscuity of Network Culture*, 94.

75. James McWilliams, "Our Failed Food Movement," *Pacific Standard*, June 17, 2015, http://www.psmag.com/health-and-behavior/our-failed-food-movement.

76. James McWilliams, "The Dangerous Psychology of Factory Farming," *The Atlantic*, August 24, 2011, http://www.theatlantic.com/health/archive/2011/08/the-dangerous-psychology-of-factory-farming/244063/.

77. Eric Schlosser, "Why Being a Foodie Isn't 'Elitist,'" *Washington Post*, April 29, 2011, https://www.washingtonpost.com/opinions/why-being-a-foodie-isnt-elitist/2011/04/27/AFeWsnFF_story.html. See also Chad Lavin, "Factory Farms in a Consumer Society," *American Studies* 50, no. 1/2 (2009): 71–92.

78. Carol J. Adams, *The Sexual Politics of Meat: A Feminist-Vegetarian Critical Theory* (New York: Continuum, 1990).

79. Belle Cushing, "Is Chipotle's New Hulu Series Art, Advertising, or Something Else?" *Bon Appétit*, February 17, 2014, http://www.bonappetit.com/entertaining-style/trends-news/article/chipotle-hulu.

80. Megan Boler, "Introduction," in *Digital Media and the Democracy: Tactics in Hard Times* (Cambridge, MA: MIT Press, 2008), 15.

81. Adrian Del Caro, *Grounding the Nietzsche Rhetoric of Earth* (Berlin: Walter de Gruyter, 2004), 77. The notion of an advertising ecosystem comes from Mark Crumpacker, Chipotle's chief marketing and development officer.

82. Beltrone, "Willie"; Lucas Kavner, "Willie Nelson's Coldplay Cover For Chipotle," *Huffington Post*, August 31, 2011, http://www.huffingtonpost.com/2011/08/31/willie-nelson-coldplay-the-scientist_n_943420.html.

83. "The Scarecrow," *Chipotle Mexican Grill*, http://www.scarecrowgame.com/film.html.

84. "Ads Worth Spreading," TED, https://www.ted.com/about/programs-initiatives/ads-worth-spreading.

85. Elizabeth Olson, "An Animated Ad With a Plot Line and a Moral," *New York Times*, February 9, 2012, http://www.nytimes.com/2012/02/10/business/media/chipotle-ad-promotes-sustainable-farming.html?_r=0.

86. Kathleen Davis, "Chipotle's Brilliant Branding Move," *Entrepreneur*, September 13, 2013, https://www.entrepreneur.com/article/228389.

87. Samantha Grossman, "Beautiful Chipotle Promo Video Goes Viral," *Time*, September 16, 2013, http://newsfeed.time.com/2013/09/16/beautiful-chipotle-promo-video-goes-viral/.

88. James Hamblin, "Popular Dystopian Burrito Campaign Implies Chipotle Is Not a Giant Fast-Food Chain," *Atlantic*, September 16, 2013, http://www.theatlantic.com/health/archive/2013/09/popular-dystopian-burrito-campaign-implies-chipotle-is-not-a-giant-fast-food-chain/279675/.

89. Kirsten Ostherr, *Cinematic Prophylaxis: Globalization and Contagion in the Discourse of World Health* (Durham, NC: Duke University Press, 2005), 194. See also Thierry Bardini, *Junkware* (Minneapolis: University of Minnesota Press, 2011), 169–206.

90. See Dan Charles, "Why We Can't Take Chipotle's GMO Announcement All That Seriously," *NPR*, April 30, 2015, http://www.npr.org/sections/thesalt/2015/04/30/403249551/why-we-cant-take-chipotles-gmo-announcement-all-that-seriously; Greenaway, "What's Eating"; Dana Liebelson, "Behind the Burrito: 5 Things Chipotle's Ads Don't Tell You," *Mother Jones*, September 25, 2013, http://www.motherjones.com/politics/2013/09/chipotle-commercial-sustainable-food-truth; Deena Shanker, "9 Disappointing Facts About Chipotle," June 25, 2014, http://www.buzzfeed.com/deenashanker/disappointed-in-chipotle; and Weiss, "What Does 'The Scarecrow' Tell Us." See also Chipotle's response at https://www.chipotle.com/2015incidents.

91. Noam Cohen, "Chipotle Blurs Lines With a Satirical Series About Industrial Farming," *New York Times*, January 27, 2014, http://www.nytimes.com/2014/01/27/business/media/chipotle-blurs-lines-with-a-satirical-series-about-industrial-farming.html. See also Elizabeth Weiss, "Chipotle's Seasoned Première," *New Yorker*, February 13, 2014, http://www.newyorker.com/business/currency/chipotles-seasoned-premire; Matthew Yglesias, "*Farmed and Dangerous*: A Satirical New Show on Hulu Created by Chipotle," *Slate*, January 27, 2014, http://www.slate.com/blogs/moneybox/2014/01/27/chipotle_s_farmed_and_dangerous_chipotle_satirizes_factory_farming.html; and others.

92. Polis, "Chipotle 'Scarecrow' Ad."

93. This is another term used by Crumpacker.

94. Peter Weber, "WATCH: Fiona Apple and Chipotle Channel Willy Wonka to Slam Factory Farming," *The Week*, September 13, 2013, http://theweek.com/articles/460038/watch-fiona-apple-chipotle-channel-willy-wonka-slam-factory-farming; Brad Tuttle, "Chipotle's 'Farmed and Dangerous': Entertainment or Marketing Propaganda?" *Time*, February 18, 2014, http://business.time.com/2014/02/18/chipotles-farmed-and-dangerous-entertainment-or-marketing-propaganda/.

95. Sontag, *Illness as Metaphor*, 51, 73.

96. Sara Ahmed, "Collective Feelings Or, the Impressions Left by Others," *Theory, Culture and Society* 21, no. 2 (2004): 25–42.

97. See Peters, *Speaking into the Air*, 62.

98. Susan Berfield, "Inside Chipotle's Contamination Crisis," *Bloomberg*, December 22, 2015, http://www.bloomberg.com/features/2015-chipotle-food-safety-crisis/.

99. Julie Jargon and Jesse Newman, "Fresh Ingredients Came Back to Haunt Chipotle,"

Wall Street Journal, February 2, 2016, http://www.wsj.com/articles/fresh-ingredients-came-back-to-haunt-chipotle-1454463065.

100. William Leiss, Stephen Kline, and Sut Jhally, *Social Communication in Advertising*, 2nd ed. (New York: Routledge, 1997), 1.

101. Leiss, Kline, and Jhally, *Social Communication in Advertising*, 1; Payne, *Promiscuity of Network Culture*, 59.

6

Isocratean *Tropos* and Mediated Multiplicity

Rosa A. Eberly and Jeremy David Johnson

Even before President Bill Clinton proposed that "it depends upon what the meaning of the word 'is' is," in answer to questions about the status of his relationship with White House intern Monica Lewinsky, the forty-second president of the United States had been accused of political "spin." In a 1996 "On Language" column in the *New York Times*, the former public relations executive, speechwriter for President Richard Nixon, and *conservateur* of written style, William Safire reported receiving a correction to his own definition of "spin"—and its origins in sport—from Clinton's then-press secretary Michael McCurry.[1] "When a pitcher tosses a curveball . . . he affects the physics of the motion so that the ball actually 'breaks,'" Safire reports McCurry to have written to him. "It does more than 'appear' to go in a slightly different direction—unless, of course, the curve 'hangs' like tortured spin falling on deaf ears." Safire describes more of the correspondence from McCurry, whom he calls "the Clintonian Newtonian," in a way that emphasizes the very thin line between effective spin and entropic dissolution: "The spinner's 'English' on a story is designed to move the story in a more favorable direction. That said, the pitch had better cross the plate. Too many wild spins, and someone sooner or later sends you to the showers."

Declining to weigh in on the proximity of the Clinton presidency to being sent "to the showers," this chapter instead takes its point of departure from the observation that, unlike his intellectual step-grandbaby Aristotle, ancient protopragmatist Isocrates did not theorize what is usually translated "character" as *ēthos*.[2] What Isocrates's English translators Norlin, Van Hook, and Mirhady and Too have rendered as "character" in Isocrates is not *ēthos*—among the most durable concepts from antiquity—but almost always *tropos*.[3] Further, whereas *ēthos* suggests dwelling and consistency, *tropos* suggests turning and mutability: conceivably, and with enough velocity, *tropos* might suggest dizzying "spin."[4] Indeed, taken to its spinning limits, *tropos* troubles not only the line between what a founder of American pragmatism, Ralph Waldo Emerson, described as the "me" and the "not-me";[5]

tropos sings the promise and peril of exhausting finally even the fumes of the liberal autonomous subject, admitting with John Nesselroade, primary among developmental methodologists, that evidence for "intraindividual variability" across the lifespan far outweighs evidence that we substantially are "the same person" from our births, through the twists and turns of our lives,[6] to—and perhaps beyond—our deaths, in memoria if not (e)utopia.

Indeed, the theoretical and practical chasms between Aristotelian *ēthos* and Isocratean *tropos* span a profound duplicity at the ancient heart of Western conceptions of self: Are we one? Are we many? Beyond theoretical conceptions of identity, how does actually existing identity-in-time—longitude—complicate even further how we make sense of ourselves and how others make sense of us? How can we trace the spaces between the turning places of multi-ply mediated identity? Decades after social constructionist theories became commonplace in the humanities as well as the social sciences, how might an Isocratean conception of identity-as-*tropos* offer explanatory power for analyzing the rhetorical practices of distributed, digitized selves? Might rhetorical acts have the capacity to integrate (or disintegrate) identity, however ephemerally?

In a postmodern context plump with possibilities for anonymous and pseudonymous engagement via virtual platforms, the notions of "self" and "identity" have become increasingly troubled.[7] If Lester Faigley is correct in asserting that "because the subject is the locus of overlapping and competing discourses, it is a temporary stitching together of a series of often contradictory subject positions," then the explanatory force of rhetoric should be apparent; after all, stitching together subject positions is a thoroughly rhetorical act, characterized by negotiation through discourse.[8] Using Aristotle's core term for character— *ēthos*—often elucidates the dynamics of perception, giving weight to an audience's view of a rhetor's moral and ethical dimensions. Yet in developing explanations for networked identity, *ēthos* seems to fall short. User names, handles, avatars, and other digital representations of "real names" undoubtedly involve perceptual processes, with audiences assessing a given rhetor's character. Over time and across multiple iterations, possibly in multiple media, character descriptions based on dwelling or constancy become even less accurate: The rhetorical dynamics of digital identity go well beyond the confines of perception. Because our chapter explores how multiplicious identities are developed and negotiated, our turn is from *ēthos* to *tropos*.

In this chapter, we identify multiple theoretical and critical exigences for a description of Isocratean *tropos*. We then sketch the implications of Isocratean *tropos* as a useful concept for describing processual, variously mediated selves in two disparate cases. In so doing, this chapter uses ancient rhetoric productively to relocate, reinvent, and rearticulate contemporary theoretical and practical problems surrounding identity, character, and agency. This chapter thus practices what Eberly has been for more than two decades calling "ancient rhetoric for contemporary purposes." Of course, the historical and philological work itself has value: It is work upon which additional scholarship might be based. Still, demonstrating to dubious colleagues or puzzled publics the pertinence of ancient ideas to contemporary practices is a capacity humanist scholars can no longer demur to develop and deploy.

After contextualizing our chapter in pre-Platonic conceptions of identity and character, we focus the balance of our chapter on two contemporary figures as illustrations of the descriptive power of Isocratean *tropos*: multimedia artist Harry Shearer and Reddit provocateur Michael Brutsch. Two of Shearer's projects—his six-episode 2013–2014 television series *Nixon's the One* and his long-running weekly radio program, *Le Show*—illustrate several senses of Isocratean *tropos* and offer material for theorizing *tropos* as periodic, echoing one of Isocrates's innovations in ancient Greek prose and, thus, thought: the periodic sentence.[9] Michael Brutsch, linked to the username Violentacrez as creator and operator of several questionable Reddit subforums such as /r/jailbait, demonstrates *tropos* in attempting to separate his corporeal person from his performances on Reddit. The disparate figures of Shearer and Brutsch underscore the complexity of negotiating identity in contemporary life, in various mediums, and with various performative purposes. Taken together, Shearer and Brutsch exemplify the power of Isocratean *tropos* to explain the dynamics of identity in contemporary multiply mediated rhetorical practice.

Isocratean Tropos

Jane S. Sutton and Mari Lee Mifsud's *A Revolution in Tropes: Alloiostrophic Rhetoric* attempts "to create a rhetoric that can sustain the ground for difference."[10] *Alloi*—"others" in ancient Greek—are the central concern of Sutton and Mifsud's project, the goal of which is to "revolutionize rhetoric" by "playing with tropes" to find "new ways of relating with the other."[11] In full support of Sutton and Mifsud's project, we here restrain our conception of "the other," following Nesselroade's vocabulary, to intraindividual otherness. Further, our sources are entirely pre-Aristotelian: In attempting to come to terms with contemporary multi-ply mediated selves, we eschew Platonic and Aristotelian modernisms and rely entirely on the grounded, protopragmatic lexicon of Isocrates, who wrote and practiced what he called *logos* without the consolations of theorized *ēthos*.

A significant rhetorical antecedent[12] to our conception of Isocratean *tropos* is Homeric *polutropos*. The Homer-poet describes both Odysseus and Achilles as *polutropos*, and meanings range from "well-traveled" and "wily" to "cunning" and "mendacious." Following Pietro Pucci,[13] Jeffrey Barnouw points out that "the epithet *polutropos*, 'of many turns'—used to introduce the hero in the first line of the poem, while his name is withheld until line 21—has been taken to mean 'much traveled' or 'versatile' but also to suggest devious duplicity"; Circe also, Barnouw adds, calls Odysseus *polutropos*.[14] Suggesting the ambi-valent nature of *tropos* as descriptor of character, Barnouw concludes that Odysseus *polutropos* "should have 'single-minded' added to his epithets. He is single-minded not in the sense of guileless or straightforward let alone ingenuous or simple-minded . . . but for his unwavering focus on his ultimate aim and his ability to integrate other aims into its eventual accomplishment."[15] It is this interplay between mutability and constancy in relation to identity—how far-flung the poles of the antithesis, before the forces that bind it give way?—that we intend to entertain in the balance of this chapter.

Classicist Christopher Gill has spent much of his career investigating ancient notions of identity and self. His books *Personality in Greek Epic, Tragedy, and Philosophy: The Self in Dialogue* and *The Structured Self in Hellenistic and Roman Thought* offer detailed insights into ancient language used to explore identity.[16] We echo Gill's claim that "there are a number of (more or less well-defined) debates in Greek philosophy which can be seen as analogous to those just outlined in modern theory."[17] Gill outlines a few key points of contention in ancient Greek thought, noting that much of the discourse is wrapped up in metaphysical questions about the mind, the psyche, and the body. Pondering the differences between ancient and modern thinking, Gill offers, "One possible point of difference is that these three strands in ancient theory, while overlapping and interrelated, are not explicitly regimented under a single heading, whereas, in recent years at least, this is the case with the modern theory of personhood."[18] This modern theory of personhood thus seems ill-suited to characterizing networked identity. Rhetorical theory needs an other-than-metaphysical way of understanding the interplay of what Gill describes as "personal identity and social identity," undercutting the "pervasive idea in modern western thought" that "our personal identity (or our 'real self') is *fundamentally* different from our social identity."[19]

In all of Isocrates's works, *ēthos*—in Norlin translated as "true character" —is used 6 times and *tropos* 117 times, seemingly with several meanings. *Liddell and Scott's Greek-English Lexicon* lists these meanings of *tropos*: turn, direction, way; manner, fashion, mode; "guise"; "way of life, habit, custom"; "character, temper"; "mode," "style," "mood."[20] Significantly for this discussion, the lexicon also describes a verb form of *tropos*, *tropé*, which comes to English in the cartographic concept of "tropic": "When the sun appears to turn his [grammatical gender excused] course," or, also significantly, when one who is facing defeat is able to turn the tables and "rout an enemy."[21] Finally, the lexicon explicitly articulates another nominative form, *tropikos*, with a slightly later invention, *rhētorikē*: "of or like a turn or turning . . . in Rhetoric, tropical or figurative."[22]

Specific to Isocrates, juxtapositions between *ēthos* and *tropos*, and *tropos* and other words in his corpus, help explain what roles *tropos* plays in his own rhetorical practice. Of Isocrates's writings that are perhaps better known to scholars of rhetoric (Isocrates remains generally ignored by philosophers), *tropos* occurs in all of them: *Against the Sophists*, *Antidosis*, *Areopagiticus*, *Helen*, *Nicocles*, *On the Peace*, and *Panathenaicus*. *Ēthos* occurs once in *On the Peace*, *Panathenaicus*, and *Antidosis* and twice in *Panegyricus*.

In the introduction to Mirhady and Too's 2000 translation of some of Isocrates's works, the translators and editors reflect on possible reasons Isocrates has been seen as marginal or "devalued"[23] since his death and departure from his contemporaneous community: "Another feature of Isocrates' rhetorical teaching . . . is that he challenges the common perception of discourse as predominantly oral. Indeed, he explicitly rejects the spoken word as a political medium, claiming that a 'small voice' and lack of courage prevent him from speaking in public."[24] Mirhady and Too suggest that Isocrates' account "may best be understood as the product of a paradoxical self-fashioning, since in claiming to be unable to speak in public, Isocrates conveniently excuses himself from his contemporary culture of public oratory. . . . For him, the written word was the basis of interactions with the larger Hellenic community."[25] Writing rather than speaking was how Isocrates interacted directly with his community. Isocrates's "paradoxical self-fashioning" vis-à-vis his community thus provides a final point of departure for applying the concept and practice of *tropos* to contemporary multi-ply mediated identities, in an age where literacy and orality (primary, secondary, tertiary, residual) once again seem equal forces in public discourse.[26]

In no work more so than *Antidosis* does Isocrates set out to defend against what he views as public misperception of his thought, his teaching, and his very character. "When my eyes were opened, as I have said, to the fact that a greater number than I supposed had mistaken ideas about me, I began to ponder how I could show to them and to posterity the truth about my character [*tropon*], my life, and the education to which I am devoted."[27] Throughout *Antidosis*—a written discourse in the form of a delivered speech that adopts the fiction of a trial—and indeed throughout his *oeuvre*, Isocrates seems to see *tropos* as character that can be, or can be perceived as, otherwise. Isocrates's longest work, *Antidosis* uses the relatively new rhetorical technology of prose to build recurring bouts of momentum that culminate in extended arguments about the nature of the speaker himself: "If, therefore, I were to agree with my accuser and concede his claim that I am the 'cleverest' of men and that I have never had an equal as a writer of the kind of speeches which are offensive to you, it would be much more just to give me credit for being an honest man than to punish me; for when a man has superior talents whether for speech or for action, one cannot fairly charge it to anything but fortune, but when a man makes good and temperate use of the power which nature has given him, as in my own case, all the world ought in justice to commend his character [*tropon*]."[28] These repeatedly culminating arguments take the form of periodic sentences, sentences in which the main subject and predicate are syntactically and thus chronologically distant from one another. Periodic sentences thus have longitude. Richard Lanham stresses the centrality of chronological time to the periodic style: "The mind shows itself after it has reasoned on the event; after it has sorted by concept and categorized by size; after it has imposed on the temporal flow the shapes through which that flow takes on a beginning, a middle, and an end."[29] Together, these intermittent long periods accrue over the considerable span of *Antidosis* to leave readers with an indelible sense of the character of the rhetor-in-writing, or at least his estimation of his character—all with the reminder that because of *tropos*—because of turning rather than dwelling—character can indeed be otherwise.

Harry Shearer's Plastiduditude

On the FOX network television show *The Simpsons*, Harry Shearer has for more than six hundred episodes over twenty-eight seasons voiced in excess of one hundred characters, including C. Montgomery Burns, Waylon Smithers, Ned Flanders, Principal Skinner, Reverend Lovejoy, and Kent Brockman.[30] During the summer of 2015, it appeared once again that *The Simpsons'* previous season would be the franchise's final one, with Shearer (some of whose castmates have called him—admiringly, he claims—"the turd in the punchbowl"[31]) holding out for two months before signing a three-year contract through 2018.[32] The single page on Shearer's website devoted to *The Simpsons* (figure 6.1) suggests Shearer's ambi-valence about the franchise: "Can a nation have a collective throat?"

However, to understand Shearer as merely the voice behind innumerable characters on *The Simpsons* is reductive at best. Shearer—multimedia artist, musician, actor, author, film director and producer, and satirist—exemplifies the power of the liberal arts in practice. The son of eastern European par-

Figure 6.1. Screenshot of Harry Shearer's website, accessed July 14, 2016.

ents who escaped the Holocaust by moving to the United States, Shearer was a child actor, making his film debut in 1953 in *Abbott and Costello Go to Mars* and working for eight years on *The Jack Benny Show*.[33] His parents, Shearer reports, ended his child acting career while he was working with Benny so that he could have "a more normal childhood." Shearer has said that his mother told him, "use all your talents"; his work suggests he has listened to his mother.[34] Shearer's work spans the second half of the twentieth century, and he remains a cunning media innovator well into the second decade of the twenty-first.

Shearer studied political science and practiced print journalism at UCLA and went on to cover the Watts riots in Los Angeles and the first moon landing, from the Jet Propulsion Laboratory in Pasadena, California, for *Newsweek*. After working as a journalist and a teacher, Shearer returned to performing as a member of the radio comedy group "The Credibility Gap." Shearer was twice part of the cast of *Saturday Night Live* but left the show before the end of his contract each time. In addition to playing Derek Smalls in *This Is Spinal Tap* and Mark Shubb in *A Mighty Wind* (Shearer as Shubb closes the film by dressing as a woman but still singing bass), Shearer co-wrote the film *Real Life* with Albert Brooks and appeared in *The Truman Show*. These are just the better known of his individual and collaborative projects.

Any assumption that Shearer is merely funny would meet a sober challenge in his information-dense 2010 documentary, *The Big Uneasy*, which explained why New Orleans flooded after Hurricane Katrina. More recently, Shearer's "found comedy" video series, *Nixon's The One*, in which he "plays" President Richard Nixon verbatim from the White House tapes, aired on SkyArts in the United Kingdom during the fall of 2013 and debuted in the United States via YouTube in October 2014. Also in 2014 Shearer appeared in London's Theatre Royal Haymarket in Oliver Cotton's haunting Holocaust memory play *Daytona* with actress Maureen Lipman and the playwright, Cotton. Shearer's work on stage and screen provide copious material for any student or scholar of identity, character, voice, and embodied rhetorics. But wait: There's more.

Since December 3, 1983, Shearer has broadcast a live weekly radio program called, for most of its history, *Le Show*. A snapshot of the inside of his brain while Shearer broadcasts *Le Show* likely resembles the full score of a symphonic opera, complete with libretto. Shearer has an incredible vocal instrument lodged proximate to an incredibly bright and busy brain. The polyphonic satire on *Le Show*, taking its point of departure from each week's political and cultural news, suggests Shearer's vast intelligence. Shearer also serves as the DJ on *Le Show*, inserting among the recurring features three cuts from various musical genres. In other words, Harry Shearer's *Le Show*, like the best rhetorical artifacts across history, teaches as well as delights. Shearer's many voices and voice-overs—and the many characters he has played—make him a natural for thinking-through conceptions of identity that assume plurality, polyphony, and change rather than unanimity, consistency, and permanence. Shearer's rhetorical mastery—his copious practice in the humanistic-cum-transhuman tradition of Isocrates, Erasmus, Milton, Lanham—exemplifies the power of Isocratean *tropos* to explain the dynamics of character, voice, and identity in contemporary multi-ply mediated rhetorical practice.

In what follows, we hold Shearer's *Nixon's The One* as one among myriad examples of Shearer "playing" a particular character as a point of contrast to his *tropoi* in *Le Show*. Shearer and Nixon are clearly separate entities; the former is "playing" the latter. Nixon did not write but spoke words later transcribed into text that Shearer memorized and animated. In *Nixon's the One*, Shearer and Nixon can be understood provisionally as two parallel lines. *Le Show*, alternatively, presents Shearer in a different geometry with *tropos* itself as his animating force. Two approaches, one looking closely at an excerpt of a particular episode of *Le Show* and the other delineating different kinds of *tropoi* in *Le Show*, provide a clear though in no way exhaustive description of Isocratean *tropos*.

The September 20, 2009, *Le Show*, broadcast from New Orleans opens with an update on Shearer's previous interview with US Army Corps of Engineers whistleblower Maria Garzino and a news story about the US Homeland Security Council suggesting a change to the color-coded post–9/11 public advisory system, reducing the colors from five to three.[35] While the tone is clearly satirical, the news items are deadly serious. In general, the contrast between the seriousness of the news Shearer reads and the satiric tone of *Le Show* creates rhetorico-meteorological conditions for the *tropoi* to come. More specifically, between 19:02 and 25:05 and following, several senses of *tropos* can be heard unrolling and spinning in time.

Shearer follows the first musical selection with more news items, including one about pharmaceutical giant Pfizer being fined $2.3 billion to settle criminal and civil accusations of illegal marketing of its painkiller Bextra. Shearer reads news items extemporaneously—Shearer does not practice reading the items before he broadcasts *Le Show*—and this one, titled "Pfizer pays a record pfine" on Shearer's website, sets up the segment of *Le Show* that follows it, a satirical sketch, "'Believia' spot," which Shearer wrote, voiced, and edited earlier in the week and subsequently played via CD-to-broadcast player on the September 20, 2009, episode of *Le Show*. Here, find excerpts of the "Pfizer pfine" news item followed by Shearer's satirical sketch in full:

> In the last couple of weeks Pfizer [two-second pause], a drug company so big it has a silent P in its name [three-second pause], agreed to pay a 2.3 billion-dollar [pause] billion [pause] fine to settle civil and criminal allegations that it had allegedly marketed its painkiller Bextra. [three-second pause] Bextra! [two-second pause] Who came up with that name? Well, it's like "extra," with a B. The extra B is for extra [two-second pause] Wheeeee! [three-second pause] Bextra has been withdrawn [two-second pause]. Oh. Uh, well? Maybe invite some friends over. [pause] Oh, I see. It was the largest health care fraud settlement and the largest criminal fine of any kind, ever—the largest criminal fine of any kind, ever, just paid by Pfizer. . . . The government charged that executives and sales representatives throughout Pfizer's ranks planned and executed schemes to illegally market not only Bextra but also Geodon, an antipsychotic, Zyvox, an antibiotic, and Lyrica [two-second pause] Lyrica! [pause] which treats nerve

pain. Oh, I thought it treated people who couldn't remember the words [pause]. Lyrica. While the government said the fine was a record sum, the 2.3 billion-dollar fine amounts to less than three weeks of Pfizer's sales. . . . A former Pfizer sales rep whose complaint helped prompt the government's Bextra case—you can read all about it: Bextra, Bextra—said the company managers told him and others to dismiss concerns about [an earlier case of Pfizer illegally marketing its drugs] while pushing them to undertake similar illegal efforts on behalf of Bextra. "The whole culture of Pfizer is driven by sales, and if you didn't sell drugs illegally, you were not seen as a team player," said the [whistle-blower], John Kopchinski. . . . Consumers should ask their doctors whether the medication being prescribed is FDA-approved for their condition, said the editor of *Consumer Reports' Best Buy Drugs*. Almost every major drug maker has been accused in recent years of giving kickbacks to doctors or shortchanging federal programs. Prosecutors say they are so alarmed by the growing criminality in the pharmaceutical industry they have begun increasing fines into the billions of dollars and would more vigorously prosecute doctors as well. . . . To its credit, Pfizer has promised to not engage in [more criminal] activities but instead to develop promising new medications.[36]

[sounds of birdsong]

[Shearer doing female-sounding voice, vaguely like NPR's Susan Stamberg, circa 1980:] I thought just because I was eating right and getting plenty of exercise, my mind and body would retain their youthful vigor and innocence. But then my doctor told me I was getting age-related skeptocemia, a chronic inability to absorb and believe messages about products that might be able to help me. What's more, he said it gets worse as I get older. That's when he told me about once-a-month Believia. He said it re-establishes brain pathways that aid in credulity. And unlike other medications which have to be taken daily, Believia does its work with just one pill a month. I can handle that.

[Piano music under Shearer doing announcer voice:] Do not use or handle Believia if you are operating machinery, becoming pregnant, or operating pregnancy machinery. Common side effects include headache, dizziness, memory loss, temporary night-blindness, and chronic gullibility. Ask your doctor if Believia is right for you. And believe him.

[Shearer doing same female-sounding voice, under birdsong:] I have this one life to live. And I don't want to miss any products that could help make that life stronger, better, or just more expensive. That's why I'm relying on once-a-month Believia. So the message gets through.

Shearer then plays Peter Yarrow's "The Great Mandala," sung by Peter, Paul, and Mary, which begins, "So I told him / that he'd better / shut his mouth / and do his job like a man," quite a song to turn to after mentioning two whistleblowers who fought injustices in their workplaces.

Several additional kinds of *tropoi* are prominent in this six-minute excerpt of *Le Show*: Suspended between the voices of news-reader and commentator are a heteroglossic *tropos*, a vacillation between the official and the critical or satiric. Also between those two is a sonic *tropos* in which the sounds either underscore or overtake sense—"a company so big it has a silent P in its name"—and generate new language forms. Yet sound does not deafen Shearer to sense: Reading the word "withdrawn" *tropes* his rhetoric psychologically, urging him to suggest that Bextra "invite some friends over." The comment "read all about it—Bextra, Bextra" is a *tropos* of public memory: echoes of youth selling "extra" editions of newspaper have likely faded, but the British rock band the Who and its "Pinball Wizard" likely still echo in the memories of most of Shearer's listeners. The satiric turns in Shearer's "Believia spot" operate on similar registers, albeit within a world fully fabricated by Shearer yet nonetheless reflecting on the absurdity of an industry with so much control over so many peoples' lives and livelihoods. Additionally, chronological time itself is a *tropos*-generator: the length of utterance ("the extra B is for extra Wheeeee!") balances with the length of pauses that Shearer uses in *Le Show*. Shearer has described the influence radio commentator Paul Harvey had on his delivery.[37] For listeners unfamiliar with Shearer's *Le Show*, the surprisingly long pauses can lead to concerns about radio reception or internet connection. For Shearer, however, the pauses leave time for turns that facilitate familiar listeners' reactions. Finally, because he lives in Los Angeles, London, and New Orleans, another of Shearer's *tropoi* derives from the dynamics of his physical location when he broadcasts *Le Show*. Close listeners can anticipate when he is about to travel from one to another of his home cities by the music he plays. His musical selections—one source of his cosmopolitanism—turn toward other cities, anticipating his departures and arrivals.

Shearer's character Atty the Atom, sometimes a guest during the segment of *Le Show* titled "News of Our Friend the Atom," is arguably most akin to Shearer among the many characters on *Le Show*. Atty and Harry are competing comedians, and examples of Harry "losing" Atty—of their two voices occasionally sliding into each other and almost combining, as if making a molecule—provide rich additional examples of Shearer's turn-taking on *Le Show*. Shearer sometimes even laughs quietly at Atty's extemporaneous responses, a strange turn because it is Shearer who has just invented and delivered those responses in Atty's voice. Atty also frequently comments on his own spinning, suggesting that in the world of Shearer's *Le Show*, spinning, not dwelling, is the norm.

Ultimately, all of these *tropoi* operate as particulars within the most general *tropos* of Harry Shearer's *Le Show*, the suspension between serious news and satire, between tragedy and comedy. Thus the rhetorico-meteorological context is better thought of as itself a *tropic*: between certain latitudes at certain times certain kinds of winds blow. In October 2015 Shearer broadcast comments making this tragicomic *tropic* explicit. Reading an item about Volkswagon's use of emissions test-cheating software—or "defeat devices"—in some of its diesel engines, Shearer quoted an official from the Center for Automotive Research at the University of Duisburg-Essen reflecting on news that the extent of the problem is not yet known: "'It suggests the company doesn't know the product, which is a tragedy,' [two-second pause] unquote. [pause] Or a comedy, sir, y'know: it could be a dark comedy."[38] Suspended between empirical tragedy and rhetorical comedy, light and dark, is where and how Harry Shearer's *Le Show* spins and swings.

"The Biggest Troll On the Web"

Our intervention, already illustrated in the aural plastiduditude of Harry Shearer, is also useful in theorizing networked identity. *Tropos* sheds light on Shearer's rhetorical multiplicity, his crafting of characters through mediated performance. On the internet, identity can be formed in myriad places in myriad ways—but without a centralized "person" or body to unify various performances. In this way, the person typing on a keyboard or screen is not necessarily the "true," "genuine," or "real" self. One must navigate points of performance in networked environments, possibly leaving entirely separate different identities or weaving together different performances into unified narratives. Those performances are understood by audience members—a process in which the term *ēthos* has some explanatory force—but only as one process in a larger network of rhetorical agency.[39] *Tropos* thus provides flexibility for analyzing performances of rhetorical power in networked environments by adding new dimensions of breadth and complexity to the oft-simplified "character" of *ēthos* and *persona*.

Tropos can help explain moments in which complex identity markers are stitched together, for better or worse. One such example is the case of Michael Brutsch, a programmer from Arlington, Texas, who became associated with what he thought was a separate online identity, linked to his person only via a few people he had met in person who also frequented the social media community Reddit. Before October 12, 2012, the user account Violentacrez was performed by Brutsch without any association with his corporeal self, the family man from Arlington. On that day, Adrian Chen, writer for the news and blog website Gawker, published a piece titled "Unmasking Reddit's Violentacrez, The Biggest Troll on the Web."[40] In that piece, Chen "unmasked" Violentacrez by revealing that the man "behind" Violentacrez was the forty-nine-year-old Brutsch. For many individuals, associating their physical person with a digital identity marker (such as a gamer tag or username) might be slightly embarrassing but not life changing. Increasingly, such association is the norm, particularly as "online" and "offline" fade as discrete categories. Chen's article created a firestorm because its use of "doxing" tactics (exposure of personal details as related to a digital identity) associated him with unsavory behavior, including questionable activities on Reddit, such as sharing technically legal but morally problematic photos of underage individuals. Violentacrez moderated a number of subforums, called "subreddits," including "Chokeabitch," "Niggerjailbait," "Rapebait," "Jewmerica," "Misogyny," and "Creepshots."[41] Associating Brutsch with these discursive communities tarnished his reputation—so much, in fact, that he was fired from his job just days after Chen's article was released.[42]

The case of Violentacrez/Brutsch illuminates a number of important elements in negotiating various identities. Instead of drawing praise, as most of Shearer's tropological performances might, Brutsch drew immense criticism and was met with profound negative consequences from the turn in his identity markers. But within many of the insular communities of Reddit, Violentacrez was viewed as a hero, or at the very least, the affable "creepy uncle of Reddit."[43] How might Violentacrez be a community hero yet Brutsch a community villain? Undoubtedly community values are at the core of the discrepancy. As we have explained, *tropos* is intrinsically linked to community (in understandings of character), but the term also covers a complex navigation of ways and means for expressing, performing, and being. In continuing our analysis of Violentacrez/Brutsch, we detail how *tropos* can be leveraged to understand the complexities and contradictions of Brutsch's character and community responses.

Chen's language of unmasking Violentacrez resonates with other ancient concepts, most notably *persona*. Explored in depth by Cicero, *persona* draws upon the dynamics of the theatrical stage, where actors would don masks to change character. *Persona*, then, refers to the changing dynamics of character, wherein people figuratively might don different masks in different situations, creating distinct performative *personae*: *persona* partakes of the parallel lines we discussed briefly above. Michele Kennerly further analyzes Cicero's use of *persona* through a thoughtful and intriguing investigation of *De Officiis*, arguing that Cicero develops something of a proto-dramatism.[44] But that proto-dramatism, and indeed the very notion of *persona*, does not quite encapsulate the theoretical move we wish to make here. Chen's language points to unmasking, yet the crux of Brutsch's crisis is not the mere binary of mask-on, mask-off, Brutsch-or-Violentacrez. What is most intriguing—and what requires the most departure from extant conceptualizations of rhetorical identity construction—is Brutsch's navigation of his roles. Rather than switching from performance to performance, Brutsch simultaneously wandered among multiple communities, attempting to balance his identity in front of many audiences. Compared to *persona* or *ēthos*, then, *tropos* accounts for degrees of change—a dimmer instead of a binary light switch, lines that, honoring finitude, do get nearer and/or farther apart in time. Further, whereas *ēthopoeia* is understood as the crafting of a stable character—like Shearer's Nixon, two parallel lines that rarely if ever intersect—*tropopoeia* might be a way of understanding rhetorical identity construction beyond the either/or limits of Euclidean geometry.

After being unmasked by Chen and fired from his job, Michael Brutsch was invited onto *Anderson Cooper 360*, a program on CNN. There, Brutsch explains to Cooper that what he did under the name Violentacrez was a performance, attempting to segment his activities on Reddit from his everyday life. Asked why he partook in these subreddits, Brutsch answers, "Well, because, and this goes back to how the Violentacrez character first started. My first thought was, or Violentacrez's first thought was, you know, I see those pictures on my incoming image stream all the time. I could easily create a Reddit for that and fill it up with some pictures. So I did, to outrage people. No idea where the pictures came from."[45] Brutsch seeks to define Violentacrez as something beyond him; he calls Violentacrez a "character" and suggests that Violentacrez—almost separately from Brutsch, but somehow still connected in a way—is able to think. By assigning agency to the Violentacrez username but maintaining some connection, Brutsch creates a scenario not unlike an actor becoming deeply involved in the performance of a character; in some ways, they are one and the same, but in others, the character is a separate entity. As Brutsch would have it, audiences would see him as essentially no different from an actor.

Tropos is important for explaining the contingency of Violentacrez, beyond the binary presented by Brutsch early in the Anderson Cooper interview. Other terms, such as Cicero's *persona*, might explain characters or performance, but *tropos* almost uniquely provides for situational performance, for *turning* and *being seen in a different light* within a community.[46] Brutsch also gives weight to Violentacrez in a notably rhetorical light; he notes, "I am to some degree apologizing for what I did. Again, I played to an audience of college kids. And two years ago, when all of this was at its height, the audience was appreciative and supportive of the sort of gallows humor that I put out there."[47] Here Brutsch emphasizes his place within the particular community on Reddit. For the "college kids" on Reddit, Violentacrez was a beloved community member. Exposing Violentacrez to outside communities changed the calculus for evaluating his character, adding layers of complexity.

Much like ascribing blame in epideictic rhetoric, doxing attempts to damage a person's image or reputation. Blaming in epideictic rhetoric would traditionally involve shedding light on the faults of character—in short, degrading one's *ēthos* by demonstrating poor moral character. This is undoubtedly the treatment Adrian Chen gave to Michael Brutsch; his piece unequivocally lays blame for Violentacrez before Brutsch, condemning the man as a whole entity. But in the way Brutsch discusses his identity, and in the way members of various communities on Reddit portray Violentacrez, there is nuance beyond Chen's holistic treatment of self. *Tropos* facilitates more flexible thinking about blame as related to the fragmented self. One might levy blame on a particular moment, behavior, or action, but perhaps not on a "whole" or dwelling self as imagined under *ēthos*. Framed in context, blame in *tropos* recognizes ranges of identity and performance. This is not to say Brutsch would be blameless. He very well might have some culpability for actions he made under the name Violentacrez; still, *tropos* would unseat absolutism in assessing Brutsch's behaviors.

Members of Reddit provided a framework for understanding Brutsch's performances within the Reddit community and Violentacrez's subreddits. One low-level Reddit moderator, using the handle POTATO_IN_MY_ANUS, responded to Chen's Gawker article, proclaiming, "As moderators, we feel that this type of behavior is completely intolerable. We volunteer our time on Reddit to make it a better place for the users, and should not be harassed and threatened outside of reddit for that. We should all be afraid of the threat of having our personal information investigated and spread around the internet if someone disagrees with you. Reddit prides itself on having a subreddit for everything, and no matter how much anyone may disapprove of what another user subscribes to, that is never a reason to threaten them."[48] Many of Violentacrez's supporters were in the same boat as Brutsch: their physical identities would preferably not be linked to their endeavors in morally ambiguous forums on Reddit. The tension exhibited between citizens of these Reddit communities and social justice activists such as Adrian Chen is indicative of *tropos*—character as a site of contestation, negotiation, and discourse. What POTATO_IN_MY_ANUS envisions as the ideal Reddit community is one of pluralism, wherein segmented subreddit communities are able to cater to their user demands without fear of criticism from other communities. Unfortunately for Brutsch and others in this camp, that norm has been breached by Adrian Chen and by another subreddit community, "Shit Reddit Says."

We have focused on *tropos* because of its historical resonance with figuration. As with *kosmos*, *tropos* does collective world-building work in Isocrates; hence its resonance with collective or civic identity and possibilities of change. Networked world-building is notably cosmological in its orientation, building order from among the chaos of the internet. Here that building arises from negotiation of community values both inside and outside the Reddit community. Inside the Reddit community, "Shit Reddit Says" is a board dedicated to exposing misogyny, racism, classism, and other forms of discrimination in the Reddit community. After taking over a subreddit dedicated to Violentacrez (tagged as /r/Violentacrez), one user, Police_of_Reddit, initiated a "HELLDUMP," aimed at exposing the various ways Reddit supported Violentacrez and the problematic communities he started or moderated. The user explains,

> The people who run reddit rely on volunteers to run their site because it is cheap, but in the long run the quality of the website declines and the social issues increase. These social issues reach a boiling point, or due to media attention they rise to the surface and the people who run reddit are forced to deal with it. One has to wonder if the lack of focus on the community is due to not caring, or because if they did it would require more employees and would cut into profit. When do the social issues start hitting their bottom line? Instead of properly dealing with the issues, they face the fall out of people feeling they don't value their privacy, don't value their time spent here, and don't value them. This website can't sustain itself like this while still having a policy of allowing teenagers as young as 13 allowed on the website. Without even needing an account you can on occasion view adult material. The admins and VA [Violentacrez] are now being dragged kicking and screaming by the media through the mud of their actions that shaped their community, and they have cried the entire way.[49]

The community is negotiating its own values by discussing the character of one of its most prominent users. Violentacrez is not just a user but part of the civic character of the website. In that way, *tropos* helps explain how character operates on multiple levels simultaneously. Much as Isocrates's character was indicative not just of his personal thought but of his education and of his role in the *polis*, Violentacrez's character operates on the personal level of Michael Brutsch and on the communal levels of Reddit and Arlington, Texas. Because Violentacrez's character was so detestable to so many, his *tropos* became a *topos* for initiating communal change, pushing against the sexist, racist, and pedophilic subcommunities Reddit turned a blind eye to—an action largely driven by Violentacrez's ability to bring new users to grow the community. Change in the Reddit *kosmos* became possible through this networked negotiation of *tropos*.

In *Antidosis*, Isocrates takes a tropological turn in investigating his own character—the character of a man dedicated to his community, Athens, but also exhibiting his own agency through the practices of rhetoric. In a distinctly contemporary fashion, Adrian Chen, SRS, and Violentacrez's supporters have reenacted a scene not all that dissimilar from Isocrates's reimagined trial. Wrapped up in all of the discourse is a question central to rhetoric both in Athens and in the virtual *polis*: how does speech reflect, deflect, and inflect character? For Adrian Chen, calls for "free speech" on Reddit are indicative of the tropological problems he sought to address with his Violentacrez piece: "What happened was, some do-gooder would stumble upon one of his offensive subreddits and expose it to the rest of Reddit in an outraged post. Then thousands more would vote the thing to the front page of Reddit. Cries to censor it would sound out, to be almost inevitably beaten back by cries of 'free speech!' The idea of free speech is sacred to many Reddit users, a product of the free-wheeling online message board culture from which Reddit springs. If you criticize someone else for posting something you don't like, you are a whiny fascist."[50] Juxtaposed with Isocrates's characterization of Athens, Violentacrez and other Redditors' self-vision and Chen's critiques thereof bear striking resemblance:

> I think that you are not unaware that while some of the Hellenes are hostile to you, some are extremely friendly, and rest their hopes of security upon you. These say that Athens is the only city, the others being mere villages, and that she deserves to be termed the capital of Hellas both because of her size and because of the resources which she furnishes to the rest of the world, and most of all because of the character of her inhabitants; for no people, they insist, are more kindly or more sociable, nor could anyone find any people with whom he could spend all his days in friendlier intercourse. Indeed, so extravagant are they in their praise that they do not even hesitate to say that they would rather suffer injury at the hands of an Athenian gentleman than benefit through the rudeness of people from another city. There are, on the other hand, those who scoff at this praise, and, dwelling upon the cruel and iniquitous practices of the sycophants, denounce the whole city as savage and insupportable.[51]

Isocrates's vision of his own *tropos* in relation to the *tropos* of Athens suggests a positive relationship. Brutsch's *tropos* and the *tropos* of Reddit would be similarly praised within the Reddit community, while Adrian Chen and others would scoff at the praise, denouncing the community as savage and insupportable.

Considering Duplicity: Tropos and the Prospect of Periodic Selves

Despite the glaring differences in Shearer's and Brutsch's practices and purposes, *tropos* nonetheless illuminates the multiple, complex, and changing meanings of character negotiated via networked rhetorics. For Isocrates, Shearer, and Brutsch, character is rich, dynamic, and central to variously mediated civic life. *Tropos* merits its place at the center of that conversation by providing a vocabulary for discussing the rhetorical dynamics of fragmented, dispersed, and networked selves.

Tropos points to a rhetoric-grounded conception of identity, full of twists and turns and spins in time instead of static dwelling. This is, in essence, the nature of networks as well. There are fixed points in networks, but only at particular moments in time. Networks are dynamic, even in physical form, as Jeff Rice reminds us in his assessment of the networked nature of the cityscape in *Digital Detroit*. He explains, "Network is a concept, not a thing out there."[52] We contend the same should be said of identity, especially from the perspective of rhetoric in the twenty-first century. As a concept, *tropos* sketches relationships between individuals and communities, creating tropological maps in time. Shifting over time and space, Isocratean *tropos* brings new rhetorical life to postmodern, transhuman selves, mediated and digitized, praised and blamed.

Finally, the figuration by William Safire, speechwriter to President Nixon, of Clinton Press Secretary Mike McCurry as "the Clintonian Newtonian" provides too tempting a means of articulating a potentially productive connection among geometry, physical chemistry, and Isocratean *tropos* and character more generally. Though the parallel lines of Shearer and Nixon do not intersect in *Nixon's the One*, Shearer has commented that he "didn't fully understand Nixon until I played him"; further, Shearer has observed that he habituated some of Nixon's "delicate gestures," especially an upward wave with the right hand, a habituated gesture that endured in Shearer's gestures three years after filming the series.[53] *Tropos* might provide material for crafting a new lens through which to study the means by which character "rubs off" or affects those intraindividual others proximate in space or time.

Again, there is no consistent distinction in Isocrates's corpus between *ēthos* as "true character" and *tropos*. But little if anything about Isocrates's work is systematic. Indeed, in the introduction to Mirhady and Too's volume of translations of Isocrates, the translators, collaborating with series editor Michael Gagarin and Terry Papillion, translator of the second volume of Isocrates in the same series, conclude in a manner, a way, a *tropos*, arguably uncharacteristic of classicists: "Finally, a note of caution to the reader. Interest in Isocrates has been growing rapidly in recent years, and scholars in different fields have been exploring new ways of reading him. As a result, a diversity of views is reflected in the three translators, and although we have coordinated our translations and discussed our differences among ourselves with a view to achieving consensus wherever possible, this diversity is still evident, both among the three translators and perhaps even within the translators themselves [intraindividual variability ftw!]. The reader should not always expect to find a single voice in these two volumes but will, we hope, enter into our discourse and create his or her own voice for Isocrates." Isocrates's different treatments of the Spartans in his different discourses provide another clue to Isocratean *tropos*: Sometimes the Spartans are exemplars of civic virtue, other times of basest barbarousness. Given the distinction between *homoioi* and *barbaroi* and Isocrates's phantasy of Panhellenism, another suggestion presents itself: Isocratean *tropoi* allows for alternatives to othering essentialisms. Instead, *tropos* enables malleable, contingent, pragmatic civic identities and the potential for rhetorical agency in multiple sources of self. Perhaps, as inventor of the periodic sentence, Isocrates has also given us the periodic self—a potentially molecular rather than atomic identity with the capacity to be reintegrated or disintegrated, however ephemerally, through rhetorical acts.

Notes

1. William Safire, "The Spinner Spun," *New York Times*, December 22, 1996, http://www.nytimes.com/1996/12/22/magazine/the-spinner-spun.html.

2. See Edward Schiappa, "Isocrates' *Philosophia* and Contemporary Pragmatism," in *Rhetoric, Sophistry, Pragmatism*, ed. Steven Mailloux (Cambridge, UK: Cambridge University Press, 1995), 33–60.

3. Isocrates, *Isocrates*, trans. George Norlin and Larue Van Hook (Cambridge, MA: Harvard University Press, 1980); Isocrates, *Isocrates I*, trans. David Mirhady and Yun Lee Too (Austin: University of Texas Press, 2004).

4. Craig R. Smith, "Ethos Dwells Pervasively," in *The Ethos of Rhetoric*, ed. Michael J. Hyde (Columbia: University of South Carolina Press, 2004), 3.

5. Ralph Waldo Emerson, "Nature," accessed March 22, 2015, http://www.gutenberg.org/files/29433/29433-h/29433-h.htm.

6. J. R. Nesselroade and P. C. M. Molenaar, "Emphasizing Intraindividual Variability in the Study of Development Over the Lifespan," in *The Handbook of Life-span Development: Cognition, Biology, and Methods across the Lifespan*, ed. W. F. Overton (Hoboken, NJ: Wiley, 2010), 30–54.

7. Carolyn R. Miller, "Writing in a Culture of Simulation: *Ethos* Online," in *Towards a Rhetoric of Everyday Life: New Directions in Research on Writing, Text, and Discourse*, ed. Martin Nystrand and John Duffy (Madison: University of Wisconsin Press, 2003), 74–75.

8. Lester Faigley, *Fragments of Rationality: Postmodernity and the Subject of Composition* (Pittsburgh: University of Pittsburgh Press, 1992), 9.

9. James J. Murphy, Richard A. Katula, and Michael Hoppmann, *A Synoptic History of Classical Rhetoric*, 4th ed. (New York: Routledge, 2014), 53–54.

10. Jane S. Sutton and Mari Lee Mifsud, *A Revolution in Tropes: Alloiostrophic Rhetoric* (Lanham, MD: Lexington Books, 2015).

11. Ibid., 105.

12. Gerard A. Hauser and Carole Blair, "Rhetorical Antecedents to the Public," *PRE/TEXT* 3 (1982): 139–67.

13. Pietro Pucci, *Odysseus Polutropos: Intertextual Readings in the Odyssey and the Iliad* (Ithaca, NY: Cornell University Press, 1987).

14. Jeffrey Barnouw, *Odysseus, Hero of Practical Intelligence: Deliberation and Signs in Homer's* Odyssey (Lanham, MD: University Press of America, 2004), 27.

15. Ibid., 32–33.

16. Christopher Gill, *Personality in Greek Epic, Tragedy, and Philosophy: The Self in Dialogue* (Oxford: Clarendon Press, 1998); Christopher Gill, *The Structured Self in Hellenistic and Roman Thought* (Oxford: Oxford University Press, 2009).

17. Christopher Gill, "Ancient Concepts of Personal Identity," in *The Oxford Handbook of Hellenic Studies* (Oxford: Oxford University Press, 2009), 261.

18. Ibid., 263–64.

19. Ibid., 266.

20. "τρόπος," *LSJ: The Online Liddell-Scott-Jones Greek-English Lexicon*, http://stephanus.tlg.uci.edu/lsj/#eid=108480&context=lsj&action=hw-list-click.

21. Ibid.

22. Ibid.

23. Mirhady and Too, *Isocrates I*, 5.

24. Ibid., 6.

25. Ibid., 6.

26. Walter J. Ong and John Hartley, *Orality and Literacy: 30th Anniversary Edition* (New York: Routledge, 2015).

27. Isocrates, *Antidosis*, sec. 6, in Norlin and Hook.

28. Ibid., sec. 35–36.

29. Richard A. Lanham, *Analyzing Prose* (New York: Scribner's, 1983), 54.

30. Jess Ferreras, "Simpsons' Infographic Captures Its Voice Talents' Greatness," *Huffington Post Canada* January 23, 2015, http://www.huffingtonpost.ca/2015/01/23/simpsons-infographic-voice-actors_n_6528346.html.

31. Conversation with Eberly in London, July 2014.

32. Jeremy Egner, "Harry Shearer to Return to 'The Simpsons,'" *New York Times*, July 7, 2015, http://artsbeat.blogs.nytimes.com/2015/07/07/harry-shearer-to-return-to-the-simpsons/.

33. Harry Shearer, "About," *HarryShearer.com*, accessed November 3, 2015, http://harryshearer.com/about/.

34. "In Conversation: Harry Shearer and Greg Kot," Future of Music Coalition Summit, Washington, DC, October 27, 2014, https://www.youtube.com/watch?v=xBOjQX4dUKI.

35. Harry Shearer, *Le Show*, September 20, 2009, http://harryshearer.com/le-shows/september-20-2009/.

36. The news story Shearer was reading seems to be Gardiner Harris, "Pfizer Pays $2.3 Billion to Settle Marketing Case," *New York Times*, September 2, 2009, http://www.nytimes.com/2009/09/03/business/03health.html.

37. Harry Shearer, *Le Show*, March 1, 2009, http://harryshearer.com/le-shows/march-1-2009/.

38. Harry Shearer, *Le Show*, October 25, 2015, http://harryshearer.com/le-shows/october-25-2015/

39. Cheryl Geisler, "How Ought We to Understand the Concept of Rhetorical Agency? Report from the ARS," *Rhetoric Society Quarterly* 34, no. 3 (Summer 2004): 9–17.

40. Adrian Chen, "Unmasking Reddit's Violentacrez, The Biggest Troll on the Web," Gawker, October 12, 2012, http://gawker.com/5950981/unmasking-reddits-violentacrez-the-biggest-troll-on-the-web.

41. Ibid.

42. Stephanie Marcus, "Violentacrez Fired: Michael Brutsch Loses Job after Reddit Troll Identity Exposed by Gawker," *Huffington Post*, October 15, 2012, http://www.huffingtonpost.com/2012/10/15/michael-brutsch-reddits-biggest-loses-job-identity-gawker_n_1967727.html.

43. Damon Poeter, "Shed No Tears for Violentacrez," *PC Magazine*, October 15, 2012, http://www.pcmag.com/article2/0,2817,2411001,00.asp.

44. Michele Kennerly, "The Stoic Nature of Early Dramatistic Theory," *Advances in the History of Rhetoric* 14, no. 1 (2011): 33–48.

45. Anderson Cooper, "Man behind 'Jailbait' Posts Exposed, Loses Job," *AC360*, October 19, 2012. See also David Fitzpatrick and Drew Griffin, "Man behind 'Jailbait' Posts Exposed, Loses Job," *CNN*, October 19, 2012, http://www.cnn.com/2012/10/18/us/internet-troll-apology/.

46. Here we italicize "turning" and "light" to emphasize two translations of *tropos* in Isocrates as translated by Norlin and Hook; see Isocrates, *To Philip*, sec. 45; and *Panathenaicus*, sec. 41.

47. Cooper, *AC360*.

48. POTATO_IN_MY_ANUS, "The Real Reason Why Violentacrez Deleted His Account: Adrian Chen, Gawker Media, Creepshots, PM's and Real-Life Doxxing,"

Reddit, October 10, 2012, http://www.reddit.com/r/SubredditDrama/comments/118qdg/the_real_reason_why_violentacrez_deleted_his/.

49. Police_of_Reddit, "[HELLDUMP] ViolentAcrez—A Severe Disconnect or a Cover Up. A Year in Review," r/Violentacrez on Reddit, 2012, http://www.reddit.com/r/violentacrez/comments/11qjfb/helldump_violentacrez_a_severe_disconnect_or_a/.

50. Chen, "Unmasking Reddit's Violentacrez."

51. Isocrates, *Antidosis*, sec. 299–300, in Norlin and Hook.

52. Jeff Rice, *Digital Detroit: Rhetoric and Space in the Age of the Network* (Carbondale: Southern Illinois University Press, 2012), 6; quoting Bruno Latour, *Reassembling the Social—An Introduction to Actor-Network Theory* (Oxford: Oxford University Press, 2005), 131.

53. Harry Shearer, interview on "West Coast Live," October 10, 2015, http://www.wcl.org/index.php/past-programs-2015/323-october-10–2015-live-from-the-sf-ferry-building-show-1133.

7

Plato's *Phaedrus* and the Ideology of Immersion

EKATERINA V. HASKINS AND GAINES S. HUBBELL

One does not have to be a technophile to notice the excitement surrounding new virtual reality applications. Head-mounted virtual reality display devices such as the Oculus Rift are taking over the senses and imaginations of adolescent players,[1] while three-dimensional immersive environments such as Cave Automatic Virtual Environments (CAVEs) and Ultimate Reality Theaters (YURTs) tantalize faculty at research universities. Witness, for example, Brown University's YURT (unveiled in 2015), a $2.5-million virtual reality room whose domed ceiling, curved walls, and clear acrylic floor represent a three-dimensional screen for projecting high-resolution computer graphics. On the opening day, the YURT dazzled its visitors with a display of virtual objects, from simple shapes drawn by a handheld wand and the animated language of a student's 3-D poetry to the surface of the Moon. According to the faculty involved in its assembly, future applications of the facility remain unknown, although they are hopeful that it will "accelerate science."[2]

The attraction of virtual reality, it seems, lies not in any specific intended use but in the technologically mediated effect of immersion. Scholarly and public discussions of immersion typically describe it in two ways: as a sensation of being transported to another realm or as an affectively charged experience of "being there." The two aspects are connected: The qualities of realism and "immediacy" of simulated environments are believed to enhance the overall sensation of presence and thus lead to a more embodied, authentic, and pleasurable experience. Small wonder that "immersion" has become a new buzzword and that many, including tech investors, members of the military-industrial complex, and academic administrators, embrace its promise.

This essay offers a critique of the discourse of immersion, with particular attention to immersion's educational promise, from the vantage point of one of the most fascinating texts in the history of rhetoric: Plato's dialogue *Phaedrus*. The *Phaedrus* can be seen as a paradigm of immersive learning and a metacommentary on the promises and dangers of technologized discourse. The dialogue models engagement and presence by painting a rich scene capable of hosting rigorous philosophical conversation. Scenic references in the dialogue not only foreshadow and illustrate its key lessons about love, rhetoric, and writing but also create an immersive land- and soundscape that renders the exchange between Socrates and Phaedrus into a higher form of communion. The reader is transported to an enchanted place where Phaedrus's unlikely transformation from a lover of trendy sophistic discourses to a lover of philosophical learning seems utterly plausible and even inevitable. The scene in all its sensory variety serves a scaffolding function: it gently guides and envelops the interlocutors—and the reader—and renders natural the interaction between the philosophical lover and his beloved.

If the *Phaedrus* furnishes the immersive scaffolding to promote Plato's distinct vision of a philosophical life, contemporary fascination with immersive technologies tends to celebrate the scaffolding regardless of what it might support. If Plato's text acts as a surrogate of human conversation, today's designers and scholars locate authenticity in the interaction between users and technological artifacts. They thus unreflectively embrace the ideology of authenticity exemplified by the *Phaedrus* without examining its ethical, political, and pedagogical dimensions.

The Promise of Digital Immersion

"Immersion" is touted today as a distinguishing feature of computer games and virtual reality simulations, but the ideas of immersion and immersive environments long predate contemporary digital media. From Plato's description of the world of shadows in the Allegory of the Cave and illusionistic wall paintings in Rome to modern panoramas and IMAX movies, the human desire to enter the space of a representation appears as an anthropological constant even as the technologies that feed this desire evolve. Whether the medium of representation is as old as an epic poem or as recent as multisensory virtual reality simulations, "our brains are programmed to tune into stories with an intensity that can obliterate the world around us," as narrative theorist Janet Murray surmises.[3]

Most definitions of immersion oscillate between two sides of the phenomenon: one that foregrounds the very experience of entering a different world and another that stresses the effect of absorption in an activity such as reading, watching, or playing. Numerous scholars point to the sensation of moving into a reality different from one's own as the essence of immersion. Highlighting the metaphorical nature of the term, Murray compares it to "a plunge in the ocean or swimming pool": "the sensation of being surrounded by a completely other reality, as different as water is from air."[4] Griffiths describes "the sense of being present in a scene, the cognitive dissonance that comes from feeling like you're elsewhere while knowing that you haven't moved and forgetting for a moment about the mediating effects of technology."[5] Commenting on the effect of literary fiction on readers, Ryan identifies immersion as "the experience through which a fictional world acquires the presence of an autonomous, language-independent reality populated with live human beings."[6] Immersion through virtual reality, the most multisensory simulation technologies to date, is often equated with "presence"; presence is the "extent to which a person's cognitive and perceptual systems are tricked into believing they are somewhere other than their physical location."[7]

Immersion is also frequently associated with the *intensity* of participating in the simulated reality. Here, the accent is placed on the psychological process that accompanies the feeling of being "elsewhere." Murray states that an immersive experience "takes over all of our attention, our whole perceptual apparatus."[8] In addition to mental absorption, Grau notes "diminishing critical distance to what is shown" and "increasing emotional involvement."[9] In her discussion of reading as immersion, Ryan contends that the "experience of being transported into the narrative scene is so intense and demanding on the imagination that it cannot be sustained for a very long time; an important aspect of narrative art consists, therefore, of varying the distance, just as a sophisticated movie will vary the focal length of the camera lens."[10] Video-game researchers isolate several levels of immersion, ranging from engagement to engrossment to total immersion.[11] These studies liken the effect of immersion to Csikszentmihalyi's notion of "flow," defined as a state of effortless absorption in a performance or a task.[12] In this sense, immersion is equated with pleasurable concentration.

Whether the stress falls on the effect of being transported to a different realm or the intensity of one's participation in this process, immersion is overwhelmingly regarded as an intrinsically valuable reward for one's attention. Many popular and academic discussions do not question the merits of immersion but instead debate the technological and artistic means of enhancing the experience of immersion. Not surprisingly, entertainment industry equates greater immersion with visually stunning environments that possess a high degree of representational and spatial realism (consider the rise of 3-D movies). In video-game design, this "immersive fallacy" is manifested by preference for increased photorealism and greater levels of aural authenticity.[13] However, academics too have embraced the "immersive fallacy" in their quest to identify factors that generate the most credible illusion. Virtual reality researchers believe that adding other modalities to the visual and aural experience, such as haptic feedback, augment the feeling of presence and participation in the simulated environment.[14] The more sophisticated the technology, the argument goes, the better the sensation of being there.

To be sure, many scholars regard verisimilitude and sensory stimulation as less crucial than other mechanisms of engaging players and readers in a particular narrative. Some point to the player's identification with characters in the game world as such a mechanism.[15] Others argue that players "must have a non-trivial impact on the environment."[16] Similarly, Dede contrasts "sensory immersion," the kind created by a simulated three-dimensional space, with "actional" and "symbolic" types: Actional immersion empowers the participant "to initiate actions impossible in the real world that have novel, intriguing consequences" (like flying), and symbolic immersion "involves triggering powerful semantic, psychological associations by means of the content of an experience."[17] In the case of literary texts especially, immersion is the product of the reader's imagination, as reading converts "codes into contents."[18] "The function of language in this activity," writes Ryan, "is to pick objects in the textual world, to link them with properties to animate characters and setting—in short to conjure their presence to the imagination."[19] Mimetic texts—those that represent "states of affairs involving individual existents in space and time"—are therefore more apt to stimulate immersion than "texts that deal exclusively with universals, abstract ideas, and atemporal categories."[20] While often critical of the technophilia of the industry-driven rhetoric of immersion, these accounts still privilege formal features of an artifact that enable immersion.

Immersion has become a god-term of sorts, and its ascendance in discussions of learning environments especially indicates a preference for technological solutions to educational problems at the expense of considering how contemporary classrooms can foster critical thinking and broader cultural awareness. Rather than pondering the ideological work of immersion, designers and scholars tend to ask a more instrumental question, "What can be done with immersion?" When not considered an end in itself (the way the entertainment industry views it), immersion becomes an instrument for other ends. Early versions of virtual reality were widely employed in military training thanks to the technology's ability to simulate flight conditions and combat situations. The remarkable popularity and variety of commercial first-person shooter games suggest, however, that gratifications of virtual combat are not purely instrumental; there is, after all, a worldview that one's participation in the game world cultivates.[21]

Eager to find more socially acceptable uses for immersive technologies, education researchers have poured much effort into identifying instructional applications that would foster so-called twenty-first-century skills. The fact that immersion can promote attention and concentration has been noted by scholars, who see this aspect of video games as "extremely relevant to learning."[22] According to numerous studies, one of the selling points of educational games is that they can monopolize attention, making one oblivious to distractions in the physical surroundings and the passage of time, an important consideration in teaching "digital natives," who purportedly prefer video games to books.[23] The guiding concern in much of this research is how best to integrate educational "content" with game play so that students not only stay engaged but also internalize and retain embedded lessons.[24] Game play and immersion are thus seen as a delivery device (because pleasurable and absorbing) for some form of educationally sound "content." In a way, students are tricked into learning particular lessons through immersion but are not encouraged to step outside and examine the game's immersive mechanism. In this scenario, immersion becomes an instrumental good as a component of a digitally and technologically mediated educational process, the goodness of which is largely taken for granted. As a result, educational software—not the actual classroom—is articulated as the main site of instruction and learning, with teachers and students playing roles determined for them by accredited designers and institutions. Intended to ease the pressure on teachers, who must engage students in knowledge acquisition and creative problem-solving while conforming to bureaucratically imposed standards, the wide adoption of "new" learning tools is premised on immersion fetishism.

Our brains may indeed be "programmed" for immersion, as Murray maintains, but technologies that allow us to feel present in the made-up world of a story or game are not ethically or politically neutral.[25] As art historian Oliver Grau argues, "Immersive art often molds propagandistic messages, conveyed by its images, thus working specifically against distanced and critical reflection. . . . The recurrent model follows the utopian notion of relocating the observer *in the image*, removing the distance to the image space, intensifying the illusion, and renewing the power exerted over the audience—an idea that has consistently driven constitutive dynamics in the development of new media of illusion. For, in essence, all socially relevant new image media, from classical antiquity to the revolution of digital images, have advanced to serve the interests of maintaining power and control or maximizing profits."[26] For Grau, the loss of reflective distance marks the experience of all historically known forms of immersive media, yet the spectator's susceptibility to illusion depends on previous media exposure. Today we are far less likely to be thrilled by "Masaccio's fresco of the Holy Trinity, the Lumieres' approaching train, or a panorama-landscape"[27] than then-contemporary observers of these attractions, because we can easily see their artifice. At the same time, we should not become complacent about our media savvy, since new image technologies have been perfecting the "dissolution of the interface" between the observer and the image so that "the medium becomes invisible."[28] "Like Plato's prisoners in the cave," urges Grau, "what we need to do is turn toward the light, to face the new, and armed with our knowledge, confront it squarely."[29] He admits, however, that "there is no way out of the cave," as "there are only old and new media, old and new attempts to create illusions"; we must therefore "engage critically with their history and their future development."[30]

Following Grau's call to examine immersive media historically, we present Plato's dialogue *Phaedrus* as an example of ancient educational technology that models both immersion and critical distance. In the *Phaedrus*, the reader witnesses an intimate scene of exchange between Socrates and his young companion as the two stroll in the countryside outside the walls of Athens and discuss love, rhetoric, and writing. The dialogue's examination of poetic and rhetorical genres and writing constructs a critical perspective from which the interlocutors judge the speeches showcased in the course of the conversation. Yet the *Phaedrus* enacts its own kind of immersive seduction that disguises the artifice of the dialogic form.

Immersion and Reflection in the Phaedrus

At first glance, Plato hardly seems a likely source for a discussion of immersion in ancient rhetoric: His distrust of the senses and appearances is a leitmotif in many of his dialogues. The Allegory of the Cave in the *Republic* teaches that the philosopher must turn his back on the world of ordinary appearances, which are mere shadows of the Real, in order to ascend to the radiant realm of eternal Forms.[31] Moreover, Plato's Socrates often criticizes or ridicules the enchanting effect of performed poetry and rhetoric on their audiences, as in his mock-serious admission in the *Menexenus* to have become transported to the Islands of the Blessed upon hearing an Athenian funeral oration.[32] At the same time, Plato constructs his philosophical vision through dialogues that include dramatic settings and actions in addition to characters' utterances. In this way he invites his readers to place themselves within the discussion rather than merely attend to the dialogues' doctrinal content.[33] Nor does he avoid using vivid imagery to spellbind the reader. For example, while the *Republic* banishes poets from the ideal *polis*, it also employs "highly visual, physical language" to conjure the realms of being and becoming and "to spirit the reader to new and unknown regions."[34] Andrea Nightingale calls this "the rhetoric of estrangement," which works by inviting the reader "to enter into the perspective of the philosophic *theoros*."[35] Although the goals of Plato's verbal artistry are governed by his distinct metaphysical and ethical assumptions, the rhetorical mechanism of estrangement is similar to other historical forms of immersion: It conveys us to a realm that is separate from our own world and effects a transformation from one mental state to another.

Plato's *Phaedrus*, the evocative dialogue about love and rhetoric, is a particularly striking example of immersion. In it, Plato depicts Phaedrus's transformation from an omnivorous lover of rhetorical speeches to a student of philosophy through a conversation with Socrates amid a lush natural backdrop. Socrates does not openly proselytize the virtues of philosophical life over and against other intellectual and artistic trends of his time; instead, they emerge—seemingly spontaneously—as a result of a leisurely stroll in the countryside.[36] We not only follow the drama of this organically developing conversation and thereby glimpse the tenets of Platonic philosophy, but we also become present in the scene of the dialogue itself. Once established, the scenic backdrop never completely fades from the text, since its sensory impact and symbolism continue to be invoked by the characters. In other words, the lessons of the dialogue are inseparable from the immersive experience, both for the characters and for the reader.

The dialogue's first lines quickly establish its setting and rhythm, dropping us in the middle of an encounter between Phaedrus and Socrates. "Wither and wherefrom," asks Socrates, signaling the peripatetic quality of the conversation about to begin. One could almost hear Socrates catching his breath as he tries to keep up with his younger companion. Phaedrus admits that he spent the morning's early hours listening to the Sophist Lysias and is now on his way to walk outside the city wall. He adds that in his choice of country roads over dusty city streets he follows the advice of their mutual friend Acumenus the doctor. We do not get to hear how Socrates spent *his* morning, because he is eager to quiz Phaedrus about Lysias and his speeches. Phaedrus invites Socrates to walk along and listen, if he has "leisure" (*scholē*). Feigning fascination with hearing about the erotic speech of Lysias, Socrates promises to follow Phaedrus "all the way to Megara" and jokingly likens his appetite for discourses to that of a voracious beast: "As people lead hungry animals by shaking in front of them a branch of leaves or some fruit, just so, I think, you, by holding before me discourses in books, will lead me all over Attica and wherever else you please."[37] Socrates's hunch—that Phaedrus is actually harboring Lysias' speech under his cloak—proves correct as the two approach the bank of the Ilisus River.

The opening scene simultaneously draws the reader into the time and place of the action (late morning of a warm summer day outside of Athens), portrays Phaedrus as a connoisseur of trendy discourses, presents Socrates as a playful (and possibly sexually aroused) conversationalist, and forecasts the topics of the dialogue (love, rhetoric, and writing). "Leisure" is also an important term here, as it signifies not only spare time but also freedom from mundane responsibilities, the mark of a "leisure" class to which both Phaedrus and Socrates belong. Phaedrus subscribes to fashionable notions of leisure, represented by sophistic rhetoric and medical advice, both of which are set up to be questioned later on.[38]

We soon discover that Socrates's notion of leisure is at odds with current intellectual trends. In the next scene, as the characters walk along the Ilisus with their bare feet in the cool water and approach the shady plane tree, we find the first major display of Socrates's peculiar preoccupation with self-knowledge. The passage is sensuous in its evocation of nature that greets and envelops the interlocutors, and it is conjured entirely through the characters' remarks about their setting. Impatient to display his erudition, Phaedrus probes Socrates about the story of Oreithyia's abduction by Boreas connected to the place, as the location is surrounded by a rich mythological lore. Socrates shows that he knows both the mythic and "rational" explanations of it. Although he does not disavow this rational account, he contends that such explanations "in accordance with probability" require a great deal of leisure *(scholē)*, and that he, Socrates, has no leisure for these things: "The reason, my friend, is this: I am not able, as the Delphic description has it, to know myself; so it seems to me ridiculous, when I do not yet know that, to investigate irrelevant things. And so I dismiss these matters and accepting the customary belief about them, as I was saying just now, I investigate not these things, but myself."[39] Socrates finds the plane tree and the area around it enchanting: The plane tree is "spreading and lofty"; the shady willow is covered with fragrant blossoms; the water in the stream is pleasantly cool; the grass on the gentle slope is thick enough for reclining; and the air is breezy and filled with the summer music of the chorus of cicadas. Moreover, this seems to be "a sacred place of some nymphs and of Achelous judging by the figurines and statues."[40] Socratic expression of wonder and delight implies a contrast between the artifice and sterility of the rational theories about mythology and nature's sensuous beauty.

The description also helps to spotlight Socrates's unique ways of seeking knowledge. Phaedrus remarks that Socrates appears more like a stranger than a native—he calls Socrates *atopotatos*, the most "out of place, extraordinary."[41] Phaedrus finds it odd that his companion so eagerly embraces the beauty of the countryside, given that he rarely leaves the city walls. Socrates agrees that as a lover of learning he prefers to spend his time with the people in the city but concedes that Phaedrus has found the charm to lead him out. Socrates seems to be equally motivated by the natural scenery, with its promise of rest and inspiration, and by his desire to hear a speech of Lysias, a discourse that, we suspect, he knows to be deficient and inferior. Yet it is precisely the tension between authenticity and pretense, inspiration and sobriety, created by the exchange about the scene of the dialogue that sets up our expectation of the unfolding drama. The three speeches that form the heart of the *Phaedrus*—and the characters' discussions of them—showcase a progression from a conventional approach to human relationships and forms of discourse to a preferred philosophical view that transcends their errors and limitations. It is not, however, a mere progression of arguments. What we witness is also a transformation of what begins as a friendly conversation into an inspired form of communion between the philosophical lover and his beloved. And each step in this development is signaled and motivated by the elements of the scene.

As Socrates lies down on the soft grass by the plane tree, Phaedrus begins to read the speech of Lysias on the superiority of a nonlover to a lover. It characterizes the relationship between an older suitor and his beloved as if it was a business contract and uses an argumentative structure to match the content. This rhetorical showpiece is marked by sterile rationality and a transactional approach to social relationships. It is akin to a finished piece of joinery; later on Socrates and Phaedrus discuss it with regard to writing, noting that unlike a living dialogic exchange, this speech-artifact is not responsive to its audience. As such, it is similar to the rational theories about mythological creatures whose contrived skepticism Socrates found so tedious and irrelevant. These are the kinds of trendy rationalism that Socrates works against.

Meanwhile, the magic of the place has already begun to change Phaedrus from a connoisseur of discourses into an enthusiast, a transformation highlighted by Socrates's mocking praise of Phaedrus's performance: "I am quite overcome by it. And this is due to you, Phaedrus, because as I looked at you, I saw that you were delighted by the speech as you read. So, thinking that you know more than I about such matters, I followed in your train and joined you in the divine frenzy."[42] But the enthusiasm Phaedrus exhibits is tinged with (erotic) violence: He seems to have internalized the *ēthos* of competition that goes along with the rationalism of sophistic rhetoric. Challenging Socrates to a verbal contest with Lysias, Phaedrus invokes the location and half-jokingly threatens physical force unless Socrates speaks voluntarily ("We are alone in a solitary spot, and I am stronger and younger than you"). He then swears on the plane tree that unless Socrates produces a discourse in the very presence of this tree, he will never read to him another speech. Similar to an ominous musical score layered over a tranquil cinematic landscape, scenic references here subtly color our experience of the shady spot under the plane tree: A site of friendly banter about clever speeches has become a scene of confrontation.

Socrates acquiesces, but because he is uttering his discourse under compulsion, he conceals himself under a cloak. This dramatic gesture indicates to the reader that Socrates wishes to distance himself from the utterance while performing it. Socrates's speech is an improvement, given its better organization and poeticized style. Though technically superior, his discourse shows that the love of the lying lover, however artfully disguised, is just as possessive and selfish—hardly a departure from the arrangement praised by Lysias. If the *erōs* of Lysias is cold and nonresponsive, its counterpart is hot and predatory. "As the wolf loves the lamb so the lover loves his beloved," Socrates suggests provocatively.

To disavow his authorship, Socrates claims being possessed by the nymphs of the place. This madness is preferable to artful sobriety of Lysias, but it is still inferior to inspired poetry and true wisdom. Halfway through, Socrates interrupts himself and remarks ironically: "Well, my dear Phaedrus, does it seem to you, as it does to me, that I am inspired? . . . Then listen to me in silence; for truly the place seems filled with a divine presence; so do not be surprised if I often seem to be in a frenzy as my discourse progresses, for I am already almost uttering dithyrambics."[43] As he ends—prematurely in his friend's opinion—Socrates declares that he'd better stop because he is already speaking in hexameters and if he does not he shall "sure be possessed of the nymphs" to whom Phaedrus exposed him. Socrates's ironic reference to divine frenzy underscores the immersive power of the place even as he tries to distance himself from it.[44]

At the end of his first speech, Socrates threatens to cross the stream to escape further compulsion from Phaedrus—an eloquent spatial detail that shows Socrates attempting to extract himself physically from the scene. As he tries to do so, he admits that his *daimon*, the inner voice that prevents him from doing something wrong, "forbade my going away before clearing his conscience, as if he had committed some sin against deity."[45] Socrates goes on to claim that the two speeches committed the sin of impiety against Erōs, the god of love, and that recantation is in order. He cites the lines from the poet Stesichorus and announces that he will offer his own recantation to Erōs, now with his head bare. By the time Socrates presents his recantation, the mania that infuses the setting of the dialogue now becomes central to the argument. His declaration—that "the greatest of blessings come to us through madness"—applies both to love and to discourse.[46] In the course of the speech, so different from the preceding two in content and form, Socrates blends the elements of his dialectic—the classification of madness, love, and parts of the soul—with the discursive conventions of myth and poetry.[47] (It should be noted that natural scenery is again introduced, but this time it functions metaphorically to describe the plane of truth as "the fitting pasturage for the best part of the soul.")[48] It is also the most earnest of the speeches in the dialogue: There are no ironic asides or other indications that the speaker is not serious. Indeed, many commentators regard the second speech of Socrates (also known as his "palinode") as a direct statement of Plato's doctrine of the soul.

The imagery Socrates employs is striking in its sensory specificity, given the metaphysical nature of the story of the soul's transformation (*metabolē*) in the presence of earthly beauty. Indeed, it creates another immersive landscape on top of the one surrounding the plane tree. When the philosophical lover beholds the beautiful youth, he is immediately transported to the divine realm and remembers true beauty. Far from being deceptive and untrustworthy, vision and its beautiful object aid in the philosophical journey of the soul, allowing it to become winged. Despite the instantaneous effect of the gaze, however, the soul's process of change is an excruciating fusion of pleasure and pain, as the language depicts bodily sensations that accompany the growth of feathers. First, there is a sensation of heat and moisture that nourish the roots of the feathers. Yet as the feathers begin to grow, the soul suffers from fever and itching, to be relieved only through the sight of the beloved.[49] These two mingled sensations produce the effects of befuddlement (*aporia*) and disorientation (*atopia*), identified by the very terms that capture the rigors of the philosophical journey to the realm of truth, justice, and beauty.[50] The story both sets before the eyes a complex psychological process and conveys major lessons of Platonic philosophy, all in the name of Erōs, to whom Socrates offers a prayer at the end of the speech.

Socrates has unquestionably bested Lysias in this performance, as Phaedrus confirms, but the dialogue does not end there; Socrates and Phaedrus are only halfway through. They continue their conversation, now focused explicitly on the art of rhetoric, in the shade of the plane tree, accompanied by the slumber-inducing chorus of cicadas. Importantly, the lore of the cicadas—insects whose idle singing the poets contrast with the industriousness of the ant—is reinterpreted by Socrates to connote a higher form of leisure.[51] Cicadas become the beneficiaries of the Muses, in whose honor they sing tirelessly, without respite or nourishment. In this new version, "the prattling of the poet/cicada/rural conversationalist is not only acceptable but preferable to other alternatives. In the context of the dialogue, this updated fable encourages us to stay in the country and engage in the sort of manic, erotic, and philosophical dialogue that occupies Socrates and Phaedrus, better that than returning to the city and the ant-like practical rhetoric of Lysias and his ilk."[52] According to Socrates, this is the kind of leisure that yields knowledge—of the soul, love, and true rhetoric—instead of opinion. Socrates contrasts his own alertness and that of his companion in the midst of cicadas' drowsy song with the slavishness of those who allow themselves to be lulled into sleep.[53] This aside pointedly signals the distinction between "base" rhetoric, the kind that makes its hearers slaves to opinion, and philosophical discourse, an ongoing conversation addressed to the soul and aimed at higher truths. We can also view it as an allegory of the relationship between immersion and critical distance—one remains alert despite the seductive power of the scene.

The remaining discussion of the speeches from the first part of the dialogue demonstrates the advantages of critical reflection over uncritical recitation of written texts. Focusing now on the medium, Socrates begins by mounting an attack on writing as a means of *psuchagōgia* ("soul-leading"). He again combines his dialectical analysis with mythological lore—namely, the Theuth and Thamus story about the invention of writing—to press the difference between the inauthentic craftsmanship of rhetoricians and the living, breathing discourse of philosophers. A written speech is not only an artifact devoid of liveliness but also deaf to any questions that might be directed to it. In fact, it discourages questioning and instead promotes a numbing repetition of itself. Writing is, at best, an art of reminding (*hypomnēsis*) rather than memory. Phaedrus's earlier uncritical recitation of the speech by Lysias—the speech that, Socrates suspects, Phaedrus had intended to memorize word-for-word—exemplifies the kind of practice that is antithetical to philosophical rhetoric. Socratic performances, by contrast, are supposedly animated by nature, nymphs, and the Muses. Moreover, they are also rejoinders, rather than set pieces, since they respond to previously uttered speech, whether it is direct, like the one by Lysias, or indirect, such as the various mythological or poetic references.

Socrates qualifies his assault on writing by describing the philosophical *logos* as "the word that is written with intelligence in the mind of the learner, which is able to defend itself and knows to whom it should speak, and before whom to be silent."[54] Natural reference once again helps to distinguish Socratic authenticity and nobility of purpose from the frivolity of other kinds of writing. One can plant the seeds in the heat of summer in the garden of Adonis, where the seeds become blossoms within a mere eight days. This practice, however amusing—and occasionally indulged in by Socrates himself—is a far cry from true philosophy. Philosophers follow a responsible type of husbandry, planting their words/seeds in a fitting soil/soul, thereby providing for a continuous chain of intellectual influence/reproduction.[55]

The ending of the dialogue suggests that the interlocutors have achieved what Kenneth Burke would call "consubstantiality."[56] Phaedrus *seems* to have been transformed into a fitting recipient of Socratic *logoi*, judging by his agreement with Socrates on all matters pertaining to the topics under discussion.[57] The heat has subsided, and it is time to return to the city. Socrates suggests that they offer a prayer to Pan and the other gods of the place, one final invitation to the reader to refocus attention on the significance of the location. Phaedrus eagerly joins him, declaring that "friends have all things in common."[58] His assent, importantly, is not to a particular intellectual proposition but to the principle of friendship, *philia*. The notion of *philia* applies to a relationship between free equals, and its invocation serves as a fitting closure to a conversation about better and worse ways of spending one's free time, *scholē*.

The boundaries of the dialogue are determined not by the finality of the analysis or the topics but by the natural rhythm and duration of a leisurely chat. The dramatic structure is circular, beginning with the characters' departure from the city and ending with their return. The entire exchange between the two interlocutors happens in the countryside, outside the city walls. It is an unlikely setting for rhetoric, the art that traffics in opinion, on which, as Socrates' second speech points out, the souls of nonphilosophers feed. Yet, we should also recall that Socrates has spent his life in the noisy, dusty environs of the *polis*, criticizing rhetoricians and poets alike and trying to win over the hearts and minds of the city's promising youths. Venturing outside is for Socrates a rare opportunity to persuade his young companion to turn away from rhetoric and pursue philosophy. The change in the scenery invites the juxtaposition of "city" discourses with mythical and poetic lore. It is important that Phaedrus is the one who leads Socrates out, even though his reasons for doing so may be far from philosophical: He is too fond of hearing and reciting trendy speeches and arguments rather than questioning them, as Socrates does.

So Socrates works with what he is dealt and takes the route of playful iconoclasm rather than direct refutation. In so doing, he demonstrates not only a superior rhetorical skill (the way it was commonly understood) but also a concern for the soul of his audience (something with which rhetoricians, and especially writers of speeches, allegedly did not bother). Over and over, through demonstration and analysis, mimicry and reflection, Socrates nudges his friend to reconsider his habitual ways of speaking and listening and become a more discerning judge of both artistic technique and moral purpose. Socrates plays a game of *psychagōgia* with Phaedrus, pretending to be fascinated with Lysianic rhetoric and miming its technique at the beginning, only to oppose this rhetoric through his own highly poetic performance and, later, critical analysis. Phaedrus and Socrates take turns imitating rhetorical and poetic genres as well as discuss how their parts fit together, a set of tasks that today's technology designers would likely applaud as a good combination of immersive and critical game strategies. Phaedrus is being converted to philosophy through a combination of immersion and critical reflection.

The critical distinctions between different types of leisured pursuits, social relationships, and forms of discourse are not forced upon Phaedrus or the reader in a dogmatic fashion but are revealed gradually and organically in the course of the journey to the plane tree and back. The depiction of the natural setting and the invocation of the cultural lore associated with it work as scaffolding—or "infrastructure," as John Durham Peters would call it—that lends authenticity to the exchange between Socrates and Phaedrus. "Infrastructure," notes Peters, "is often defined by being off the radar, below notice, or off stage," even as it is indispensable to human action. "Technology," on the other hand, "is a concept biased toward newness."[59] Although today we would consider writing as such a taken-for-granted element of our communicative infrastructure, for Plato writing—especially in the form of easily portable scrolls—represents a potentially troublesome new technology. At the same time, the infrastructure—the countryside, the plane tree, the stream, the cicadas, and of course, the privileged condition of leisure—plays a supportive role in the dialogue's drama.

The setting immerses us in the fiction of the dialogue and veils the artifice of the text we are reading. If Plato objectifies poetic and rhetorical discourses through implicit and explicit distancing and draws attention to their persuasive mechanism, the dialogic form itself and its infrastructural conditions do not receive the same degree of scrutiny. To be sure, Socrates tries to impress on Phaedrus the distinction between noble and base forms of leisure, but neither interlocutor questions whether the freedom to stroll outside the city walls and engage in unhurried conversation about written texts in a beautiful natural setting is his prerogative.

The moment of distancing occurs in the context of the discussion of good and bad writing. Writing does not exist without readers, and the commentary on writing as a "reminder" rather than "memory" can be taken as a reflection about reading as a cultural and educational practice. We are encouraged to think upon how texts figure in the process of pedagogical *psychagōgia*, a rhetorical process during which teachers, students, and texts come together. Reading texts can be an amusing and engaging adventure, since it can temporarily transform one into a character in a story. Yet a more lasting qualitative transformation comes only as a result of repetitive exposure to texts and their questioning with other learners. Indirectly, then, the dialogue asks us to view it as a prompt, rather than as a hermetically sealed world. Despite its charms as an immersive text, the *Phaedrus* can only provoke—not replace—an ongoing intellectual conversation about ancient Greek rhetoric and philosophy, because such conversation, unlike the drama depicted by the text, cannot be finalized.

Immersion and Teaching: From the *Phaedrus* to the Twenty-First-Century Classroom

The *Phaedrus* does not define the nature of philosophical discourse directly and unambiguously, relying instead on the spatial and narrative dynamic to give shape to the philosopher's pedagogical vision. The text's immersive qualities lend the dialogue form a sense of naturalness and elevate it above other competing genres and models of teaching, just as Socrates holds his manic discourse on love above Lysianic rhetoric and by extension all rhetorical instruction. Presenting the interaction between Socrates and Phaedrus as a playful and inspired conversation, Plato promotes his pedagogical model over and against the likes of Lysias and Isocrates, who is the object of sarcastic praise at the end of the dialogue and quite possibly its major target. The *Phaedrus* was likely read both inside and outside Plato's Academy, and it probably functioned as a demonstration of the virtues of Plato's pedagogical approach in the competitive intellectual milieu of fourth-century BCE Athens.[60] While his rivals for the minds and hearts of ambitious young men also disseminated their textual displays of rhetorical virtuosity as a form of self-advertisement, Plato's text offered more than a demonstration of its author's superior rhetorical skill: Through artful effacement of its own textuality it created an enduring appeal to authenticity.

Students of rhetoric have learned to be wary of Plato's charms, and they may find this appeal to authenticity suspect for at least a couple of reasons. The first one is the dialogue's promise to transcend the artistic and ethical limitations of all other discourses by subordinating them to a seemingly apolitical form of leisure dramatized by the dialogue. This vision promises an escape from the constraints of political interdependence that marks the relationship of rhetors and audiences in the city. With the pressures of politics and society thus lifted, language is seemingly freed from its previous habitation and can be deployed in pursuit of higher knowledge and a more genuine human happiness. The *Phaedrus* nurtures a troublesome assumption that rhetoric, understood as an address to an interlocutor's soul, can be freed from the bonds of conventional politics.[61]

Another problem with the ideology of dialogic authenticity is its arguably reactionary notion of knowledge transmission. Particularly in its image of insemination through speech, the *Phaedrus* offers a "patriarchal vision" by promoting "philosophical instruction as a kind of reproduction without women."[62] As well as an expression of an "archaic set of anxieties" about legitimate paternity, writes Peters, it also constitutes an enduring theory of good and bad forms of communication: "Socrates faces two kinds of 'AI': the artificial intelligence of the written text, which simulates a caring teacher, and the artificial insemination of its distribution, which makes paternity undecidable. Writing for Socrates is something like an intellectual sperm bank: conception can occur between anonymous partners whose junction can be manipulated across great distances of space and time."[63] On this view, Plato's elevation of dialogue pathologizes dissemination, even though, according to Peters, the latter is "far friendlier to the weirdly diverse practices we signifying animals engage in."[64]

Plato's *Phaedrus* may indeed be offering us an impossible dream of authenticity and mutuality, the dream that was—and still is—at odds with most of the diverse practices of signifying animals. Yet Plato's vision of the dialogue as a way of questioning the cultural and psychological effects of discourses is not entirely obsolete. One does not need to share Plato's metaphysics and cultural elitism to see how the *Phaedrus* can illuminate the process of learning as an intrinsically good form of leisure.

As we have seen in our earlier discussion, video games and virtual reality are often defended on the grounds that they deliver an inherently satisfying experience because the experience appears to be protected from the pressures of the outside world. In a game, one is bound by the game's rules and procedures as well as its constructed fantasy world. But the player's actions, along with the simulated reality of the game, create the effect of immersion and thereby authenticate the process of playing. This activity, claims game designer and scholar Eric Zimmerman, needs no external social validation, since games "are beautiful."[65] To extend the erotic analogy of the *Phaedrus*, video games offer their users a form of autoeroticism, as each player creates a distinct fantasy from the menu of encoded options. Like Phaedrus prior to his unplanned rendezvous with Socrates, excitedly clutching Lysias's speech under his cloak, today's game players seek gratification in a human-digital text interaction.

The autoeroticism of game play is doubly valuable if it can be leveraged for education—after all, the English word "school" is a direct descendant of Greek *scholē*. Because users perceive them as a form of leisure—no matter how demanding they may be in actuality—games have been hailed as the "new" frontier in educational technology. So, someone like Zimmerman can assert that games possess both aesthetic appeal and educational value: games can foster "understanding [of] how the parts of a system fit together" because "as we play, we think about thinking and we learn to act in new ways."[66] Not all games, of course, provide this kind of intellectual exercise, as some scholars have pointed out. But the debate regarding educational uses of games and game-like environments is preoccupied with how to design games with the right ratio of immersion to reflection, not with how the adoption of increasingly sophisticated immersive technologies alters the situation of the learner and the entire educational infrastructure.

This is precisely the question that the *Phaedrus* invites us to ponder. Given that the principal characters of the dialogue meet outside the city walls by sheer chance, we might wonder: What would Phaedrus get out of Lysias's speech had he not run into Socrates? Socrates's mention of his own voraciousness for discourses seems a fitting description of Phaedrus, whose weakness for *logoi* is well established in the dialogue. It is very likely that Phaedrus would have taken delight in reciting the speech of Lysias to himself while strolling the country roads, since he is shown to be quite immersed in its reading in the presence of Socrates. He would be likely to internalize the commonplaces of praise and blame that structure the encomium as well as its organizational and stylistic features. He may even reflect on how Lysias's composition may be different from other speeches of its genre, given its clever way of praising the nonlover over the lover, and decide to emulate this paradoxical strategy in the future. Indeed, Socrates's first speech, intended as a parodic "improvement" over the original encomium, may be the kind of discourse Lysias would be able to produce in the absence of guidance from Socrates. All of that he could do entirely on his own. But he would not have progressed much further intellectually or ethically, remaining in the same state of pseudo-erudition as he was at the time of leaving the city.

Socrates adapts his pedagogical approach to Phaedrus's complex and omnivorous soul and succeeds, possibly, in leading him away from Lysias's rhetoric and toward philosophy. Setting aside the disciplinary and political reservations about Plato's agenda, we cannot help admiring the spontaneity and ingenuity with which Socrates anticipates and answers his interlocutor's moves. This is what a good classroom conversation looks like, according to David Bromwich's polemic against massive open online courses, or MOOCs. Experienced teachers, observes Bromwich, learn to recognize the look "of the participant who is almost ready to say something interesting or surprising," which can indicate "a half-baked idea that is well lost," but "at other times it may mark the beginning of a perception that ought to be encouraged and followed up."[67] In this kind of conversation, argues Bromwich, "you learn a good deal that can't be quantified, packaged, or transmitted by an efficient impersonal medium, no matter how up-to-date, no matter how well engineered."[68] Bromwich's point resonates with this essay's criticism of new educational technologies that are praised for their unprecedented ability to engage students in learning through immersive experiences. Such technologies may be better than a bad teacher or no teacher at all (as may be the case when schools cannot afford hiring teachers to offer certain subjects), but they cannot replace a good teacher. Moreover, the majority of immersive technologies offer preprogrammed experiences that can teach a set of skills but cannot lead students to ask questions for which there are no easy answers.

Even the most innovative games that are meant to teach students critical thinking are likely to fail in their mission without the often neglected infrastructure of a classroom environment. As a case in point, Bogost, Ferrari, and Schweizer point to Molleindustria's *McDonald's Videogame*, an ironic game that asks players to control a single McDonald's vertical conglomerate from the South American pastures to the meat packing factory to the storefront.[69] The game's rules corral the player into performing a series of environmentally and socially disastrous actions in order to meet monthly profit goals: clear cutting, bribing public officials, and creating unhealthy livestock conditions. When the player's actions reach a height, environmental activists or regulatory agencies levy fines. By design, the fines and protests by social justice advocates are supposed to make clear the irony of the game's promotion of the McDonald's brand. Yet "many players . . . find themselves lamenting the difficult job of McDonald's executives rather than incensed by their corrupt corporate policies."[70] *McDonald's Videogame* cannot teach critical consumption to players without the help of schooling in such skills as appreciation of irony and parody, skills that take time and the guidance of experienced instructors to develop.

We have argued that contemporary studies of immersion have focused on educational technologies such as games and virtual reality at the expense of the conversation about education as a process. In their search for a more "authentic" context of education, designers, scholars, and university administrators have unwittingly fetishized so-called immersive technologies as a way to engage, motivate, and instruct digital natives. The praise of the pedagogical value of immersion (even when it is accompanied by its counterpart, critical distance) seamlessly fits into the capitalist model of education as the delivery and consumption of products, not the cultivation of the life of the mind or civic identity. We need pedagogies that combine immersion and critical reflection on the nature of that immersion, not merely new shiny fixes for age-old problems of boredom and distraction.

Notes

1. The cover of the August 17, 2015, issue of *Time* magazine features Palmer Luckey, founder of Oculus VR (the company that created the Oculus Rift) along with the title, "The Surprising Joy of Virtual Reality and Why It's About to Change the World." See Joel Stein, "Inside the Box," *Time* magazine, August 17, 2015, 40–49.

2. "Boston University Unveils Virtual Reality Room," *Boston Globe*, June 20, 2015, https://www.bostonglobe.com/lifestyle/style/2015/06/19/brown-university-unveils-virtual-reality-room/QoTOOp66NpPZeGMFobapjO/story.html.

3. Janet H. Murray, *Hamlet on the Holodeck: The Future of Narrative in Cyberspace* (Cambridge, MA: MIT Press, 1998).

4. Ibid., 98.

5. Allison Griffiths, *Shivers Down Your Spine: Cinema, Museums, and the Immersive View* (New York: Columbia University Press, 2008), 4.

6. Marie Ryan, *Narrative as Virtual Reality: Immersion and Interactivity in Literature and Electronic Media* (Baltimore, MD: Johns Hopkins University Press, 2001), 14–15.

7. Emily Brown and Paul Cairns, "A Grounded Investigation of Game Immersion." Paper presented at the *Annual Meeting of the Computer-Human Interaction Conference*, Vienna, Austria, 2004, 1297. Scholars of rhetoric are perhaps more familiar with presence as defined by Perelman and Olbrechts-Tyteca: "One of the preoccupations of a speaker is to make present, by verbal magic alone, what is actually absent but what he considers important to his argument or, by making them more present, to enhance the value of some of the elements of which one has actually been made conscious"; in *The New Rhetoric: A Treatise on Argumentation* (Notre Dame, IN: University of Notre Dame Press, 1991), 117.

8. Murray, *Hamlet on the Holodeck*, 98.

9. Oliver Grau, *Virtual Art: From Illusion to Immersion* (Cambridge, MA: MIT Press, 2003), 13.

10. Ryan, *Narrative as Virtual Reality*, 139.

11. See especially Brown and Cairns, "A Grounded Investigation of Game Immersion," 1298.

12. Mihaly Csikszentmihalyi, *Flow: The Psychology of Optimal Experience* (New York: Harper, 1990), 76.

13. For criticisms of immersive fallacy, see Frank Lantz, "Something Called the Immersive Fallacy." Presentation at the annual meeting for the Game Developers Conference, San Francisco, California, 2006, http://www.gdcvault.com/play/1013249/Burn-Baby-Burn-Game-Developers; Robert Yang, "On Focalization and Against Convenient Understandings of Immersion/Flow," *Radiator Design Blog*, May 15, 2013, http://www.blog.radiator.debacle.us/2013/05/on-focalization-and-against-convenient.html.

14. Grigore C. Burdea, "Keynote Address: Haptic Feedback for Virtual Reality," *Proceedings of International Workshop on Virtual Prototyping* (1999), 87; Richard Moss, "Haptic Technology: The Next Frontier in Video Games, Wearables, Virtual Reality, and Mobile Electronics," *GizMag*, January 15, 2015, http://www.gizmag.com/haptic-tech-vr-wearables-games-sightlence/35616/.

15. James Newman, "The Myth of the Ergodic Videogame: Some Thoughts on Player-Character Relationships in Videogames," *Game Studies* 2, no. 1 (2002), accessed July 19, 2015, http://www.gamestudies.org/0102/newman/; Jan Van Looy, "Uneasy Lies the Head that Wears a Crown: Interactivity and Signification in *Head Over Heels*," *Game Studies* 3, no. 2 (2003), Accessed July 19, 2015, http://www.gamestudies.org/0302/vanlooy/.

16. For example, Alison McMahan, "Immersion, Engagement, and Presence: A

New Method for Analyzing 3-D Video Games," in *The Video Game Theory Reader*, ed. Mark J. P. Wolf and Bernard Perron (New York: Routledge, 2003), 69.

17. Chris Dede, "Immersive Interfaces for Engagement and Learning," *Science* 323 (2009): 66. This description resonates with Kenneth Burke's discussion of the psychology of the audience and his definition of form as "the creation of an appetite in the mind of the auditor, and the adequate satisfying of that appetite"; see *Counter-Statement* (Berkeley: University of California Press, 1968), 31.

18. Sven Birkets, *The Gutenberg Elegies: The Fate of Reading in an Electronic Age* (New York: Fawcett Columbine, 1994), 97.

19. Ryan, *Narrative as Virtual Reality*, 92.

20. Ibid., 92.

21. See, for example, Aaron Hess, "'You Don't Play, You Volunteer': Narrative Public Memory Construction in Medal of Honor; Rising Sun," *Critical Studies in Media Communication* 24, no. 4 (2007): 339–56.

22. Ricardo Rosas, Miguel Nussbaum, Patricio Cumsille, Vladimir Marlanov, Mónica Correa, and Patricia Flores, "Beyond Nintendo: Design and Assessment of Educational Games for First and Second Grade Students," *Computers & Education* 40, no.1 (2003): 76.

23. See, for example, Mary Jo Dondlinger, "Educational Video Game Design: A Review of the Literature," *Journal of Applied Educational Technology* 4, no. 1 (2007): 21–31; Glenda A. Gunter, Robert F. Kenny, and Erik H. Vick, "Taking Educational Games Seriously: Using the RETAIN Model to Design Endogenous Fantasy into Standalone Educational Games," *Educational Technology Research and Development* 56, no. 5 (2008): 511–37.

24. Gunter, Kenny, and Vick, "Taking Educational Games Seriously," 512.

25. Murray, *Hamlet on the Holodeck*, 98.

26. Grau, *Virtual Art*, 339.

27. Ibid., 341.

28. Ibid., 203, 340.

29. Ibid., 346.

30. Ibid., 346.

31. Plato, *Republic*, book VII.

32. Plato, *Plato VII: Timaeus, Critias, Cleitophon, Menexenus, Epistles*, trans. R. G. Bury (Cambridge, MA: Harvard University Press, 1914), 235c.

33. Many scholars have argued that Plato's dialogues are dramatic enactments—that their theoretical import does not reside in any specific utterance but instead permeates the entire exchange. See, for example, Albert Cook, "Dialectic, Irony, and Myth in Plato's *Phaedrus*," *American Journal of Philology* 106, no. 4 (1985): 427–41; Rosemarie Desjardins, "Why Dialogues? Plato's Serious Play," in *Platonic Writings, Platonic Readings*, ed. Charles Griswold (London: Routledge, 1988), 110–25; Francisco Gonzalez, ed. *The Third Way: New Directions in Platonic Studies* (Lanham, MD: Rowman and Littlefield, 1995); and Gerald A. Press, ed., *Plato's Dialogues: New Studies and Interpretations* (Lanham, MD: Rowman and Littlefield, 1993).

34. Andrea Wilson Nightingale, *Spectacles of Truth in Classical Greek Philoso-*

phy: Theory in Its Cultural Context (Cambridge, UK: Cambridge University Press, 2004), 96.

35. Ibid., 96.

36. We are by no means the first to emphasize the setting's dramatic importance in the *Phaedrus*. G. R. F. Ferrari's *Listening to the Cicadas: A Study of Plato's* Phaedrus (Cambridge, UK: Cambridge University Press, 1987) is perhaps the most extensive and astute of the treatments of the dialogue. However, Ferrari draws our attention to the dialogue's scenery to highlight its role in the characterization of Socrates. We agree with Ferrari's insights but suggest that the setting lends authenticity not only to Socrates's idiosyncratic persona but to the dialogue form itself.

37. Plato, *Euthyphro, Apology, Crito, Phaedo, Phaedrus*, trans. H. N. Fowler (Cambridge, MA: Harvard University Press, [1914] 1995), 230d–e.

38. In classical Greece, leisure was an attribute of the elites, who did not have to work for a living. Leisured pursuits thus came to be identified with nobility and to be contrasted with merely "useful" occupations. Aristotle's *Politics* illustrates the classification of educational activities according to their suitability to noble leisure. See Aristotle, *Politics*, trans. H. Rackam (Cambridge, MA: Harvard University Press, 1977). At Plato, *Phaedrus*, 276d, we arrive at a meditation on noble and base ways of passing time.

39. Plato, *Phaedrus*, 229e–230a.

40. Ibid., 230b–c.

41. Ibid., 230c. In Plato's dialogues, the term *atopos* often describes the philosopher's condition of disorientation. As Nightingale explains, "the detachment from the world and contemplation of the Forms radically transforms the philosopher. . . . When he returns to the human realm, he sees (and is seen) very differently. . . . The philosopher has become a sort of stranger (*atopos*) in his own land," *Spectacles of Truth*, 97.

42. Plato, *Phaedrus*, 234d.

43. Ibid., 238c.

44. Ferrari notes that Socrates "is shaken by how things have developed; for he has felt how susceptible he is to this poisonous drug"; *Listening to the Cicadas*, 112.

45. Plato, *Phaedrus*, 242c.

46. Ibid., 244a.

47. On Plato's appropriation of Greek poetry and rhetoric in the *Phaedrus*, see Andrea Wilson Nightingale, *Genres in Dialogue: Plato and the Construct of Philosophy* (Cambridge, UK: Cambridge University Press, 1995), 157–59.

48. Plato, *Phaedrus*, 248b.

49. Sappho's erotic poetry is evoked in this passage to describe the lover's pathos. See Nightingale, *Genres in Dialogue*, 158.

50. Plato, *Republic*, book VII, 515d–e.

51. On the lore of the cicadas in Greek culture, see Malcolm Davies and Jeyareney Kathirithamby, *Greek Insects* (New York: Oxford University Press, 1986), and J. C. B. Petropoulos, *Heat and Lust: Hesiod's Midsummer Festival Scene Revisited* (Lanham, MD: Rowman and Littlefield, 1994).

52. David J. Schenker, "The Strangeness of the Phaedrus," *American Journal of Philology* 127, no. 1 (2006): 67–88.

53. Plato, *Phaedrus*, 259a–b.

54. *Ibid.*, 276a.

55. *Ibid.*, 276b–277a.

56. Kenneth Burke, *A Rhetoric of Motives* (Berkeley: University of California Press, 1969), 20–21.

57. The two characters promise each other that upon their return they each will take their conclusions about rhetoric and philosophy to their friends, Phaedrus to Lysias and Socrates to Isocrates "the fair." The reference to Isocrates is, of course, laden with irony—at the writing of the dialogue, Isocrates not only competed with Plato for students but also threatened to usurp the very term *philosophia*. Socrates's effort, we suspect, is therefore likely to be wasted.

58. Plato, *Phaedrus*, 279C.

59. John Durham Peters, *The Marvelous Clouds: Toward a Philosophy of Elemental Media* (Chicago: University of Chicago Press, 2015), 36.

60. Based on textual and intertextual evidence, Murphy concludes that Plato's dialogues found sophisticated readers outside of Academic circles. Socratic dialogue was a popular genre in the fourth century BCE, and several authors besides Plato composed in the genre. See David J. Murphy, "Isocrates and the Dialogue," *Classical World* 106, no. 3 (2013): 311–53.

61. See Russell Bentley, "On Plato's Phaedrus: Politics Beyond the City Walls," *Polis: The Journal of the Society for Greek Political Thought* 22, no. 2 (2005): 230–49.

62. John Durham Peters, *Speaking into the Air: A History of the Idea of Communication* (Chicago: University of Chicago Press, 2001), 48–49.

63. Ibid., 49.

64. Ibid., 62. Peters's critique of dialogue as a cultural norm is premised on the idea that sometimes life lived on the surface of things can be valuable and ethical. Still, in his new book Peters cautions, "we should beware of any platform of communication that pretends there is no difference between interaction online and face to face"; *The Marvelous Clouds*, 275–76.

65. Eric Zimmerman, "Manifesto for a Ludic Century," accessed July 19, 2015, http://ericzimmerman.com/files/texts/Manifesto_for_a_Ludic_Century.pdf.

66. Ibid.

67. David Bromwich, "Trapped in the Virtual Classroom," *New York Review of Books*, July 9, 2015, 16.

68. Ibid., 16.

69. Ian Bogost, Simon Ferrari, and Bobby Schweizer, *Newsgames: Journalism at Play* (Cambridge, MA: MIT Press, 2010).

70. Ibid., 27.

8

Genre in Ancient and Networked Media

Carolyn R. Miller

Aei pherei ti ē Libuē kainon. (Always something new out of Libya).
—Aristotle, *Historia Animalium* 606b20

In 2007, half of the top 10 best-selling novels in Japan were cellphone novels, originally composed, circulated, and read on the smartphones of, mostly, young women fans who craved recognition of these works as a new literary genre.[1] Wikipedia's article about video-game genres lists ten major genre categories (including an "other" category) for a total of fifty-six genres, plus alternate modes of classification.[2] In 2014, Microsoft spent $2.5 billion to acquire the game Minecraft in order "to create a new genre of gaming," for use with its Microsoft HoloLens augmented reality headset, according to company CEO Satya Nadella.[3] Shortly after the iPhone 4 was released in 2010 with the front-facing camera that made self-photography simple and ubiquitous, a new genre of photography was recognized, and "selfie" was chosen as the Oxford Dictionaries' word of the year for 2013.[4] YouTube made possible the posting and compulsive watching of "drunk videos," which include a series of subgenres: drunk-shaming videos, intervention videos, drunk animal videos, and the like. Noting that one such video of a pop star had been viewed three million times, Virginia Heffernan asked, "Why are videos of drunken people so compelling? . . . Why do we watch this stuff?"[5] Heffernan has been a dogged observer of what she calls the "mayhem and trivia" of the internet world, and she points out that the success of YouTube has depended on the enthusiasm of amateurs. "What's surprising," she says, "is how little the homemade videos resemble the pro goods. Sure, there are parodies of mainstream clips here and there, but mostly the amateurs are off on their own, hatching new genres," such as "haul" videos and "fail" videos. Such videos "often attract 100,000 views or more—and no one had even imagined such genres until recently."[6]

Examples like these could be multiplied many times over. What they have in common are two features of interest here: that the traditional, stabilizing concept of genre is being applied to a media landscape that seems constantly in motion, and that this constant motion is understood to be producing new genres. The proliferation of new media in the past thirty years has been accompanied by a proliferation of interest in genre, both in ver-

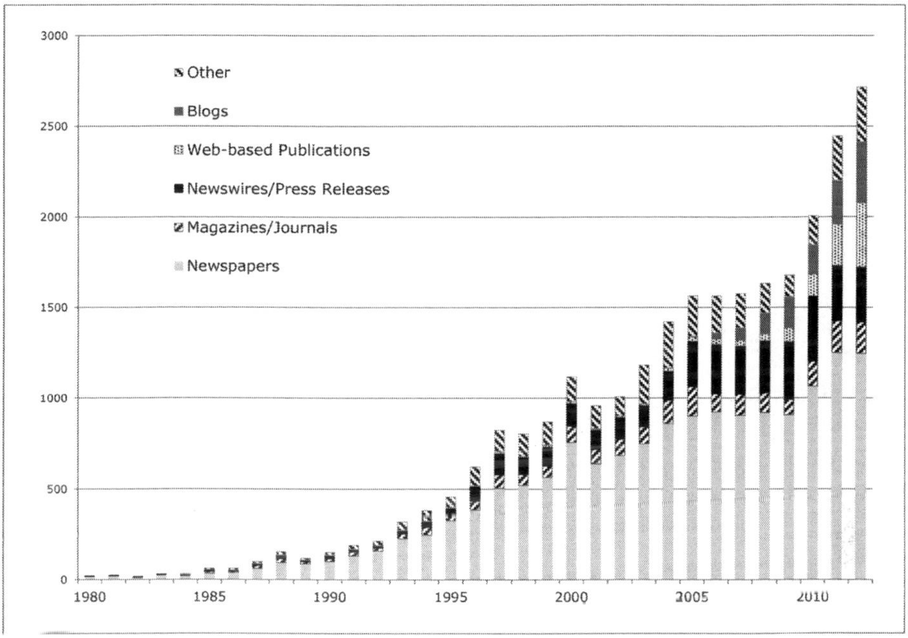

Figure 8.1. Occurrence of the phrase "new genre" in LexisNexis Academic database, 1980–2012, by type of source (data retrieved February 16, 2013).

nacular discussions about everyday life and in scholarly examinations of the media environment. The increasing vernacular interest in genre can be demonstrated by searching news databases, which are a treasure trove of how people are talking, and being urged to talk (and think), about genres. These databases show increasing use of the term "genre" over time and increasing use of the expression "new genre." For example, the *New York Times* Proquest Historical Archive shows that the proportion of articles mentioning the word "genre" doubled in the twenty-five years between 1982 and 2007 (the last year available), from .01 to .02, and the proportion mentioning the phrase "new genre" about tripled over the same period, from .0001 to .0003.[7] Over the same twenty-five years, Google's N-gram viewer for the contents of its scanned books shows that the relative frequency of the word "genre" in the English database of scanned books increased by a factor of 1.5, as did the relative frequency of the expression "new genre." The LexisNexis Academic database of print and digital news sources shows a dramatic increase in the use of the expression "new genre" between 1980 and 2012 (see figure 8.1; however, these data can't be controlled for the size of the database, so the results can only be suggestive).

In his book on genre, John Frow emphasizes that "reflection on genres and the distinctions between them is built deep into ordinary talk and writing" as we observe conventions, organize interactions, and interpret everyday linguistic phenomena.[8] We're surrounded by genres: They are "embedded in filing systems, in the organisation of books and journals in libraries, in school syllabuses, in the instructions for filling out tax returns or playing a game, in Powerpoint templates, and in online discussions of list etiquette."[9] But on the evidence from the databases mentioned above, we're also surrounded by what I call "genre-talk." Frow is discussing how genre operates in the background; what we see in genre-talk is an overt consciousness of genre. If genres are cultural categories, genre *itself* has become a cultural category, in addition to being a critical construct.

Genre matters. How much it matters has been illustrated by a number of works testing the distinction between the memoir and the novel, genres that help us sort personal history from fiction as rhetorical actions. The most notorious of these tests was James Frey's *A Million Little Pieces*, which was published in 2003, marketed as a memoir, and turned into a best seller in fall 2005 by its selection for Oprah Winfrey's television book club. It was subsequently challenged by the *Smoking Gun* website for including details about Frey's drug and alcohol addiction and recovery that were "wholly fabricated or wildly embellished," and an extensive and very public controversy ensued.[10] Other examples include Margaret Jones's book *Love and Consequences*, published in 2008 to great critical acclaim and described as "a heart-wrenching memoir" by the *New York Times* reviewer,[11] but which was shortly thereafter discovered to be "entirely fabricated."[12] The publisher recalled all 19,000 published copies and canceled the author's book tour. At about the same time, an international best-selling memoir about Holocaust survival, *Misha: A Mémoire of the Holocaust Years*, originally published in 1997, was also revealed as a fake: Belgian author Misha Defonseca is not Jewish, did not walk from Belgium to Ukraine during the war, and was not befriended by wolves.[13] The difference between memoir and fiction not only has had financial consequences for publishers and authors but also has stirred public reflection on authenticity, credibility, truth, and trust—all issues related to genre expectations.[14] As Frow notes, "Genre classifications are real. . . . They are embedded in material infrastructures and in the recurrent practices of classifying and differentiating kinds of symbolic action. And they bind abstruse and delicate negotiations of meaning to the social situations in which they occur."[15] Genre scandals like these are useful because, as Harold Garfinkel showed, it is when social norms are violated that we become newly aware of their existence.[16] Thus, we learn something important about the genre of the blog from the angry debate about the Kaycee Nicole cancer hoax, in which a healthy adult blogger posed as a teenage leukemia victim in 2001,[17] and from several instances when an ostensibly personal blog was revealed to be a form of unacknowledged corporate advertising aimed at viral dissemination. Consumer outrage at such attempts by Sony, Wal-Mart, and McDonald's became public-relations disasters.[18] Once again, we see the strength and importance of genre expectations.

Beyond these violations, we find evidence of vernacular consciousness of and concern for genre identity—as it manifests in form, content, and rhetorical action—in activities such as conversations among Wikipedians about what to include in various entries, the rankings and ratings and blogrolling of blogs by bloggers, the proliferation of parodies on YouTube and elsewhere. In addition, the very act of naming new genres exhibits a widespread care for categories: thus, we have genres known as aristocrunk, steampunk, torture porn, lolcats, fan-fic, mockumentary, kiddie noir, dirtbag sitcom, and chillwave, a list that spools out seemingly endlessly. As Heffernan concludes, YouTube and other new media platforms are "incubating novel forms of creative expression and cultivating new audiences."[19] The point is not whether these shards of contemporary culture are or are not "really" genres but rather that so many people want to talk about them as genres. Genres are, after all, not natural or theoretical categories but "sociocognitive" categories, that is, shared recognitions and mutual expectations.[20]

Much vernacular genre-talk concerns commodities in the culture industry—distinctions among musical trends, literary themes, television programming, and the like—and many isolated claims about various "new genres" we might write off as marketing hype. But such talk also reveals a concern for understanding and ordering our relationships to the symbolic world, to ballooning amounts of information, and to our multiply mediated environment. And while much of this genre-talk concerns "old media," it is the affordances of digital media, in particular, that have made genre newly important: the ease of replication and variation, the rapidity and extent of dissemination, and the blurring roles of producer and consumer have made it possible to do new things, to create new relationships, and to discover new pleasures. How different and severe the world of Aristotelian genres looks, with its short and stable lists: in poetics, the epic, drama, lyric, and (much later) the novel; and in rhetoric, the forensic, deliberative, and epideictic genres. What accounts for this difference? Is it the cultural conditions, or the theoretical basis for defining genre? Is it fair or even meaningful to compare ancient Greek genres with contemporary "mayhem and trivia"?

Contemporary rhetorical genre theory has understood genres as typified rhetorical actions based in recurrent rhetorical situations. As actions, genres are seen as pragmatic, rather than purely formal or substantive, functioning as macro-speech-acts and constituted by recognizable forms and thematic content.[21] As typifications, genres are context dependent and culturally significant, and they mediate between private intentions and socially objectified needs (exigences). Because genres mark cultural categories, genres and genre systems[22] collectively help to "constitute" society, in Anthony Giddens's sense.[23] Thus, genres that have vernacular names are culturally salient and significant, because the name signifies the shared recognitions and expectations of the community of use.[24] Furthermore, as constituents of culture, genres and their systems change over time, coming into being, transforming, fading away: As Berkenkotter and Huckin put it, "Genres must be flexible and dynamic, capable of modification according to the rhetorical exigencies of the situation. At the same time, though, they must be stable enough to capture those aspects of situations that tend to recur. This tension between stability and change lies at the heart of genre use."[25] Or, as Catherine F. Schryer has it, "Genres come from somewhere and are transforming into something else."[26]

In the era of digital networks, genre studies has been reinvigorated by the burgeoning discursive and communicative activity of the internet, with the new media platforms, new audiences and producers, new communicative interactions, new exigences, and new genres it makes possible. In the words of Giltrow and Stein, "The Internet enables a new communication setting which reconfigures the conditions to which pragmatic features of language respond."[27] We find ourselves using terms like "proliferation," "velocity," "remix," "modularity," and "chaos" to characterize the constant and rapid change that accompanies digital media, which the examples offered above can only hint at. The digital environment, then, disrupts the tension between stability and change, enabling and provoking change, attenuating the possibility of stabilization, and thus bringing the very possibility of genres into question. And yet, as we've seen, genres and genre-talk seem to abound.

Invention and Imitation of Genres

The so-called digital age poses two challenges for genre theory, both centered on this tension between stability and change. One challenge is to account for change, given the primary nature of genre as typification based in recurrence and its function as a basis of familiarity and social recognition; these features tempt us to essentialize, or reify, genres, to see them as defined by specifiable criteria and related in a logic of mutual exclusion. But long before the digital age, the dynamism of genres was recognized, leading to debates over genre definitions in the Renaissance, attempts to regulate genres during the eighteenth century, and neoclassical stultifications in both literary and rhetorical studies in the nineteenth and twentieth centuries. The recent literature includes a number of case studies of genre change, both predigital and digital, that describe the facts of change over time and explore ways to explain it: Charles Bazerman's history of the experimental article[28] and Berkenkotter and Huckin's follow-up study;[29] Sharon D. Downey's chronological survey of the apologia;[30] JoAnne Yates's study of the emergence and evolution of the business memo in the nineteenth and early twentieth centuries;[31] and Jason Mittell's work on the origins and cultural meanings of various television genres.[32] Other studies have taken up the evolution of digital genres, for example in Wikipedia articles,[33] in blogs,[34] in online components of the scientific journal article,[35] in the online mathematics dictionary and the online newspaper.[36] In these studies and others, genres are seen to emerge from technological change and media transformation, from marketing and commercial constraints, the conditions of production, intellectual explanatory ambitions, antecedent genres, sociocultural and psychosocial change, influential individuals and their goals and commitments, critical appraisal and deliberate recombination, play and experimentation, and institutional and political change.[37]

The second challenge, perhaps the more crucial one, is to account for how the constant flux of digital media is stabilized into what we call genres. In Schryer's formulation, genres are "stabilized-enough" or "stabilized-for-now," but how does stabilization occur given the dizzying volatility of the internet?[38] When change occurs at internet speed, and variation is clickably easy, what remains stable? In some situations, stabilization is produced through various mechanisms of regulation. Most work in rhetoric, writing studies, and applied linguistics, for example, has focused on structured contexts such as intellectual disciplines and the professions. Berkenkotter and Huckin focus their inquiry on "disciplinary communication," where genres are "the intellectual scaffolds on which community-based knowledge is constructed,"[39] and Yates and Orlikowski examine bureaucratic organizations such as business corporations, where "genres of organizational communication . . . influence a wide range of organizational phenomena."[40] These are regulated contexts in which genre stabilization can be achieved through administrative control and authoritative sanction. The new setting of the internet is regulated differently than traditional academic disciplines and corporate or government bureaucracies. On the internet we find voluntary activity, user-generated content, emergent communities of use, and robust experimentation and play. The achievement of stabilization is likely to be a much different process in this environment, one that relies on what Schryer has called "regularization," the frequently tacit self-regulation (or co-regulation) of a group or community.[41]

These twin challenges of accounting for genre change and accounting for genre stabilization in digital networks lead us directly to *inventio* and *imitatio*, rhetoric's drivers of change and stabilization, innovation and tradition. While some of the difficulties posed by these processes have been suggested above, looked at another way, these processes might seem easily accounted for in purely technological terms: imitation is made easy and ubiquitous by the stern governance of code (numerical representation) and its qualities of modularity and automation; invention proliferates by code's variability (and modularity) and capacity for transcoding, to use Lev Manovich's terms.[42] But although these technical qualities might show us *how* imitation and invention occur, they don't explain *why*, or more importantly how new genres result from these two processes together.

What could genre invention mean within digital networks? Are we inventing a genre or inventing within the context of a genre? In the first instance, we are asking how new genres come into being, and in the second, we are asking how an established genre changes over time. And it is not obvious whether these are two different processes or the same. If we adopt an evolutionary view, in which variation and selection of features in successive texts produce change both in a corpus and, more importantly, in the perceptions and expectations of those who engage specific texts, then a new genre can only be a combination and recombination of previous genres. "From where do genres come?" asks Tzvetan Todorov: "Why, quite simply, from other genres. A new genre is always the transformation of one or several old genres: by inversion, by displacement, by combination."[43] Kathleen Jamieson's discussion of the constraining influence of antecedent genres reinforces this view that we are, after all, dealing with only one process.[44] Whether we set out to invent a text that eloquently adapts a genre to a particular situation or, like the genre entrepreneurs we encountered earlier, to invent "a whole new genre," invention takes place in a complexly historical situation, informed and constrained by antecedent genres as well as by technical affordances and socioeconomic conditions, and it would seem that these processes would be pretty much the same, although case studies would have to explore this conjecture. This evolutionary view rules out radical novelty, such as the kind of "spontaneous cybergenres" hypothesized by Shepherd and Watters as those with "no counterpart in other media."[45] If we accept that genres are, first and foremost, social recognitions, an evolutionary view seems justified, since without antecedents there would be no way that such recognitions could occur and become shared. *Inventio*, in other words, is always conditioned and constrained, informed and enabled, by its historical location and rhetorical positioning with respect to audiences, conventions, styles, values, media, traditions. *Inventio*, at least in the case of genre, we might conclude, is thus intimately tied to *imitatio*.

As noted above, imitation seems obvious in the context of digital networks: much of what constitutes internetworked communication involves copying, embedding, quoting, forwarding, parodying, remixing, and so on—all ways of reproducing what has come before. These are certainly ways of reproducing texts, specific instances of generic patterns. But what kind of imitation informs the perpetuation and transformation of genres? Richard McKeon's extensive discussion of imitation in antiquity suggests that there are three primary types of objects, or models, of imitation: specific texts or producers of texts (as in pedagogical apprenticeship), human actions (as in Aristotelian poetics), and ideals (as with Platonic forms).[46] The imitation of a genre potentially involves all three of these objects of imitation. The imitation of specific texts produces new texts that reproduce at least selected features from the model; such features are often the most salient identifiers of a text's genre, though of course situation and exigence are also important and may also be susceptible of imitative replication. Thus, corporate and university home pages have many recurrent features, such as navigation tools and prominent content themes that have become conventionalized; similarly, shreds videos are identified by the combination of video of live music performance by a recognized performer or group with overdubbed audio of extremely low quality.[47] The imitation of rhetorical action ensures that the new text will be taken up as a genre with a particular illocutionary force: Thus, the university home page is now understood as a combination (often a vexed combination) of institutional promotion plus informative question-answering, and the shreds video is understood as cultural ridicule. And finally, the imitation of a genre involves imitating an ideal in that the genre itself does not exist in particular texts or a corpus of texts but in the socially shared recognitions and expectations of a community of use. It is these that must be reproduced to one extent or another, reproduced by being reevoked, for the genre to function as genre. We might think of this as reproduction of effect, which John Muckelbauer discusses in the context of Aristotelian poetic imitation of actions rather than the philosophical imitation of an ideal form: The poet "must attempt . . . to reproduce the effect of the model."[48] The distinction between these two types of imitation is difficult to maintain, however, because with genres we are necessarily dealing both with the "ideal" (that is, the shared recognitions) and with an action that presupposes that ideal, and only secondarily with specific texts, which are themselves, with respect to genres, always imitations.

How does imitation occur in the context of genre? Muckelbauer's discussion of the ancient tradition is helpful, particularly because he emphasizes that imitation and invention are tied together in multiple ways in all three contexts in which imitation is of interest: the philosophical context of Plato, the poetic context of Aristotle, and the pedagogical context that is the shared concern of most ancient rhetoricians, Greek and Roman. Although we think of imitation as concerning repetition and reproduction and invention as concerning variation and novelty, the ancients generally understood them in close relation, with invention as adaptive replication rather than as innovation. Muckelbauer argues that imitation and invention implicate and even require each other, their seeming opposites. Imitation produces inventive variation even when "identical reproduction" is the goal, through the use of multiple models, through imperfections in the process of reproduction, through differential capacities of the imitator. With digital genres, the example of "email forwardables"[49] would seem to demonstrate effortless identical reproduction, but each forwarding person, each recipient, and the current and prior contents of each recipient's inbox offers potential for variation in force and uptake, even if not in form or thematic content; the more recent "viral" media are similar in that "reproduction" is minimal, with one text (image, video, webpage) being loaded and reloaded as a link to it circulates on social media; here again, though, the context of reception varies.[50] It is the very extent of distribution ("sharing") of such forwardables and virals that gives them recognizable genre status, as they emerge from their non-digital antecedents in chain letters, petitions, fund-raising campaigns, advertising, and the like.

In contrast to identical reproduction, inventive variation, or what Muckelbauer calls "differential reproduction," can also be the goal of imitation.[51] The primary example here is the Aristotelian representation of human action in drama, which imitates life in a way that is necessarily different from life; in other words, "invention occurs within the act of repetition," a process that constitutes "the very possibility of artistic *mimesis*," which aims at reproducing not the model but "the *effect of the model*."[52] Platonic imitation of ideal forms in the actualized world is necessarily repetition as difference, since such imitation can never be identical and is therefore always inferior. In the pedagogical tradition, repetition with difference occurs as "emulation," or an agonistic attempt to surpass the model, to retain recognizable qualities of the model yet succeed in a new way.[53] In the emergence and perpetuation of digital genres, agonistic emulation is salient: today, for example, rhetors compete to produce memes that are wittier or more timely variations on some trending theme, such as "Trump your cat" (photos of cats with Donald Trump–like hairpieces as Trump gained momentum in the 2016 Republican presidential primaries) or Pluto-themed messages (images combining the 2015 NASA flyby of the planet with the Disney character) or the more persistently relevant "Overly Attached Girlfriend" image of a manically beaming young woman with captions demonstrating various modes of intrusive attachment.[54] Likewise, in the example of university home pages, the promotional aims put them in competition with each other while at the same time they benefit from the generic conventions of navigation and screen design that make them imperfect copies of each other.

Muckelbauer also discusses what he calls imitative inspiration, in which the model provokes "responsiveness itself."[55] What is reproduced in this case is not a text but the receptive act of production, an imitative invention. We might see this in the expansion of blogging from the early use of blogs as personal diaries to their later use for many different shared purposes: mommy blogs, travel blogs, quitter blogs, journalism blogs, classroom blogs, and the like. Or in the case of the image-meme, its dissemination as a game anyone can play. In these examples, what is reproduced is the very idea of imitating in order to invent: of putting blogging technology to a new use, or of wanting to produce a new meme type.

Another way of looking at the processes of genre change is offered by Alastair Fowler in his discussion of literary kinds.[56] He suggests that genre transformation occurs by the same inventive strategies that produce any kind of literary change (strategies that resemble rhetorical figures and *topoi*), and there is no reason these strategies would be restricted to literary genres. Several of these strategies, for example, involve ways of combining one genre with another to produce a new one: aggregation of multiple works, combination of features, inclusion or embedding, mixtures and hybrids (a relatively equal mixture); in rhetorical terms, we can think of these as forms of *amplificatio* and parenthesis. The university website, for example, includes by combining and embedding what used to be distinct genres of documents: the faculty directory, the recruiting brochure, the course catalog, the press release, tenure regulations, curriculum requirements, fund-raising requests, and so on, adding more and more of these functions over time. Another cluster of strategies involve changes of scale or function. Twitter, for example, has introduced new genres through a reduction of scale, or *brachylogia*, such as Twitter microfiction.[57] Change of function is seen in the frequent parodic or satirical appropriation of previously produced material, of which the shreds video is one example, and slash fiction another. Finally, Fowler suggests transformation by counterstatement, which we can think of as *antithesis* or invention by contraries, another strategy that can produce parodic "antigenres."[58]

All of these strategies rely on imitation as well as invention, on some combination of reproduction and variation. We might better understand imitation and invention as complementary, not as oppositional, movements and thus see the processes of genre stabilization and transformation as likewise complementary. Without transformation, stable genres die, as sociotechnical change makes them obsolete. Without stabilization, the flux of mediated experience becomes all detail and no pattern, all instance and no recurrence, fragmenting social life and dissipating social meaning.

Classical Genre Theory

Given this picture of genres and genre theory in the contemporary digital world, and the relevance of the ancient concepts of invention and imitation, it's worth looking in more detail at the ancient world's thinking about genres and genre change, and the conceptual tensions they demonstrate. The Greek terms that we now understand as designating "genre" are *genos* and *eidos*. These terms often seem to be used as effective synonyms, though they point in different semantic directions, and the contrast between them is suggestive about some of the issues we've examined above.

The earliest discussion of genre seems to be Plato's analysis of literary production in the *Republic*, an analysis that probably influenced Aristotle's somewhat similar approach in the *Poetics*.[59] In Plato's well-known discussion of what kinds of speech and storytelling are appropriate for the *polis*, several distinctions are at work. First, Socrates distinguishes two species (*eidē*) of stories, true and false.[60] Then, given that poetry is understood as imitation, he distinguishes imitation (as "modes of enunciation"[61]), based on whether the performer is narrating in his or her own voice (dithyramb) or is imitating the voice of another (drama, i.e., tragedy and comedy); a third mode mixes these two (epic) (*Republic*, 394c). Although this passage provides the basis for a long tradition of genre distinctions, Plato does not use either *eidos* or *genos* here. He does go on to make a third relevant distinction, based on the character of the poet: "There is a form (*eidos*) of diction and narrative in which the really good and true man would narrate anything that he had to say, and another form (*eidos*) unlike this to which the man of the opposite birth and breeding would cleave and in which he would tell his story."[62] This line of argument provides the rationale for Socrates to exclude all poets from his city, except for "the more austere and less delightful" who "imitate the diction of the good man."[63] Plato's goal here is not so much to define literary production or its kinds as to evaluate its suitability for the ideal state and the education of the young, and perhaps for that reason these three lines of argument are not reconciled.[64]

Aristotle's approach to genre in the *Poetics* is, characteristically, more systematic. He begins by laying out both the essence and the several species (*eidē*) of poetics, naming several examples, epic, tragedy, comedy, dithyramb, and "most flute-playing and harp-playing," which are all imitations (*mimēseis*), the defining essence of poetics. The species differ from each other in three ways: the means of representation (tune, rhythm, and language), the objects represented (good or inferior men and their actions), and the manner of representation (Plato's distinction between narration and dramatization).[65] Presumably, one could use these three criteria in combination to mark out all possible genres, but Aristotle does not do this, focusing rather on tragedy, with brief excursions to comedy and epic—in other words, on that literature already taken to be most culturally serious and significant. There is here no mention of audience, effect, or occasion. "Poetry's fate," according to Jeffrey Walker, "was to be conceptualized . . . in ways that obscured or diminished its rhetoricity or that constrained its rhetorical possibilities. And . . . this reduced and fundamentally grammatical conception [was] to become the dominant version of poetics transmitted to the literary culture of the modern West."[66] Walker's characterization invokes a static, essentialized version of genre, one that becomes exacerbated in later appropriations of the classical tradition.

Gerard Genette draws our attention to the many forms of contemporaneous Greek literature that Aristotle does not discuss: lyric, all kinds of prose including Plato's dialogues, scientific treatises in meter (such as those of Empedocles), the mimes of Syracuse (farcical sketches), or any work in meter that does not conform to a conventional combination of topic and mode of presentation.[67] Aristotle does mention explicitly that there are no common names for some anomalous types: "The art which employs words either in bare prose or in meters . . . happens up to the present day to have no name."[68] So there is some slippage here between the stabilizations offered by the theory and the variable practice it purportedly describes, and indeed, Joseph Farrell makes the point that in the ancient world, poetic practice never did conform to theoretical strictures about genre, pointing for example to the "weird essay in genre theory" by Horace, the *Ars Poetica*, which, he says, exemplifies a practice that violates the principles of generic purity that it articulates.[69]

Turning to rhetoric, Aristotle's classification of deliberative, forensic, and epideictic genres is one of the most familiar parts of his *Rhetoric*. George Kennedy notes that although a general classification of the sort we are familiar with predated Aristotle, earlier writers apparently did not provide a conceptual basis for the genres.[70] This Aristotle does, describing the essence of rhetoric, a *dunamis* of seeing the available means of persuasion,[71] and its three kinds (*eidē*), "for such is the number (of classes) to which the hearers of speeches belong," the hearer being the objective (*telos*) of speech.[72] With this, Aristotle identifies the audience as the basis for discerning the different classes, with the relationship between the speech, the subject treated, and the audience leading to distinctive features such as the characteristic goal, time, context, and *topoi* of each genre. Although Aristotle typically uses *genos* as the larger category and *eidos* as what we would call "species," in this passage, he uses both terms to refer to the three types of rhetoric, *eidos* at the beginning and *genos* at the end.[73]

Here again, Aristotle's systematizing and conceptualizing impulses can conflict with his empirical predilections. He tells us, for example, that because of the limited number of functions an auditor can perform, there are "necessarily" (*anankē* is repeated twice, at 1.3.2 and 1.3.3) three genres of rhetoric. But he tells us elsewhere that "the greatest and most important of all things in an ability to persuade and give good advice is to grasp an understanding of all forms of constitution and to distinguish the customs and legal usages and advantages of each."[74] Presumably, different states with different constitutions might sustain different genres. Thus, empirical variety might challenge his conceptual system; practice might complicate theoretical "necessity." However, this delineation of the rhetorical genres persisted, becoming codified in the work of Cicero,[75] in the influential *Rhetorica ad Herennium*[76] and elsewhere.

Some of the ancients did recognize empirical variety in their discursive environment and value it over theoretical neatness, Isocrates being the exemplar here. Ekaterina Haskins has contrasted Isocrates's approach to genre with that of Aristotle, characterizing Aristotle's approach as a primarily "content-based classification" that stabilizes and reifies genres and Isocrates' approach as based in *kairos*, emphasizing adaptation to specific occasions and offering a "nuanced understanding of discourse types" akin to Bakhtin's notion of performative speech genres.[77] Isocrates himself rejects rhetorical pedagogy based in "hard and fast rules," favoring a view of rhetoric as "a creative process."[78] He uses both *eidos* and *idea* to refer to categories of discourse but seems to reserve *genos* for references to races or tribes of people.[79] In cataloging Isocrates's various uses of *eidos* and *idea*, Robert Sullivan concludes that one of the major uses relevant to rhetoric is to refer to categories of discourse, a usage common in writers contemporary with Isocrates.[80] He finds five such uses, of a total of twenty-three in the corpus. For example, in this passage in the *Antidosis*, Isocrates describes the vast diversity of discourse: "It would be no slight task to attempt to enumerate all of the forms of prose (*pasas tas ideas tōn logōn*)," forms that include the genealogies of the demigods, studies of the poets, histories of wars, and dialogues.[81] As Haskins emphasizes, Isocrates's abiding interest is in the particularities of situations rather than in their similarities.

Thus, in his consideration of genre, Quintilian describes the "dispute" about whether there are three genres or possibly a number "almost past calculation" (*paene innumerabilia*): "On what kind of oratory," he asks, "are we to consider ourselves to be employed, when we complain, console, pacify, excite, terrify, encourage, instruct, explain obscurities, narrate, plead for mercy, thank, congratulate, reproach, abuse, describe, command, retract, express our desires and opinions, to mention no other of the many possibilities? . . . What was the reason that led earlier writers to restrict a subject of such variety to such narrow bounds[?]" He suggests that "the old writers of textbooks only included those kinds of oratory which were most in vogue." However, after a copious defense of this alternative view, he finally concludes that "the safest and most rational course seems to be to follow the authority of the majority," that is, Aristotle was correct that there are (only) three genres of oratory.[82]

The power of the Aristotelian system is strong. As Farrell says regarding poetics, "Classical genre theory was a powerfully essentializing discourse": It reinforced the notion that genre was a defining and determining quality or essence inherent in a poem (or speech, presumably), "not putty in the hands of an inventive poet, and not a discursive tool to be invoked at will by critics for the sake of the argument."[83] Indeed, for both Plato and Aristotle, determining the essence of a subject is the basic strategy of definition and the first step in inquiry. In both the *Poetics* and the *Rhetoric*, Aristotle seems to assume the existence of the genres and then investigate what makes them what they are rather than examining examples and deriving the genres from observation. His assumptions rely on prior opinion together with terminology already in use; both of these would reflect dominant cultural values of the time. The emphasis is on genres as stable, definable entities without variation or change over time.

This emphasis is reinforced by the semantic tonalities of the term *eidos*, which comes from the verb "to see" and carries the primary meaning of that which is seen, form, shape, or figure, and comes to mean kind, class, or sort.[84] It is the term that Plato uses for the eternal "forms" of suprasensible reality, of which transitory empirical reality is always an imperfect imitation. And while Aristotle doesn't accept this doctrine, he does seek to understand the phenomenal world in terms of stable defining "essences," the "what it is to be" something (*to ti esti*), by stating the class (*genos*) a given entity belongs to and its differentia (*diaphora*), which makes it distinct from other members of the larger class,[85] as we have seen above for the literary and rhetorical genres. The differentia is "the difference that makes a species" (*eidopoios diaphora*).[86] In his early works, Aristotle associates definitional essence with *eidos*, for it is the *eidos* that is subject to definition, and "a definition is an account that signifies an essence."[87] Moreover, it is only *eidē* that have essences.[88] What we define is not an individual entity or a broad class but a species, that is, an *eidos*, and what the definition provides is its essence.

The essentializing power of classical genres extended far beyond the ancient cultures of Greece and Rome, reappearing in Renaissance translations and adaptations of the classical sources. Aristotelian poetic genres maintained a strong hold on poetic theory during the Renaissance and neoclassicism of the eighteenth century, with increasing emphasis on poetic meter and other aspects of literary form, and occasional debate about the possibility and advisability of mixed or hybrid genres, but general canonization—even ossification—of what were mistakenly understood to be the classical triad of epic, lyric, and dramatic genres.[89] The rhetorical genres rarely served as major organizing concepts until they were resuscitated for the twentieth-century academy in the neo-Aristotelianism inaugurated by Herbert Wichelns in 1925 and codified in the teaching tradition of the new departments of speech communication. This movement, which arose as scholars of oratory separated themselves from departments of English, was motivated by the aim to formulate a criticism appropriate to the study of oratory as distinct from literature. As Richard B. Gregg notes, Wichelns and Hoyt Hudson made a case for a strong separation between rhetoric and poetics, in a "struggle to affirm their scholarly identity as a separate and legitimate field of study";[90] a set of distinct essentialized genres would offer an advantage in this struggle. Edwin Black's influential critique of this movement in 1965 identified the classification of rhetoric into forensic, deliberative, and epideictic genres first among "the primary and identifying ideas of neo-Aristotelianism."[91] Black's critique, coupled with other events such as the National Developmental Project on Rhetoric,[92] had the effect of weakening neo-Aristotelianism and enabling a fresh approach to genre, represented by the work of Karlyn Campbell and Kathleen Jamieson beginning in the late 1970s.[93] There followed in speech communication an "explosion of interest in genre studies"[94] that began to conceptualize genre as a functional social dynamic rather than a static essence. Somewhat later, in the adjacent fields of composition, professional writing, and applied linguistics, a more sustained research effort developed that has become known as "rhetorical genre studies,"[95] including the work on genre change and stabilization discussed earlier.

The countercurrent to the essentializing predilections of classical genre theory, discernible in the brief passages from Aristotle and Quintilian above, acknowledges the multiplicity of discursive forms and functions and their changing configurations over time. This countercurrent is present in the semantic tonalities of the alternate term for genre, *genos*, which is derived from the verb "to come into being," "to be born," "to happen," a verb of generation, incorporating both a sense of change over time and a sense of familial relatedness; it is used in contexts where we would use race, family, clan, tribe, breed (as Isocrates does). By settling on *eidos/idea* as his terms for genre, Isocrates highlights the static, rather than the dynamic, dimension of genre, formal pattern rather than diachronic relatedness. He thus reinforces the Aristotelian essentialization of genre while rejecting it. As noted earlier, Plato uses *genos* as a general synonym for *eidos* (but not when he's talking about transcendent forms). Aristotle usually reserves *genos* for a larger category subdivided into species, *eidē*.[96] He defines *genos* with respect to *eidos*: "A genus [*genos*] is that which is predicated in the category of essence of several things which differ in kind [*eidos*],"[97] meaning that the genus must identify an essence that is shared by its several species. In other words, the *genos* specifies a basis of relatedness and, in this sense, cannot be a semantic equivalent to *eidos*, which identifies distinctive difference. Elsewhere, Aristotle identifies *genos* with the material (as opposed to the formal), using as examples the continuous generation of a race or tribe of people and their derivation from a common ancestor, or more generally types that share the same material.[98] *Genos*, then, as James G. Lennox concludes, "does not represent an abstract category, but refers to a shared or common aspect of the nature of individuals, serving as the substratum of differentiation."[99]

Aristotle's thinking about *genos*, *eidos*, and genus-species relationships changed over time, from the *Topics* and *Categories*, through the *Metaphysics*, and then at a later stage with the multiple works on biology.[100] Scholars don't agree on the exact nature of those changes, though it does seem clear that in the works on logic Aristotle uses examples from biology to explain problems in logic and that in the early biological works he is struggling to make biological categories comprehensible in logical terms, while in the later ones, he begins to develop a more biologically inflected logic.[101] It is instructive, then, to examine the biological works, where we see him applying these same terms and methods to very different problems and as a result thinking differently.

Aristotle's Biology

In his effort to account for animal life, Aristotle is faced with the burgeoning diversity of the natural world, drawing as he does not only from his own observations during his time on Lesbos after he left Athens but also on reports of travelers to India, Egypt, Libya, and elsewhere.[102] In the lengthy *History of Animals*, he offers a factual survey of the information at hand to understand the many differences that exist among the various animals, and in the *Parts of Animals* and other works, he attempts to explain the facts he has gathered, their formal, material, efficient, and final causes. To do this, he thinks long and hard about what his unit of analysis (as we would call it) should be, ultimately deciding that it would be most efficient and informative to deal with groups of similar animals, even when they "lack a popular name."[103] The word he uses here for groups is *genos*.[104] The problem, then, is to establish the proper groupings, and he decides, on the basis of the complexity and multiplicity of factual information he has gathered, that the approach used in other sciences or in his earlier works on philosophical method will not suffice. As Lennox points, out, Aristotle's work on astronomy and meteorology do not include a preliminary fact-gathering treatise such as we find in *History of Animals*.[105] These works concern natural phenomena that are not subject to the processes of generation and decay, phenomena that are more worthy but about which we know less "because we do not live among them."[106] The study of biology, however, involves not only the generation and decay of specific organisms but also newly discovered organisms and newly inventive ways to observe and study them.[107] As Aristotle remarks when he is recounting variations in the size of animals in different regions, "In Libya the snakes develop a monstrous size, so it is said," adding, "in fact there is a proverb that Libya ever bears something new.[108] For it is thought that owing to the lack of rain animals meet at the water-holes and mate with each other even if not of the same breed."[109]

A science concerned with variation and change over time, with the possibility of hybridization, requires methods different from those appropriate to the sciences of the seemingly immutable. In *Parts of Animals* Aristotle addresses these methods, rejecting dichotomous division, so prevalent in Plato (and often attributed to Aristotle within the scholastic tradition), and instructing us that "the proper course is to endeavour to take the animals according to their groups [*genē*], following the lead of the bulk of mankind, who have marked off the group of Birds and the group of Fishes. Each of these groups is marked off by many differentiae, not by means of dichotomy."[110] He acknowledges that popular usage often correctly identifies the *genē* (644b2), using multiple differentia at once. Given Aristotle's emphasis on multiple differentia, James Franklin argues that Aristotle is at least open to the notion that species are not discrete and immutable, citing this passage as evidence: "The Ascidians [sea-squirts] differ very little in their nature from plants, but they are more akin to animals than the Sponges are, which are completely plants. Nature passes in a continuous gradation from lifeless things to animals, and on the way there are living things which are not actually animals, with the result that one class[111] is so close to the next that the difference seems infinitesimal."[112] Aristotle, it seems, is not prepared to essentialize the species of the natural world. In arguing this point, D. M. Balme claims that although Aristotle uses *eidos* to refer both to essence and to species, for him species is not the same as essence but is a generalization over instances that include many accidental (nonessential) features. *Eidos*, Balme suggests, "indicates merely that a subject-matter is to be considered in the formal mode."[113]

The more general argument of the *Parts of Animals*, according to Balme, is that biology must be understood teleologically, that the diversity of plants and animals is determined by the relationship between inherited form (that is, the tendency of an organism to resemble the parent) and functional advantage: "The fact that individuals develop in their most advantageous way in given circumstances, within the limits of the form inherited from their parents, is enough to explain the existence of species in Aristotle's sense of it."[114] The species, the phenomenal units of the biological world, are constituted by essential features (those explainable by teleology) as well as by nonessential accidents and qualities: essence, then, has to do with final causes and *eidos* with formal causes. But the groups themselves, the *genē*, are to be understood as coming about through the influence of both form and function.

Here is Aristotle telling us how to identify the *genē*: "Now it is practically by resemblance of the shapes [*schēmasi]* of their parts, or of their whole body, that the groups [*genē*] are marked off from each other: as e.g. the groups Birds, Fishes, Cephalopods, Testacea [bivalves such as oysters]. Within each of these groups, the parts do not differ so far that they correspond only by analogy . . . ; that is, they differ not structurally, but only in respect of bodily qualities . . . to put it generally, they differ 'by the more and less.'"[115] A group or *genos* is defined by similarity relations of its members, "resemblance of shapes" (and note the use of *schēma* here, like *eidos* a highly visual word); between groups, there are structural differences that "mark off" one group from another by "shapes" and thereby make divisions recognizable.

Whether Aristotle's application of *genos* and *eidos* to his extensive descriptions of many types of animals actually produced a classification is in dispute; many subsequent scholars read him as trying but failing to produce one. Armand Marie Leroi, however, credits Aristotle not only with the first classification of the animals but also with two methodological insights regarding this exasperatingly complex task. One was his rejection of dichotomous division, noted earlier: The use of multiple differentiae simultaneously means that classification must be based on central tendencies rather than on mutually exclusive categories defined by essential features. The other was his point that there are degrees of difference, that animals differ from each other at different scales or levels on a hierarchy.[116] The greatest differences, at the highest levels, the *megista genē*,[117] are the differences by analogy, noted just above, that is, differences of architecture and geometry (blooded vs. bloodless animals, quadrupeds vs. bipeds); animals in different *megista genē* (jellyfish and horses, for example) have analogous parts such as digestive and reproductive functional systems, but these are shaped much differently from each other.[118] Differences between *genē* are differences in kind, whereas differences between *eidē* are differences in degree.

Although Aristotle frequently seems to focus on form, we should also note that in *Parts of Animals* when he inquires why an animal has a particular part or feature in the first place, he turns again and again to functional explanation: It is useful (*chrēsimos*). For example, shrimp have a tail because tails are useful to swimmers, but crabs do not because they live near the land (684a3); octopuses have long feet, which are useful for walking as well as for swimming (685a28); the joints of the forelimbs in humans bend in the opposite direction from those of quadrupeds because they are put to different uses, feeding and running (687b29); in humans and some quadrupeds the lower jaw moves sideways as well as up and down but in fishes and birds it moves only up and down because the different motions are useful for the different kinds of food they eat (691b1). Thus, the *genos*-group is marked off by perceptible form, but it is a *genos* because of similar function.

Form and Function in Genre

Regrettably, Aristotle did not apply to the cultural phenomena of rhetoric and poetics the same subtlety of thinking and sensitivity to change that he did to the organic phenomena of biology. But those virtues did not actually influence subsequent thinking in biology, either. Aristotelian thought—especially his earlier work in logic and method—was taken to authorize an essentialism expressed as formalism, which pervaded thinking about the natural phenomena of biology, as well as about the cultural phenomena of rhetoric and poetics. Biologists, including both Linnaeus and Darwin, as well as generations of rhetorical and literary critics, labored under the influence of essentialism, which made it difficult to ask questions about variation and change over time. But the recent reevaluation of Aristotle's thought in light of the development of evolutionary theory has revived interest in his biology and in his functional-formal analysis of the diverse and changeable organic world.[119]

What changed biology was the piling up of evidence from geographic exploration and colonization, evidence that natural philosophers in the eighteenth and nineteenth centuries couldn't fit into the essentialist framework. Likewise, what challenges rhetoric and poetics is the piling up of incontrovertible evidence of variation and change, starting with the crisis in neo-Aristotelianism in the 1960s, when rhetorical studies was challenged to account for new forms of rhetorical activity produced by social protest movements. It is, perhaps, no coincidence that genres and genre theory become salient and controversial during times of cultural change, when new social relations, new technologies, new economic forces, enable (and sometimes produce) new patterns of social action; the European Renaissance was one such period and the current era seems to be another.[120] Perhaps the ancient world—over its many centuries—presented others. As evidence of the preoccupation with genre, today we have the ubiquitous digital evidence of what I earlier called "genre-talk" that genres are both useful and troublesome. Evidence from the Renaissance includes multiple debates and controversies over what "kinds" (or genres) of literary production were acceptable, as well as theoretical treatises such as Scaliger's catalog of over one hundred genres and subgenres, the *Poetices libri septem*, published in Lyon in 1561.[121] Genre-talk from the ancient world is of course much more difficult to access, but it surely includes the discussions I have reviewed above and, importantly, the clouds of semantic experimentation surrounding the terms *eidos* and *genos*. It's not that Aristotle was insightful in using *genos*, or that Isocrates was obtuse in using *eidos*. It's that neither of these terms is sufficient in itself, that genre is (or has become) a concept most useful when it includes both semantic dimensions, static patterning, and functional relatedness.

We are still learning how to ask questions about variation and change: How can we explain a stable pattern in a world of flux? How can we account for change in a world of recognizable (and therefore at least somewhat stable) forms?[122] The recent turn to functional-formal understandings of genre—the turn that Aristotle made so long ago in biology—helps us examine these questions about rhetoric in its social, instrumental, interactional dimensions. After all, Aristotle did begin by telling us that "rhetoric is useful [*chrēsimos*]."[123] Aristotle's approach to the questions about stability and change in biology is useful for us, as well: Identifying genres involves using multiple differentia and family resemblances, rather than logical dichotomies; it can be informed by the wisdom of vernacular naming systems; it is better served by observation than by theoretical principles; and it requires attention to both form and function, as well as to their interactions. In fact, on this last point, it has been argued that rhetorical genre studies has gone too far in reaction to the centuries of essentialist-formalism, overlooking the importance of form in its eagerness to make function central.[124]

Form, of course, is necessary for genre to function. As Aristotle says, form is what makes it possible to discern similarity and difference, to discriminate one genre from another. Form recurs and can be replicated, reproduced, imitated. It both allows the recognition of recurrence and constitutes the terms in which recognition is accomplished. Like sonnets, memes are recognized as such in large part by their formal features, as are shreds videos. And as noted earlier, the numerical form underlying digital networks enables automated recognition and reproduction by computational processes. Jamieson suggested that the reproducibility of form gives it a "momentum" that can sometimes overwhelm judgments about functional appropriateness and effectiveness, as in the counterproductive use of classical Latin and imperial protocols in the papal encyclical.[125] I have elsewhere suggested that Kenneth Burke's discussion of "piety" provides another way to think about the power of form: piety can be seen as the motivation for imitation, "the engine of recurrence,"[126] because it derives from "the yearning to conform with the 'sources of one's being,'"[127] that is, our need to belong, our need for the familiar. Piety makes form functional. Form has yet another enabling effect on recurrence, and this derives from the satisfactions—the pleasures—that form offers: the pleasures of recognitions made and expectations met or surprised; the pleasures of rhythm, repetition, and variation; the pleasures of putting a familiar form to an unfamiliar use, of merging form and function into a satisfying whole. Because so much of the rhetorical activity in digital networks is voluntary rather than required, regularized rather than regulated,[128] it seems likely that these pleasures of form and function together serve to motivate both imitation and invention in digital networks. And this conclusion leads to another: that rhetoric needs to turn its attention to the formal and the aesthetic just as poetics needs to acknowledge the social and the functional.

The ancient distinction between rhetoric and poetics haunts genre theory, even as there are many reasons to question it now. If the reified Aristotelian genres rely on this distinction, a formal-functional approach to genres can get along without it, except as a historical artifact that describes current culture only vaguely. In his effort to create an academic space for rhetoric in the early twentieth century, Hudson acknowledged that the distinction was not leakproof in Aristotle or elsewhere: that the poet may trespass into the field of rhetoric, and the rhetorician into the realm of poetry. Nevertheless, Hudson characterizes rhetoric and poetics as occupying "opposite poles" and concludes that "poetry is for the sake of expression; . . . rhetoric is for the sake of impression."[129] To follow this aphorism is for rhetoricians to lose sight of form, textuality, and aesthetics and for literary scholars to ignore readers, occasion, and effect—and for a time, this is exactly what happened. But in stating the implications this way, I hope to have reminded you that this situation no longer obtains. In literary study, we have come a long way from formalist New Criticism, through reader-response, new historicist,

Marxist, and feminist criticism, all concerned with effect. Terry Eagleton's 1983 *Literary Theory* offers rhetoric as "the oldest form of 'literary criticism' in the world," concerned with "the way discourses are constructed in order to achieve certain effects."[130] In rhetorical studies, we have had a new interest in form, aesthetics, and close reading, as shown in Michael Leff's work on close reading, Bradford Vivian's argument about aesthetics as shared sentiment, Robert Hariman's study of political style, and Barry Brummett's work on style as global aesthetic.[131] The distinction between poetics and rhetoric, inaugurated by Aristotle himself according to John Kirby, may be an aspect of ancient thinking that we no longer need.[132] But genre theory can be richer and more persuasive to the extent that it finds ways to learn from both rhetoric and poetics, ancient and modern, as it struggles to understand the relationships between form and function, between imitation and invention, and the twin challenges of stability and change.

Notes

1. Norimitsu Onishi, "Thumbs Race as Japan's Best Sellers Go Cellular," *New York Times*, January 20, 2008, http://www.nytimes.com/2008/01/20/world/asia/20japan.html.

2. "List of Video Game Genres," Wikipedia, accessed July 6, 2015.

3. "Microsoft HoloLens Augmented Reality For Developers, Enterprises First Says CEO Satya Nadella," *Franchise Herald*, July 21, 2015, http://www.franchiseherald.com/articles/32752/20150721/microsoft-hololens-augmented-reality.htm.

4. "The Oxford Dictionaries Word of the Year Is . . . ," *OxfordWords Blog*, November 18, 2013, http://blog.oxforddictionaries.com/2013/11/word-of-the-year-2013-winner/.

5. Virginia Heffernan, "In Vino Veritas," *New York Times Magazine*, February 10, 2008, http://www.nytimes.com/2008/02/10/magazine/10wwln-medium-t.html.

6. Virginia Heffernan, "Uploading the Avant-Garde," *New York Times Magazine*, September 6, 2009, http://www.nytimes.com/2009/09/06/magazine/06FOB-medium-t.html.

7. This is the only news database I could find for which the results could be normalized to the size of the database in a given year.

8. John Frow, *Genre: The New Critical Idiom*, 2nd ed. (London: Routledge, 2014), 12.

9. *Ibid.*, 13.

10. "A Million Little Lies," *Smoking Gun*, January 4, 2006, http://www.thesmokinggun.com/documents/celebrity/million-little-lies.

11. Mimi Read, "A Refugee from Gangland," *New York Times*, February 28, 2008.

12. Motoko Rich, "Gang Memoir, Turning Page, Is Pure Fiction," *New York Times*, March 4, 2008.

13. "Holocaust Memoir Turns Out to Be Fiction," *New York Times*, March 3,

2008; David Mehegan, "Author Admits Making Up Memoir of Surviving Holocaust," *Boston Globe*, February 29, 2008.

14. Blake Eskin, "Crying Wolf," *Slate*, February 28, 2008; Daniel Mendelsohn, "Stolen Suffering," *New York Times*, March 9, 2008.

15. Frow, *Genre*, 14.

16. Harold Garfinkel, *Studies in Ethnomethodology* (Cambridge: Polity Press, 1967).

17. Adam Geitgey, "The Kaycee Nicole (Swenson) FAQ." In *We've Got Blog: How Weblogs Are Changing Our Culture*, ed. John Rodzvilla (Cambridge, MA: Perseus Publishing, 2002), 89–98.

18. Noelle Weaver, "What We Should Learn from Sony's Fake Blog Fiasco," *Advertising Age*, December 18, 2009, http://adage.com/article/small-agency-diary/learn-sony-s-fake-blog-fiasco/113945/.

19. Heffernan, "Uploading the Avant-Garde."

20. Carol Berkenkotter and Thomas N. Huckin, "Rethinking Genre from a Sociocognitive Perspective," *Written Communication* 10, no. 4 (1993): 475–509.

21. Karlyn Kohrs Campbell and Kathleen Hall Jamieson, "Form and Genre in Rhetorical Criticism: An Introduction," in *Form and Genre: Shaping Rhetorical Action*, ed. Karlyn Kohrs Campbell and Kathleen Hall Jamieson (Falls Church, VA: Speech Communication Association, 1978), 9–32; Berkenkotter and Huckin, "Rethinking Genre"; Frow, *Genre*; Carolyn R. Miller, "Genre as Social Action," *Quarterly Journal of Speech* 70, no. 2 (1984): 151–76; for a full review of rhetorical genre theory, see Anis S. Bawarshi and Mary Jo Reiff, *Genre: An Introduction to History, Theory, Research, and Pedagogy* (West Lafayette, IN: Parlor Press, 2010).

22. Charles Bazerman, "Systems of Genres and the Enactment of Social Intentions," in *Genre and the New Rhetoric*, ed. Aviva Freedman and Peter Medway (London: Taylor and Francis, 1994), 79–101.

23. Anthony Giddens, *The Constitution of Society: Outline of the Theory of Structuration* (Berkeley: University of California Press, 1984).

24. Miller, "Genre as Social Action," 155. Another approach sees genres as constructions of critics rather than of the communities of production and reception. Rhetorical studies generally favors the latter position; Rosmarin has advocated the former for literary studies (Adena Rosmarin, *The Power of Genre* [Minneapolis: University of Minnesota Press, 1985]); see also Todorov's related discussion of theoretical vs. historical approaches to genre (Tzvetan Todorov, *The Fantastic: A Structural Approach to a Literary Genre*, trans. Richard Howard [Ithaca, NY: Cornell University Press, (1970) 1975]).

25. Berkenkotter and Huckin, "Rethinking Genre," 501.

26. Catherine F. Schryer, "Records as Genre," *Written Communication* 10, no. 2 (1993): 208.

27. Janet Giltrow and Dieter Stein, "Genres in the Internet: Innovation, Evolution, and Genre Theory," in *Genres in the Internet: Issues in the Theory of Genre*, ed. Janet Giltrow and Dieter Stein (Amsterdam: John Benjamins, 2009), 9.

28. Charles Bazerman, "Reporting the Experiment: The Changing Account of

Scientific Doings in the Philosophical Transactions of the Royal Society, 1665–1800," in his *Shaping Written Knowledge: The Genre and Activity of the Experimental Article in Science* (Madison: University of Wisconsin Press, 1988), 59–79.

29. Carol Berkenkotter and Thomas N. Huckin, "News Value in Scientific Journal Articles," in their *Genre Knowledge in Disciplinary Communication: Cognition/Culture/Power* (Hillsdale, NJ: Lawrence Erlbaum Associates, 1995), 27–44.

30. Sharon D. Downey, "The Evolution of the Rhetorical Genre of Apologia," *Western Journal of Communication* 57 (1993): 42–64.

31. JoAnne Yates, "The Emergence of the Memo as a Managerial Genre," *Management Communication Quarterly* 2, no. 4 (1989): 485–510.

32. Jason Mittell, *Genre and Television: From Cop Shows to Cartoons in American Culture* (New York: Routledge, 2004).

33. Malcolm Clark, Ian Ruthven, and Patrik O'Brian Holt, "The Evolution of Genre in *Wikipedia*," *Journal for Language Technology and Computational Linguistics* 24, no. 1 (2009): 1–22.

34. Carolyn R. Miller and Dawn Shepherd, "Questions for Genre Theory from the Blogosphere," in *Genres in the Internet: Issues in the Theory of Genre*, ed. Janet Giltrow and Dieter Stein (Amsterdam: John Benjamins, 2009), 263–90.

35. Gael Ayers, "The Evolutionary Nature of Genre: An Investigation of the Short Texts Accompanying Research Articles in the Scientific Journal Nature," *English for Specific Purposes* 27, no. 1 (2008): 22–41; see also essays in Alan G. Gross and Jonathan Buehl, eds., *Science and the Internet: Communicating Knowledge in a Digital Age* (Amityville, NY: Baywood, 2016).

36. Michael Shepherd and Carolyn Watters, "The Evolution of Cybergenres," in *31st Hawaii International Conference on System Sciences*, ed. Ralph H. Sprague Jr. (Maui: IEEE Computer Society Press, 1998), 97–109.

37. For a more thorough review of this literature, see Carolyn R. Miller, "Where Do Genres Come From?" in *Emerging Genres in New Media Environments*, ed. Carolyn R. Miller and Ashley R. Kelly (London: Palgrave Macmillan, 2017), 1–34.

38. Schryer, "Records as Genre," 208.

39. Berkenkotter and Huckin, *Genre Knowledge*, 24.

40. JoAnne Yates and Wanda Orlikowski, "Genres of Organizational Communication: A Structurational Approach to Studying Communication and Media," *Academy of Management Review* 17, no. 2 (1992): 321.

41. Catherine F. Schryer and Philippa Spoel, "Genre Theory, Health-Care Discourse, and Professional Identity Formation," *Journal of Business and Technical Communication* 19, no. 3 (2005): 249–78; it should be noted that regularization is not restricted to digital networks but also operates in face-to-face and other environments.

42. Manovich, *The Language of New Media* (Cambridge, MA: MIT Press, 2001), 27–49.

43. Todorov, "The Origin of Genres," *New Literary History* 8, no. 1 (1976): 161.

44. Jamieson, "Antecedent Genre as Rhetorical Constraint," *Quarterly Journal of Speech* 61, no. 4 (1975): 406–15.

45. Shepherd and Watters, "Evolution of Cybergenres," 99.

46. McKeon, "Literary Criticism and the Concept of Imitation in Antiquity," *Modern Philology* 34, no. 1 (1936): 1–35, doi: 10.2307/434493. See also the discussion of McKeon in Edward P. J. Corbett, "The Theory and Practice of Imitation in Classical Rhetoric," *College Composition and Communication* 22, no. 3 (1971): 243–50, doi: 10.2307/356450, and John Muckelbauer, "Imitation and Invention in Antiquity: An Historical-Theoretical Revision," *Rhetorica* 21, no. 2 (2003): 61–88.

47. Jörgen Skågeby, "Dismantling the Guitar Hero? A Case of Prodused Parody and Disarmed Subversion," *Convergence: The International Journal of Research into New Media Technologies* 19, no. 1 (2013): 63–76, doi: 10.1177/1354856512456791; also see http://knowyourmeme.com/memes/x-shreds.

48. Muckelbauer, "Imitation and Invention," 79. It is the necessity of both these aspects of generic functioning—pragmatic force and audience effect, or uptake—that underlies Freadman's insistence that genre applies to pairs of texts, rather than to single texts: a text "is contrived to secure a certain class of uptakes, and the . . . uptake text confirms its generic status by conforming itself to this contrivance"; Anne Freadman, "Uptake," in *The Rhetoric and Ideology of Genre: Strategies for Stability and Change*, ed. Richard M. Coe, Lorelei Lingard, and Tatiana Teslenko (Cresskill, NJ: Hampton Press, 2002), 40.

49. Marjorie D. Kibby, "Email Forwardables: Folklore in the Age of the Internet," *New Media and Society* 7, no. 6 (2005): 770–90.

50. Limor Shifman, *Memes in Digital Culture* (Cambridge, MA: MIT Press, 2014).

51. Muckelbauer, "Imitation and Invention," 84.

52. Ibid., 78, 79, italics in original.

53. Ibid., 82.

54. http://knowyourmeme.com/memes/trump-your-cat; http://mashable.com/2015/07/14/pluto-memes/; http://knowyourmeme.com/memes/overly-attached-girlfriend.

55. Muckelbauer, "Imitation and Invention," 87.

56. Fowler, *Kinds of Literature: An Introduction to the Theory of Genres and Modes* (Cambridge, MA: Harvard University Press, 1982), especially chapter 10.

57. See the *Guardian*'s 2013 feature, "Twitter Fiction," accessed June 30, 2016, http://www.theguardian.com/books/series/twitter-fiction.

58. Fowler, *Kinds of Literature*, 175.

59. Plato, *Republic*, 392d–398b, in *The Collected Dialogues*, ed. Edith Hamilton and Huntington Cairns (Princeton, NJ: Princeton University Press, 1961).

60. Ibid., 376e.

61. Gerard Genette, *The Architext: An Introduction*, trans. Jane E. Lewin (Berkeley: University of California Press [1979] 1992), 61.

62. Plato, *Republic*, 396c. In all three of these classifications, we can see Plato's predilection for dichotomous division at work, a strategy for pursuing the truth about a topic by finding its essence, which can be seen as the foundation of dialectic.

63. Ibid., 398b. Subsequent literary theorists have emphasized one or more of these three classifications, but Plato has introduced plenty of room for debate here.

64. According to Peters, Plato elsewhere uses *genos* as a synonym for *eidos* in the general sense of "kind," though not to my knowledge in reference to literary or oratorical genres: F. E. Peters, *Greek Philosophical Terms: A Historical Lexicon* (New York: New York University Press, 1967), 72.

65. Aristotle, *Poetics*, trans. W. Hamilton Fyfe (Cambridge, MA: Harvard University Press, [1927] 1932), 1.1–3, 3.3. According to Joseph Farrell, Aristotle follows Plato in associating the nobility and seriousness of the object with the character of the poet: see "Classical Genre in Theory and Practice," *New Literary History* 34, no. 3 (2003): 384.

66. Walker, *Rhetoric and Poetics in Antiquity* (New York: Oxford University Press, 2000), 278.

67. Genette, *Architext*, 10.

68. Aristotle, *Poetics*, 1.7.

69. Farrell, "Classical Genre," 395.

70. Aristotle, *On Rhetoric: A Theory of Civic Discourse*, 2nd ed. and trans. George A. Kennedy (New York: Oxford University Press, [1991] 2007), 46.

71. Ibid., 1.2.1.

72. Ibid., 1.3.1.

73. Kennedy points out that Aristotle may have used *genos* at this point because in the next passage he divides each type further into subtypes: deliberative into exhortation and dissuasion, judicial into accusation and defense, and epideictic into praise and blame (Aristotle, *On Rhetoric*, 48n77). Aristotle also uses *genos* to refer to the three kinds at 2.18.2 and 3.12.6.

74. Aristotle, *On Rhetoric*, 1.8.1. Aristotle's discussion of the *topoi* for deliberative rhetoric suggests an empirical familiarity with the trajectories of actual debates.

75. Marcus Tullius Cicero, *De Inventione*, 1.9, in *De Inventione, De Optimo Genere Oratorum, Topica*, trans. H. M. Hummell (Cambridge, MA: Harvard University Press, 1949).

76. [Cicero], *Rhetorica ad Herennium*, trans. Harry Caplan (Cambridge, MA: Harvard University Press, 1981), 1.2.

77. Ekaterina Haskins, *Logos and Power in Isocrates and Aristotle* (Columbia: University of South Carolina Press, 2004), 78. Bakhtin's performative conception underlies a good deal of contemporary rhetorical genre theory, as described above.

78. Isocrates, "Against the Sophists," in *Isocrates*, vol. 2, trans. George Norlin (Cambridge, MA: Harvard University Press, 1929), 12.

79. A search for the nominative form of *genos* in the online Loeb Classical Library edition of Isocrates yields seventeen instances; a typical example uses *genos* to refer to "the race of the Hellenes" as contrasted with the barbarians: Isocrates, *Antidosis*, in *Isocrates*, vol. 2, trans. George Norlin (Cambridge, MA: Harvard University Press, 1929), 294.

80. Robert G. Sullivan, "*Eidos/idea* in Isocrates," *Philosophy & Rhetoric* 34, no. 1 (2001): 80–81, doi: 10.2307/40238081.

81. Isocrates, *Antidosis*, 46.

82. Quintilian, *Institutio Oratoria*, trans. H. E. Butler (Cambridge, MA: Harvard University Press, 1920) 3.4; Quintilian uses the Latin word "genus," which is here translated as "kind."

83. Farrell, "Classical Genre," 383, 386.

84. H. G. Liddell, *An Intermediate Greek-English Lexicon* (Oxford: Clarendon Press, [1889] 1986).

85. Robin Smith, "Aristotle's Logic," in *The Stanford Encyclopedia of Philosophy*, ed. Edward N. Zalta (Palo Alto, CA: Center for the Study of Language and Information, Stanford University, 2015) section 7.2. http://plato.stanford.edu/archives/spr2016/entries/aristotle-logic/.

86. Ibid., section 7.4, referring to Aristotle, *Topica*, 143b.

87. Aristotle, *Topica*, 102a3; trans. S. Marc Cohen, "Aristotle's Metaphysics," in *The Stanford Encyclopedia of Philosophy*, ed. Edward N. Zalta (Palo Alto, CA: Center for the Study of Language and Information, Stanford University), 2015. http://plato.stanford.edu/archives/fall2015/entries/aristotle-metaphysics/.

88. Aristotle, *Metaphysics, Books 1–9*, trans. Hugh Tredennick (Cambridge, MA: Harvard University Press, 1980), 1030a11.

89. Heather Dubrow, *Genre* (London: Methuen, 1982); Genette, *Architext*.

90. Gregg, "The Criticism of Symbolic Inducement: A Critical-Theoretical Connection," in *Speech Communication in the 20th Century*, ed. Thomas W. Benson (Carbondale: Southern Illinois University Press, 1985), 44.

91. Black, *Rhetorical Criticism: A Study in Method* (Madison: University of Wisconsin Press, [1965] 1978), 31.

92. Lloyd F. Bitzer and Edwin Black, eds. *The Prospect of Rhetoric* (Englewood Cliffs, NJ: Prentice-Hall, 1971).

93. Campbell and Jamieson, eds., *Form and Genre: Shaping Rhetorical Action* (Falls Church, VA: Speech Communication Association, 1978).

94. Stephen E. Lucas, "Genre Criticism and Historical Context: The Case of George Washington's First Inaugural Address," *Southern Speech Communication Journal* 51, no. 4 (1986): 354. It should be noted, however, that the enthusiasm within speech communication barely lasted through the 1980s.

95. Aviva Freedman, "Beyond the Text: Towards Understanding the Teaching and Learning of Genres," *TESOL Quarterly* 33, no. 4 (1999): 764–67, doi: 10.2307/3587890.

96. Peters, *Greek Philosophical Terms*, 35, 72.

97. Aristotle, *Topica*, trans. E. S. Forster (Cambridge, MA: Harvard University Press, 1960), 102a.

98. Aristotle, *Metaphysics*, 1024a–b.

99. James. G. Lennox, "Aristotle on Genera, Species, and 'the More and the Less.'" *Journal of the History of Biology* 13, no. 2 (1980): 335.

100. Aristotle's use of *eidos* also changes over time. In his early logical works, the term *eidos* means "species" (contrasted with *genos*), but in the *Metaphysics* it means "form" or "actualization" (contrasted with "matter"), and in *Physics* it refers to formal cause; Cohen, "Aristotle's Metaphysics," section 7; Peters, *Greek Philosophical Terms*, 50.

101. Compare D. M. Balme, "Aristotle's Biology Was Not Essentialist," *Archiv für Geschichte der Philosophie* 62, no. 1 (1980): 1–12; "Aristotle's Biology," Herbert Granger, "Aristotle and the Genus-Species Relation," *Southern Journal of Philosophy* 18, no. 1 (1980): 37–50, doi: 10.1111/j.2041-6962.1980.tb01362.x; and G. E. R. Lloyd, "The Development of Aristotle's Theory of the Classification of Animals," *Phronesis* 6, no. 1 (1961): 59–81.

102. Introduction to Aristotle, *History of Animals, Volume I, Books 1–3*, trans. A. L. Peck (Cambridge, MA: Harvard University Press, 1965), lxvi.

103. Ibid., 1.4, 644b1–5.

104. Like commentators on the logical and philosophical works, scientists who have studied Aristotle's biology are wary about the apparent lack of system with which Aristotle uses this term. For more on the term *genos*, see Ernst Mayr, *The Growth of Biological Thought: Diversity, Evolution, and Inheritance* (Cambridge, MA: Belknap Press, 1982); Balme, "*Γένος* and *Εἶδος* in Aristotle's Biology," *Classical Quarterly* 12, no. 1 (1962): 81–98; and Peck's "Introduction to Aristotle," *Historia Animalium*, lxiv–lxvi.

105. Lennox, "The Complexity of Aristotle's Study of Animals," in *The Oxford Handbook of Aristotle*, ed. Christopher Shields (New York: Oxford University Press, 2012), 290.

106. Aristotle, *Parts of Animals*, trans. A. L. Peck (Cambridge, MA: Harvard University Press, [1937] 1961), I.v.644b25, 644b32.

107. Lennox, "The Complexity of Aristotle's Study of Animals," 291.

108. Or, in a translation I prefer, "Always something new out of Libya"; Armand Marie Leroi, *The Lagoon: How Aristotle Invented Science* (New York: Viking, 2014), 48.

109. Aristotle, *Parts of Animals*, 606b10–20; here the Greek for "breed" is *homophulon*; *phulon* is a term similar to *genos*, meaning race, tribe, or class, whether of animals or humans.

110. Ibid., I.iii.643b10–15.

111. No direct equivalent appears in the Greek, neither *eidos* nor *genos* nor *phulon*.

112. Aristotle, *Parts of Animals*, IV.v.681a10; James Franklin, "Aristotle on Species Variation," *Philosophy* 61, no. 236 (1986): 247.

113. Balme, "Aristotle's Biology," 6. See also Peters, *Greek Philosophical Terms*: for Aristotle, *eidos* is the formal cause of things, from *Phys*. II. After an inventory of Aristotle's uses of both *genos* and *eidos* in the biological works, Balme concludes that they do not designate higher and lower taxonomic levels and are occasionally used to refer to the same group or species of animals. When there is a difference, it has to do with the root semantics of the terms: *genos* refers to kinship and *eidos* to form or shape; Balme, "*Γένος* and *Εἶδος*," 87.

114. Balme, "Aristotle's Biology," 2.

115. Aristotle, *Parts of Animals*, I.iv.644b8–15.

116. One important way Aristotle's understanding of *eidos* differs from Plato's is that in Aristotle it becomes a relative concept: *eidē* can be grouped into ever widening classes, or *genē*, each of which serves as *eidos* to the wider *megista genē*; Peters, *Greek Philosophical Terms*, 50; Introduction to Aristotle, *History of Animals*,

Volume III, Books 7–10, trans. D. M. Balme (Cambridge, MA: Harvard University Press, 1991), 16.

117. Aristotle, *History of Animals, Volume I, Books 1–3*, e.g., 1.6, 490b8; 2.15, 505b27.

118. Leroi, *The Lagoon*, 107–13.

119. Ibid.; Mayr, *Growth of Biological Thought*, 11, 191, 256–72, 304–7.

120. Carolyn R. Miller, "New Genres, Now and Then," in *Literature, Rhetoric, and Values*, ed. Shelley Hulan, Murray McArthur, and Randy Allen Harris (Newcastle upon Tyne: Cambridge Scholars Publishing, 2012), 127–49.

121. Julius Caesar Scaliger, *Poetices libri septem* (Lyon: Joannes Crispinus, 1561); accessed October 31, 2015, https://books.google.com/books?id=gddfAAAAcAAJ; and, of course, the work of many literary scholars on the topic of Renaissance genres. See Miller, "New Genres," for further discussion and citations.

122. Evolutionary biologists still struggle with the question whether species is a useful concept or an illusory stabilization; see Mayr, *Growth of Biological Thought* and Daniel C. Dennett, *Darwin's Dangerous Idea: Evolution and the Meanings of Life* (New York: Simon and Schuster, 1995).

123. Aristotle, *On Rhetoric*, I.i.12; 1355a.

124. Amy Devitt, "Re-fusing Form in Genre Study," in *Genres in the Internet: Issues in the Theory of Genre*, ed. Janet Giltrow and Dieter Stein (Amsterdam: John Benjamins, 2009), and Janet Giltrow, "Form Alone: The Supreme Court of Canada Reading Historical Treaties," in *Genre Studies around the Globe: Beyond the Three Traditions*, ed. Natasha Artemeva and Aviva Freedman (Edmonton, AB: Inkshed Publications, 2015), 207–24.

125. Jamieson, "Antecedent Genre," 410.

126. Miller and Shepherd, "Questions for Genre Theory," 285.

127. Kenneth Burke, *Permanence and Change: An Anatomy of Purpose* (Indianapolis: Bobbs-Merrill, [1935] 1965), 69.

128. Schryer and Spoel, "Genre Theory."

129. Hoyt H. Hudson, "Rhetoric and Poetry," *Quarterly Journal of Speech Education* 10, no. 2 (1924): 144, 146.

130. Terry Eagleton, *Literary Theory: An Introduction* (Minneapolis: University of Minnesota Press, 1983), 205.

131. Michael Leff, "Things Made by Words: Reflections on Textual Criticism," *Quarterly Journal of Speech* 78, no. 2 (1992): 223–31; Bradford Vivian, "Style, Rhetoric, and Postmodern Culture," *Philosophy & Rhetoric* 35, no. 3 (2002): 223–43; Robert Hariman, *Political Style: The Artistry of Power* (Chicago: University of Chicago Press, 1995); Barry Brummett, *A Rhetoric of Style* (Carbondale: Southern Illinois University Press, 2008).

132. Kirby, "The 'Great Triangle' in Early Greek Rhetoric and Poetics," *Rhetorica* 8, no. 3 (1992): 213–28; it should be noted that in the ancient world this was a distinction observed more in theory than in practice, just as ancient genre theory never accurately or fully described ancient practice.

9

Poiēsis, Genesis, Mimēsis

Toward a Less Selfish Genealogy of Memes

Michele Kennerly and Damien Smith Pfister

"Socrates may or may not have a gene or two alive in the world today," evolutionary biologist Richard Dawkins once mused, "but who cares? The meme-complexes of Socrates, Leonardo, Copernicus and Marconi are still going strong."[1] By a contrast all the more notable due to his nearly identical name, the cultural legacy of *I*socrates, the fourth-century BCE Athenian educator and political advisor, seems less mighty. Isocrates has not been inconsequential, however. He animated Cicero's *De Oratore*, Baldassare Castiglione's *The Courtier*, and John Milton's *Areopagitica*, to name but a very few major works of Western thought. Since the 1990s, rhetoricians have championed his most extensive contribution to Western culture: his system of education for the *polis*.[2] More recently, Kathleen Welch showed how Isocrates's use of the ascendant communication technology of his day, the papyrus book roll, might guide our understanding of screen-based "electric rhetoric," and David Perlmutter claimed Isocrates as "an ancient protoblogger."[3] Perhaps Isocrates's greatest fortune thus far in the second decade of the twenty-first century, however, has been to find an unlikely ally: Ryan Gosling.

The meme (figure 9.1) first appeared, unattributed, on the Rhet-Comp Ryan Gosling Tumblr, a *mimēsis* of the vastly more popular Feminist Ryan Gosling Tumblr.[4] The Feminist Ryan Gosling meme, launched by Danielle Henderson when she was a gender studies graduate student seeking an outlet for her theory overload, pairs an image of the actor with a quote from or allusion to a feminist theorist and a related and usually amorous invitation to the hailed "girl." As silly and insignificant as that meme might seem, a 2015 study of college students concluded that those who viewed Feminist Ryan Gosling memes became more feminist in their thinking thereafter. The researchers suggest that "popular internet memes can serve more than an entertainment function in that they can also serve as a persuasive device for relaying ideological information."[5] Those who made and shared the Gosling-loves-Isocrates meme likely found their thoughts turn inward toward Isocrates and outward toward spreading the good word on behalf of a

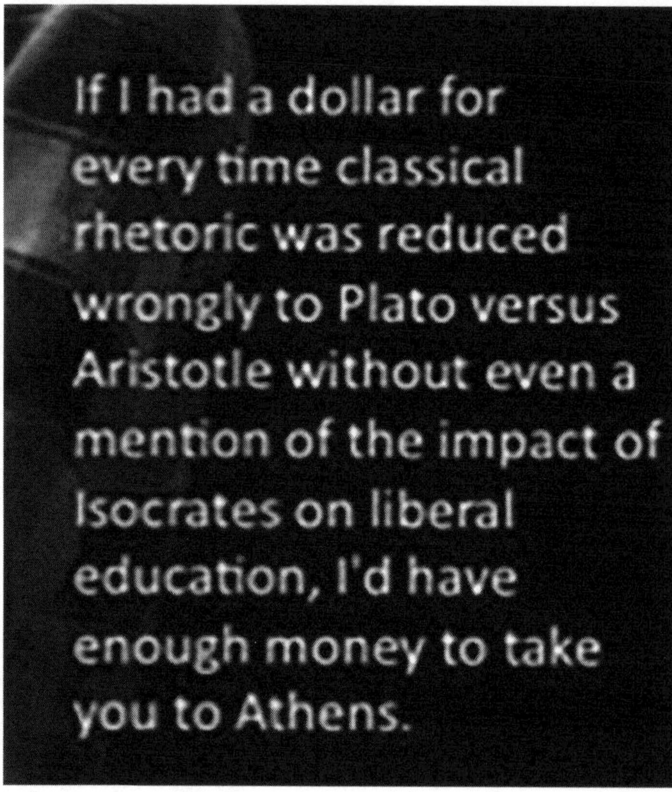

Figure 9.1. Sample of *Rhet/Comp Ryan Gosling* (November 14, 2011, image retrieved July 31, 2015).

figure marginalized even within his own tradition. Recognizing the meme as a Rhetoric-based iteration of Feminist Ryan Gosling may have brought an amusement people could not keep to themselves, as may have the implausibility that an actor—however unconventional—would throw shade at Plato and Aristotle and light on a figure mostly unknown outside of the academy.

Or should we consider what *the meme* was up to? Should we consider the formation and circulation of memes from the "meme's eye view"? What conditions did the meme exploit to replicate, mutate, and make its survival more likely? As Davi Johnson explains, "from the 'meme's eye view,' . . . humans are an effect of memes (and genes), and the memes that drive human culture are those that survive, or are selected for, in a competitive environment."[6] This meme-centric critical perspective accords with the origin of the word itself. In his 1976 surprise best seller, *The Selfish Gene*, Richard Dawkins—a quotation from whom opened this chapter—named the "meme" the cultural counterpart to the biological gene: "The new [primeval] soup is the soup of human culture. We need a name for the new replicator, a noun which conveys the idea of a unit of cultural transmission, or a unit of *imitation*. 'Mimeme' comes from a suitable Greek root, but I want a monosyllable that sounds a bit like 'gene.' I hope my classicist friends will forgive me if I abbreviate mimeme to *meme*. If it is any consolation, it could alternatively be thought of as related to 'memory', or to the French word *même*."[7] For Dawkins, memes are as selfish as genes. A meme looks after itself, assessing its environment and acting ruthlessly to ensure its survival. In an ecology of human culture, competition is for attention: "If a meme is to dominate the attention of a human brain, it must do so at the expense of 'rival' memes. Other commodities for which memes compete are radio and television time, billboard space, newspaper column-inches, and library shelf-space."[8] Dawkins calls the agency-granting conceit of the selfish meme "convenient," "fruitful," and "only a figure of speech," cautioning that no one should "get mystical about it."[9] We fear mysticism has settled in.

"Gene talk" of a selfish bent abounds in rhetorical scholarship about memes.[10] For example, Davi Johnson channels Dawkins's theory of the meme: "Memes persuade, or rather memes infect, because they 'program' people to respond in particular ways. Memes operate according to laws of selection and replication. . . . Memes use us to alter their environments so that their chances for replication are enhanced."[11] Leslie Hahner similarly writes in a sociobiological vein that "the meme is a virus that replicates by imitation, 'jumping from brain to brain'" and that "memes work by changing the behavior of the organism."[12] Sociobiological language enjoys some explanatory power due to its authorizing, vaguely scientific origins and reassuring (for some) materialism. Since Dawkins took inspiration from the word family of *mimēsis* when coining "meme," we see grounds for reconceptualizing the meme along rhetorical rather than sociobiological lines.[13] Rhetorical pedagogies following the principle of *mimēsis* guide students in the selection, critical investigation, and creative emulation of *exempla*, helping students pattern not only their words in the manner of the best rhetors but also their judgment.[14] Furthermore, mimetic pedagogies draw attention to the argumentative architecture, figurative dimensions, and contexts of emergence for a given text or genre or, today, a meme.

The consonance between memes and the many terms for repetition within rhetoric's long-spanning lexicon further invites a rhetorical approach. Such terms predominate in inventories of tropes and figures, stretching from *anaphora*, when a word or phrase is repeated at the beginning of successive clauses, to *zeugma*, in which a repetition is artfully suppressed. The *koinoi topoi* (common topics/places) are so called because of their recurrent applicability to rhetorical situations. A maxim, which Aristotle calls a shortened enthymeme, compresses social knowledge into a form easy to recall and repeat.[15] Sound bites and iconic images are repeatable and remixable. Given the centrality of repetition to rhetoric, we might say that rhetorical culture is a "culture of the copy," or, in a more processual sense, a culture of copying.[16]

Of course, following the logic of *mimēsis*, repetition and copying do not mean slavish reproduction. In 2013, Dawkins himself reflected upon the rapid changes "meme" has undergone: "The very idea of the meme has itself mutated and evolved in a new direction. An internet meme is a hijacking of the original idea. Instead of mutating by random chance before spreading by means of Darwinian selection, internet memes are altered deliberately by human creativity. In the hijacked version, mutations are designed, not random, with the full knowledge of the person doing the mutating."[17] Dawkins's observation about the "hijacking" of "meme" points to a breakdown in the analogy between genes/biology and memes/culture. Fidelity, fecundity, and longevity may fare well as criteria for the spread of genes, but variation, profundity, or timeliness (to take some loose antonyms) are as likely to explain the circulation of particular memes. The evolutionary logic embedded in Dawkins's original theory of memes was too limited to explain human culture, especially since that culture has become inflected by the digital. Dawkins's observations about the mutation of the meme concept in the context of digital culture show the inadequacy of this evolutionary logic when applied to rhetorical cultures: Of course, the concept changes with repetition over time—that is what symbols do. That memes have mutated in a way that aligns them closer to their rich rhetorical heritage in the *mimēsis* word family is a delicious irony that prompts our effort to root an alternative genealogy of the meme in ancient rhetoric.

This alternative genealogy of memes uses three germinal ancient concepts—*poiēsis*, *genesis*, and *mimēsis*—to sow three propositions: memes are made and done, memes are generative, and memes are mimetic. The common concern of *poiēsis*, *genesis*, and *mimēsis* with "bringing forth" emphasizes the productive character of rhetoric. Rhetoric is not only generative but also generous: A rhetorical product—be it a speech or an image or a meme—is an offering, a humble submission, a modest contribution, be it to pop culture, academic culture, or democratic culture.[18] The language of "sharing" attached to memes cues us to this rhetorical generosity. We note, though, that from a perspective rooted in *mimēsis*, even acts of sharing the "same" meme are variations, since the meaning of a given meme changes in context, in time, and in space.[19] For us, the original authorship of a meme—when such an origin point is even discernible—is less important than the situation that prompts it and the varied publics who assemble around it and keep it going, or not.

In thinking through the civic and communal work memes do when people craft and copy them, we draw energy from Seyla Benhabib's concepts of *jurisgenerative politics*—that is, "cases of legal and political contestation in which the meaning of rights and other fundamental principles are reposited, resignified, and reappropriated by new and excluded groups, or by the citizenry in the face of new and unprecedented hermeneutic challenges and meaning constellations"—and *democratic iterations*, which she defines as "linguistic, legal, cultural, and political repetitions-in-transformation, invocations that also are revocations. They not only change established understandings but also transform what passes as the valid or established view of an authoritative precedent."[20] For Benhabib, democratic iterations are required to negotiate differences that accompany cultural pluralism; the alternatives to communicative interchange about public controversies are an untenable relativism or an uncritical imposition of dominant values. Through democratic iterations, "the democratic people can reconstitute itself . . . so as to enable the extension of democratic voice."[21] Such extensions happen through cycles of communicative, democratic iterations, such as debates, referenda, court cases, protests, boycotts, and, now, memes.

In particular, we focus on democratic iterations sparked and sustained by a specific kind of meme: the hashtag. The power of hashtags, and hashtag activism, is accumulative, building by sheer repetition. A given hashtag, however, typically accompanies an observation, insight, or anecdote from the tweeter that can be considered on its own but is more important as part of a larger, iterating narrative. Hashtag activists contribute to rhetorically exigent conversations and deliberations, amplifying a "trending" topic or initiating a trend themselves. A tweet's own limit of 140 characters lends itself to the creation and circulation of anthems. As journalist Sasha Weiss observes, "There is something about the fact that Twitter is primarily designed for speech—for short, strong, declarative utterance—that makes it an especially powerful vehicle for activism, a place of liberation."[22] Weiss's use of "place" is evocative. The technical term for the hash or pound sign—the now ubiquitous #—is "octothorpe." In cartographic renderings of villages, an octothorpe represents a central public square surrounded by eight fields.[23] Having taken residence within the global, digital village, the octothorpe registers the public dimensions and material realities of our connectedness. While the octothorpe visually represents a common, public space where the *dēmos* deliberates, in radiating outward in many directions, it also signals ties to communities whose membership in the *dēmos* is contested and negotiated.

The contestations and negotiations mediated by hashtags in the contemporary, networked media ecology are exemplars of democratic iteration. Iteration derives from the Latin *iterare*, to do again, repeat, rehearse. The inhering humility—if at first you don't succeed, try, try again—accords with the experimental *ēthos* of democracy. Democracy itself has been called, likely first by Churchill, the worst form of government except for all the others that have been tried. Democracy is ever a work in progress, and its habits—both good and bad—are established and rerouted by repetitions, revisions, and reinventions. After a certain amount of reiteration, even memes from the margins can become commonplaces upon which established members of a *dēmos* or other community can draw during moments of alteration and extension; indeed, rhetors often seek the available memes of persuasion.[24]

Proposition One: Memes are Made and Done (*Poiēsis*)

In the context of so-called Web 2.0, technologies like blogs, wikis, social networking sites, and image- and video-sharing portals, words like "producer" and "user" seemed to lose their analytic precision. Internetworked media technologies could not so easily be understood in terms of the producer/consumer model of broadcast media. Axel Bruns suggests "produsage" as a hybrid term that reflects the open participation, hetarchal interaction, and collaboration that enable communities of meme-makers and sharers.[25] The logic that fused producer and user applies equally well to rhetor and audience, since their division tends to be especially untidy in digital spaces. Eric Jenkins, for example, observes that the "blurring of roles and continual remaking [of memes] makes it difficult" to disambiguate rhetor from audience and "to isolate origins and intentions," since there are many points of articulation along a meme's travels.[26] Rhetorical produsage—in this case, the rampant multiplicity and multiplying of memes—seems at odds with the trained tendency of rhetorical critics to treat an object of their analysis as "made" or "done," through rendering or actuating. The easy retreat is to a Heideggerian place of provocative ploce: memes meme (figure 9.2).

Given rhetoric's status as a productive art—a status whose antiquity rhetoricians commonly emphasize by using, untranslated, the ancient Greek word, *technē*—we do well to wonder what we surrender when we stop attending to our making and doing when we analyze memes. Rosa Eberly's claim that it is only through "public making and public doing" that the political project that is living together communally is sustained seems especially pointed in this context.[27] Moreover, a lack of attention to rhetorical making and doing risks endorsing the sociobiological understanding of memes as self-perpetuating and of unclear or unimportant origins. Even

Figure 9.2. Meme created by Kennerly and Pfister, quoting Jeffrey S. Librett's translation of Heidegger in *Of the Sublime*.

meme studies in rhetoric that do not use sociobiological language demonstrate a need for a richer rhetorical account of memes. For instance, Jenkins includes "techné [*sic*]" in his adoption of "modes," essentially, "manners or ways of engaging (image-)texts" into which we are habituated, as a critical term for understanding how memes work rhetorically.[28] He limits the word, though, to its most popular cognate: "technique." For instance, within his analysis of "Fail/Win" as a mode, he notes that it "first relied on the technique of appending captions 'FAIL' or 'WIN' but has since evolved, now featuring images without captions that allow viewers to vote Fail or Win."[29] Reducing *technē* to technique has never served rhetoric well. When rhetoric-as-*technē* becomes a repository of techniques or a handy toolbox for tasks, it tends to come with a separation from invention, evaluation, and judgment—in short, from thought.

So that we might understand memes to be made and done and why rhetoricians should not cede making and doing to the memes themselves, we return to *poiēsis* (making, doing); more specifically, to *poiētikē technē* (poetic, productive *technē*). A *technē* is a recognizable, replicable method for achieving an end. Ancient accounts of the processes and products of rhetoric share a vocabulary and pragmatics of making and doing based in unmistakably poetic language; in infinitive form, the concept is *poiein* in ancient Greek. While *poiein* underlies "poetics" and "poetry" etymologically, the meaning of the verb extends across domains of activity, production, and craft; indeed, *poiein* referred first to making in general.

Rhetoric's productive dimension has itself been interpreted productively: The rhetoric-curious from Socrates onward have wondered what, precisely, rhetoric produces. If rhetoric is a *poiētikē technē*, then what does it produce? Arguments? Speeches? Persuasion? Judgments? Modes? According to Jeffrey Walker, those responses have a troubling object orientation, focusing on what is produced rather than on who produces. For Walker, rhetoric produces rhetors.[30] That is the only "product" rhetoric uniquely yields. If we base our treatment of memes as productions on that axiom, then we must attend to those who produce them. Such a claim does not foreclose a focus on that which rhetors make, since rhetors are symbol-users whose attentiveness to forms and norms of communication shapes what they make and remake, when, where, and for whom. Nor does such a claim fetishize the agency of the maker: That any rhetor is but a point of articulation in a vast network of influences does not blur that point to nondistinction.

Since any utterance emerges from a dense communicative matrix, teaching and learning rhetoric is a complicated enterprise. In *Against the Sophists*, Isocrates challenges teachers who approach thinking and speaking well as a *tetagmenē technē* (strictly ordered art), as a few fixed details, tips, and techniques to memorize once and deploy often and unvaryingly.[31] Those same teachers also shill *technai* through handbooks, perpetuating the ruse that everything a student needs to know can be contained in an unchanging format that offers everyone the same fixed instructions and insights.[32] For his part, Isocrates deems thinking and speaking well to be a *poiētikon pragma* (poetic/productive deed).[33] Accordingly, he habituates students to be dynamically responsive to situations and occasions (*kairoi*), respectful of the proper (*to prepon*), and unafraid of the new (*to kainon*).[34] In emphasizing that memes self-generate, sociobiological accounts underplay the situatedness, appropriateness, and novelty that meme-makers respond to in their moment of *poiēsis*.

Coming a few decades after Isocrates's treatise, Aristotle's sense of *technē* affords additional language by which to distinguish the rhetorician's eye view from the meme's. When Aristotle categorizes rhetoric as both a *technē* and a *dunamis* (power, potentiality) in book 1 of the *Rhetoric*, he is not being sloppy but rather describing a relationship.[35] Once internalized, a *technē* becomes a *dunamis* responsive to situational promptings and the rhetor's own inclination to act. Within other works, Aristotle applies his characteristic trifold thinking to *technē*, offering a taxonomy of *theōretikē* (theoretical), *praktikē* (practical), and *poiētikē* (poetic, productive) *technai*.[36] Readers of the *Rhetoric* will find that Aristotle uses grammatical variants of all three words with some frequency, making it difficult to determine where, ultimately, he places rhetoric's *telos*. If rhetoric is a *theōretikē technē*, then its end is internal, residing in the seeing and speculating the *technē* enables; if a *praktikē technē*, then its end is also internal, residing in the activity the *technē* makes possible; if a *poiētikē technē*, then its end is external, residing in the object the *technē* produces. Containing the verb *theōrēsai*, Aristotle's definition of rhetoric points to it being the first type of *technē*; however, elsewhere he locates rhetoric's for-the-sake-of in judgment (*krisis*), pointing to the second and third types.[37] Turning Aristotle's trifold categorization of *technē* toward meme-making, we can see that the sociobiological position that "memes meme" renders meme-making a practical activity that does nothing beyond itself. The rhetorical position that meme-makers meme establishes a meme as a made thing (*poiēma*) whose making tells us important things about what the maker notices and values and wants others to notice and value.

One of the classification challenges for rhetoric, then, comes from its manifestation as an effervescent activity, in some instances, and a stabilized product, in others. Since it organizes and archives utterances in a manner that allows for scrutiny in the frenzy of real-time or the (mostly hypothetical) luxury of scholarly time, Twitter is a platform upon which rhetoric's practical (the "to do" translation of *poiein*) and productive (the "to make" translation of *poiein*) dimensions are on display. For example, in early June 2015, while speaking at a luncheon organized by South Korean women scientists and engineers during the World Conference of Science Journalists in Seoul, Nobel-winning British biochemist Tim Hunt volunteered his views on women in labs: "Let me tell you about my trouble with girls. Three things happen when they are in the lab. . . . You fall in love with them, they fall in love with you and when you criticize them, they cry."[38] He advocated for sex-separate labs as a solution to all this explosive relational chemistry. Connie St Louis—an award-winning broadcaster, science journalist, and director of the science journalism program at City, University of London—scribbled down his remarks, embarrassed and disgusted.[39] After confirming with other attendants of the luncheon the wording of Hunt's account of his "girl" troubles, St Louis took to Twitter.

St Louis proceeded cautiously, mindful of her positionality and aware that "women are vulnerable to vicious trolling on Twitter, and black women doubly so."[40] Although her caution was not unjustified, her disclosures did not meet with much vitriol. University College London asked Hunt to resign his professorship, and he subsequently resigned from a few other academic positions as well. Several big names in British science—all men—publicly defended Hunt, and thousands signed a petition to have him reinstated. Many women in science and their allies, however, followed St Louis's lead and took to Twitter. Using #distractinglysexy as their hashtag rallying cry, thousands of people tweeted pics—of themselves or of scientists like Marie Curie and Rosalind Franklin—with allusions to Hunt's remarks that were pithy, clever, and descriptive of the mode of knowledge-making depicted. For example, @ameliacervera tweeted a selfie, displaying herself in full lab gear, and the caption: "Filter mask protects me from hazardous chemicals and muffles my woman cries. Double win!" (June 11, 2015). @lorenelynn tweeted a picture of herself head-first in a hole and the caption, "Some call me a dirty girl, some call me a soil scientist" (June 12, 2015).

The resultant visibility of the range and representation of women in science, doing all kinds of science—lab work, fieldwork, computation, surgery—all over the world, served as a vibrant rejoinder to Hunt and the old guard. That is not to assert that Hunt's remarks gave rise to the arms-linked solidarity of all women in science, but #distractinglysexy did mobilize a networked coalition of those unwilling to let his remarks go unremarked upon. Moreover, this hashtag activism produced a meme that troubled the (in)visibility politics that explain why women remain underrepresented in STEM fields. The reasoning behind the common *topos* of visibility presumes that though there may be women in science, the public does not *see* or *hear* about them; therefore, few girls aspire to go into science, and there are few institutions and infrastructures to build interest and capacity, and round and round we go. Digital forms and *fora*, from TED Talks to blogs, however, are putting women of science in the public eye.[41] #distractinglysexy did tremendous visibility work by using but subverting the old adage of publicity- and profit-seeking that "sex sells."

The profusion of making undertaken with and through #distractinglysexy resulted in images of women at play while at their scientific work, and the ludic dimension undercut the usual accusation against people, particularly women, who react to sexist remarks: that they need to have a sense of humor. The Twitter inventory of #distractinglysexy is an inventory of invention. In science, that word has historically referred to a brilliant, sui generis eureka moment of insight and discovery, often operating in ways that seriously disserve/d women.[42] In rhetoric, of course, invention (*heuresis*) refers to the process of coming upon and generating material that produces trust (*pistis*) in the people who encounter it and publics that form around it. In terms of #distractinglysexy, each created and curated tweet is a made thing that teaches, delights, or moves—or all three—with an image of a woman of science, usually in situ engaging in a scientific activity, and 140 characters full of character (*ēthos*) as Aristotle himself analyzed it: virtue and virtuosity, experiential know-how, and goodwill. The hashtag #distractinglysexy spoke not only to mid-June 2015 but also to the past and present of science that did or would deny women places of parity or power. Contributors to the hashtag crafted or retweeted tweets that recognized the ridiculousness of Hunt's comments by addressing and maybe even redressing to some degree the more systemic equality problems in science of which his words were emblematic.

Proposition Two: Memes are Generative (*Genesis*)

In his *Nicomachean Ethics*, Aristotle specifies that a *technē* "deals with the bringing into existence [*peri genesin*] of things that could either exist or not."[43] In view of our proposition that memes are made and done, and thus in league with a conception of rhetoric as a *poiētikē technē*, a rhetorical theory of memes will fail to account adequately for the potent generativity of memes if it does not push beyond *their* power to autopoetically make more of themselves through self-replication. A *poiēsis*-based view of memes stipulates that *we* make memes and orients attention to what—aesthetically and ethically—underlies our seeming compulsion to generate and regenerate memes. Although rhetoricians who go to Plato to find terms or themes useful for rhetoric risk being grudged by their colleagues, we find in Socrates's retelling of Diotima's views on *erōs* in the *Symposium* a compelling account of how and why we wish to make more and more of that which quickens us. Diotima's account of generative repetition accords both with rhetoric's traditional, ethical, and vital insistence upon human agency and the increasing insistence across the theoretical humanities that things—such as memes—have vibrant lives of their own.[44]

Diotima's erotic dialogues with Socrates elaborate upon the principle of *genesis*: begetting, generation, fecundity. *Erōs*, she pronounces, "is a bringing forth [*tokos*] in beauty, both in body and in mind."[45] She explains that "all human beings are pregnant in body and in mind, and when we reach a degree of adulthood, we naturally desire to give birth [*genōntai*]."[46] She begins with the pregnant *sōma*, gesturing toward the heteroerotic compulsion that brings together two differently sexed bodies and results in the conception, gestation, and eventual expulsion of another human form.

People who are pregnant only of body find satisfaction in making more of themselves through carnal means. Diotima, though, privileges the pregnant *psuchē* (soul or spirit). Those who are pregnant of *psuchē* give birth to "wisdom and other kinds of virtue," and she names "all the poets and other craftsmen who are said to be inventive [*heuretikoi*]" as their begetters (*gennētores*).[47] She calls the poetic issuances of Homer and Hesiod their "children."[48] The "most important and finest type of wisdom," though, she deems to be that which is "connected with the organization of cities and households, which is called moderation and justice."[49] She explains that Lycurgus and Solon "fathered" laws.[50] Poets and statesmen, just like everyday people, want to be remembered, and the very best of them have cults established in their honor "as a result of children of that kind, but this has never happened as a result of human children."[51] As she asserted earlier in the conversation, "It is undying virtue and glorious fame [*eukleoūs*] . . . that motivates everyone in all they do, and the better they are, the more true this is: it is immortality that they are in love with."[52] *Erōs* is a passion for perpetuation.

Although the first, Diotima is far from the only creator and contemplator to articulate such a view of *genesis*. In later accounts, what is for Diotima an interpersonal (*sōma*) or intrapersonal (*psuchē*) response to beauty finds extension. Upon accepting her Nobel Prize for Literature in 1938, novelist Pearl S. Buck described what she called "the creative instinct" in similar terms, even, like Diotima, pairing the male pronoun with procreative language:

> The creative instinct is, in its final analysis and in its simplest terms, an enormous extra vitality, a super-energy, born inexplicably in an individual, a vitality beyond all the needs of his own living—an energy which no single life can consume. This energy consumes itself then in creating more life, in the form of music, painting, writing, or whatever is its most natural medium of expression. Nor can the individual keep himself from this process, because only by its full function is he relieved of the burden of this extra and peculiar energy—and energy at once physical and mental, so that all his senses are more alert and profound than another man's, and all his brain more sensitive and quickened to that which his senses reveal to him in such abundance that actuality overflows into imagination. It is a process proceeding from within. It is the heightened activity of every cell of his being, which sweeps not only himself, but all human life about him, or in him, in his dreams, into his circle of activity.[53]

The abundance and exuberance of *genesis* cannot be contained. It pulls others inward even as it spills outward, begetting cycles of generativity. A line of a poem might twist in its application to a situation in which one finds oneself. The hoots of owls might motivate one to compose a piece of music that hauntingly mimics them. A pitcher plant might inspire one to craft a jug that imitates its dangerous contours. One hashtag may spark an ironic hashtag (or bashtag) in response, or a subtly tweaked hashtag with layers of new meaning. An encounter with the good, true, or beautiful, no matter the form it assumes, prompts us to attempt to preserve and perpetuate it using whatever medium we have at hand, be it ink or paint or memory.[54]

So far in this account, human agency seems overpowered by this (pro)creative productivity: Our outpourings might be seen as a reflex rather than a response. While being pregnant in *sōma* and/or *psuchē* is a condition of being human, we can craft, through training and self-discipline, ways to deliver ourselves from the discomfort of being fully and frequently expectant, a process Diotima details. Humans have a zeal—a need—to generate, to put ourselves out there, be it through communicative participation in the now or projection into the future. Rhetoric directs this eagerness to project and participate toward common, collective issues and ventures.

Pivoting from Diotima's account of *erōs*, Elaine Scarry joins beauty and justice, critiquing the attitude that "beauty, by preoccupying the attention, distracts attention from wrong social arrangements. It makes us inattentive, and therefore eventually indifferent, to the project of bringing about arrangements that are just."[55] To the contrary, she argues that an attunement to beauty arises from or enables a "perceptual acuity" that heightens our powers of noticing. Beauty thus contributes to justice—and is, in some cases, indistinguishable from justice—because "beautiful things give rise to the notion of distribution, to a lifesaving reciprocity."[56] At the core of this potency lies generativity: "Beauty brings copies of itself into being," just as Diotima emphasized.[57]

In further extending Diotima's outlook and Scarry's explication of it, we might wonder whether encounters with falsity, ugliness, and injustice can compel creativity on behalf of and in furtherance of their opposites. *Genesis* prompted by beauty and justice has preservative purpose, whereas that prompted by their opposites has corrective purpose. The feeling of being prompted by an urgent imperfection has been called "the exigence" that compels a rhetor to speak.[58] The Latinate etymology of exigence—*exigere*, to drive out—recalls Diotima's theory of *erōs*, according to which our pregnant bodies and minds, under all sorts of influences and pressures, long to put something out into the world that will matter after we are gone. Audre Lorde, in "The Uses of Anger," asserted that anger, when properly distinguished from hatred, is a generative rather than destructive force. In her view, "hatred is the fury with those who do not share our goals, and its aim is death and destruction. Anger is a grief of distortions between peers, and its object is change."[59] The kind of change she envisions is "not a simple switch of positions or a temporary lessening of tensions, nor the ability to smile or feel good. I am speaking of a basic and radical alteration in those assumptions underlining our lives."[60] Anger, Lorde avers, "is loaded with information and energy." On the matter of racism, whose exposure and dissolution is her focus in that speech and much of her other work, few born-digital movements initially and continually energized and informed by anger, by "a grief of distortion between peers," have attained the visibility and cultural diffusion of #BlackLivesMatter.

Launched by Alicia Garza, Patrisse Cullors, and Opal Tometi, three activist women of color (and Garza and Cullors identify as queer), after the acquittal of the killer of Trayvon Martin in 2013, #BlackLivesMatter was a response to feeling "vulnerable, incredibly exposed and incredibly enraged."[61] Garza called it "a call to action, to make sure we are creating a world where black lives actually do matter."[62] Accordingly, the three founders adopted a narrative-based and participatory emphasis, inviting people to share stories on Tumblr and Twitter about why #BlackLivesMatter. The hashtag has trended on Twitter—and, of course, expanded far beyond social media—after the deaths of several black people, some dead due to vigilante violence but most due to state violence. The networked awareness that grew after each death made it increasingly difficult to support the claim that the deaths were unconnected, isolated incidents rather than the result of systemic, structural violence against and devaluation of black people. In a media environment in which the produser is active, the process of generating, finding, and disseminating information, which, according to Lorde, gives potency to anger, has not been difficult for the activists in the movement and their allies. As Elizabeth Day observes, "Because so many of the new generation of activists are young and smart and technologically engaged, there is a restless, inventive energy to their campaigns, which can be started with little more than an iPhone."[63] This generative impulse makes for an accumulation of evidence that, as Garza put it, "the lives of Black people—not ALL people—exist within [certain] conditions [as a] consequence of state violence."[64] Of course, that does not mean that everyone will derive the same meaning or moral compunction from that *copia*.

The generativity of #BlackLivesMatter has taken forms other than supportive reiteration. We saw its immediate negation with #WhiteLivesMatter and contestation with #AllLivesMatter.[65] We saw its extension after further black deaths: #iftheygunnedmedown, after Michael Brown; #blackwomenmatter, after Natasha McKenna; #sayhername and #ifIdieinpolicecustody, after Sandra Bland; and #allblacklivesmatter, since media focus often falls on black men, rather than black women, black queers, black trans. We saw its cooptation in the early days of the race for the Republican presidential nomination, when Ben Carson used #BlackLivesMatter to argue against abortion.[66] We even saw its ironic imitation after a wealthy dentist from the United

States killed Cecil the lion and #AllLionsMatter was born.[67] Its cultural uptake—in its original form or a transformed one—suggests that major negotiations among and within the *dēmos* are underway. For those in positions of power, that negotiation could be a recognition or confirmation that inclines toward alliance and solidarity, or it could be a form of reactive friction, as, for instance, #CopsLivesMatter, #BlueLivesMatter, and #WhiteLivesMatter seem to be. In #BlackLivesMatter, a democratic iteration continues to proliferate, pointing and pushing at the constructed border between those who are treated like citizens and those who, too often, are not.

Proposition Three: Memes are Mimetic (*Mimēsis*)

"The memetic approach does not ask, 'What do people want?' but 'What does discourse want?' or 'What do memes want?'"[68] The *mimetic* approach, on the other hand, attends to the (human) act and effect of representation, a typical translation of *mimēsis*. As we noted above, *mimēsis* has long been a core element of rhetorical pedagogy. "Within a mimetic pedagogy," Robert Terrill observes, "inspiration comes through a productive encounter between the internal talents of the rhetor and the external resources of her or his culture and discursive traditions."[69] *Mimēsis* also stands at the busy intersection of ancient rhetorical, poetic, and philosophical theory.[70] In book ten of the *Republic* especially, Plato's Socrates took issue with *mimēsis*, which for him refers to both the process of imitation and its product. There are lots of layers to his critique, the most familiar being his condemnation of representation, an iterative process of instantiation that moves us further and further from the pure form of a concept (i.e., justice) or object (i.e., table).[71] The layer of Socrates's critique of *mimēsis* that we take issue with is his claim that imitative poetry "*misrepresents* human virtue in a dangerous way."[72]

The danger pertains to imitative poetry's propagative tendencies, powered by the emulative urges it prompts in those who encounter it. This critique maps well onto virals, that is, items that are shared with a high volume but without an alteration of the content of "the original." Copy-makers who purvey content that does not appeal to our better daimons would no doubt meet with a Socratic sneer, but the truly dangerous copy-makers are ones who circulate images (still or moving) of *conspicuously* virtuous people, that is, people whose virtue is such that it can be captured in an image or short video—that is, when a recording device is around, which is not rare these days, but means that others are present and able to notice the virtue unfold in some instance—or who only appear virtuous in the moment. Or both. The occasional extraordinary act of courage may be captured on a smartphone and posted on YouTube, but everyday acts of courage—simple, humble, repetitive, not much to look at—are harder to capture and harder still to imitate, the logic goes. As Socrates worries, "the wise and peaceful character, remaining always nearly the same as itself, is neither easily imitated nor easy to understand when imitated, especially not by a motley crowd gathered at the theater."[73] In today's media environment, the analogue to the theatre crowd is individuals joined together in a network, watching separately yet together.

Videos and images of acts we would classify as civic commonly traffic in showy virtue, in bold, often one-time acts, an example of which we present and parse below. Such acts are hard if not impossible—due to their historical singularity—to duplicate oneself, which accounts for the whirl of sharing and celebration such acts can provoke, as if by sharing, we participate in what we share. Emulation is the kind of mimetic activity that most interests Plato's Socrates, and he would deem this digital sort not only shallow but also hazardous: it promotes through propagation an attitude that externalizes and sensationalizes virtue. The risk is that virtue becomes thought of as something done by extraordinary people other than oneself and those with whom one has everyday contact. As Socrates says of the wise and peaceful character, it remains "always nearly the same as itself." Virtue, as it builds within a single person, is iterative. Its circularity and continuity suggest an easiness to copy—it looks nearly the same, again and again—but it is precisely that invariance that makes true virtue *hard* to copy. It is not glamorous. It is seldom exciting. It is hard. It can be tedious.

Recognition of the phenomenon one might call the spectacle of virtue is an occasional feature of social movement literature. As Paul Rogat Loeb laments, some people hesitate to flex their muscles (including their vocal chords) in communal or civic arenas for fear they are not eloquent enough or not informed enough or brave enough to be effective.[74] Socrates's criticism of *mimēsis* hones in on the *eidōlon* (image), and we can see, too, that icons and idols of civil rights movements can inspire us without motivating us or can even seem to possess an all-encompassing kind of virtue we feel we are unable to imitate.

Because clipped sounds and images circulate so cheaply and abundantly across digital networks, an instance of civic virtue can be deemed "iconic" and undergo many of the types of reiteration distinctive of iconic images, including repetition without changes (i.e., repostings) and creative imitation, with a speed and shareability alien to earlier media.[75] We can thus see, in nearly real time, how civic virtue becomes memed. Bree Newsome's ascent of the flagpole in front of the South Carolina State House offers a poignant example. On the evening of June 17, 2015, a racial terrorist sat quietly among a Bible-study group at the Emanuel African Methodist Episcopal Church in downtown Charleston; he went on to murder nine members of the church. By the next morning, the South Carolina State House had lowered its US and state flags to half-mast, but the Confederate battle flag remained at full-mast, where it had flapped like racist lips since the decidedly-not-nineteenth-century year of 1962. The digital response included the swift rise of #TakeItDown and increased take-up of #BlackLivesMatter. Ten days later, as day dawned, Newsome, a community organizer and social justice activist, strapped on a harness and a hard hat, scaled the flagpole, removed the flag, and was promptly arrested for defacing a monument.

Within hours, the image of Newsome, looking at the camera, extending one arm from the pole, clutching the flag, the state house in the background rendered less stately by its failure to #TakeItDown, was called "iconic" and shared far and wide.[76] Quick to follow were countless renderings, shares, and retweets that presented Newsome scaling the flagpole costumed like a superhero or even equipped with wings like an avenging angel.[77] She has been gracious in her recognition of this citizen-fan art, gushing in her first public statement after her arrest that "words cannot express how deeply touched I am to see how yesterday's action inspired so many. The artwork, poems, music and memes are simply beautiful!"[78] Those iterations, though, amplify this one activist moment above the countless others in which she has been involved and transform her from a mortal woman into a supernatural and singular being. Within her statement, she offered this reminder to all those "admiring her courage in that moment": "Please remember that this is not, never has been and never should be just about one woman. This action required collective courage."

What we mean to suggest by rehearsing Socrates's objections to *mimēsis* and applying his objections to Newsome's iconic act of civic virtue is this: Ancient treatments of *mimēsis* may be usefully brought to bear on digital artifacts and experiences to help us see better what is new about them, that is, what the old fails to give us and what we do need to theorize by ourselves. In Newsome's case, it is ease of access—our access to her and her access to us. Nearly every day, she responds to dozens of supporters and detractors who tweet @BreeNewsome. Newsome's digital direct address demonstrates that activism takes work, time, and plenty of conversation and explanation. Digital platforms can display the discipline it takes to remain actively committed to and engaged in the pursuit of social justice and not to give in to the slow nature of change or hot tempers of those who do not want it. At the time of the composition of this chapter, her daily tweets reach more than 110,000 followers, and a single tweet can see upward of one thousand retweets. That abundant mimetic activity demonstrates that she provides much that one can share with others and endeavor to emulate oneself.

For his part, Aristotle's redemption of *mimēsis* from the epistemic and moral misgivings of Plato's Socrates occurs early in the *Poetics*, when Aristotle deems *mimēsis* to be a distinctive human proclivity: We enjoy making representations and encountering those made by others. Why? According to Aristotle, "Learning gives pleasure, not only to philosophers but also to others, although they share in the pleasure to a lesser degree."[79] Aristotle adds that encounters with *mimēseis* engage our syllogistic reasoning, our sense of what is what, what goes with what, and what their relationship might be.[80] Later, Aristotle enumerates the parts of poetry, and one of them, *muthos*, appears to mean something like plot or plot structure, the narrative content built upon "action and life."[81] A flair for *mimēsis* powers a poet's representational renderings. We derive from Aristotle, then, the following understanding: a *mimēsis* builds on the material of action and life but twists and tweaks it in ways that please those who make and those who encounter it, since the former have to exercise reason to make renderings and the latter have to exercise reason to recognize what has been rendered.

By way of example, we return to the fall of 2011, by which time the energy of the Occupy movement, centered in Zuccotti Park near Wall Street in New York City, had carried across the United States and beyond. Unfolding months after the democratic protests across northern Africa, in which digital media played a much-hyped role, Occupy was the first movement of US origin to flourish digitally. People aligned with Occupy posted to networked sites images of themselves holding up paper or poster-board upon which they had written their reasons for supporting or joining the movement, ranging from hope for relief from student or medical debt to hope for a living wage for hourly work. Many included #WeAreThe99Percent or #IamThe99Percent, anthems that unified and amplified diverse narratives. University students across the United States held peaceful demonstrations to keep the momentum going. At a sit-in at UC-Davis, onlooking students took video and photographs as a cop pepper sprayed a line of seated students, firing directly into the whites of their eyes. Very shortly after those images went public, people affronted by them or annoyed at all the attention they were getting made the pepper-spray cop meme, which hung in the air for some time.

Some meme-makers simply added text to one of the image frames, such as the snarky lines, "Oh, don't mind me. I'm just watering my hippies." Others imported pepper-spray cop into iconic images of good versus evil in film; one, for instance, placed him in the position of Voldemort in the Harry Potter film *Harry Potter and the Goblet of Fire*, and rendered his pepper-spray as the bolt of Voldemort's wand. Yet other memers transported pepper-spray cop into iconic images that had in their time galvanized public opinion against an entrenched or threatening power. He strolled through Paul Revere's woodcut of the Boston Massacre of 1770, spraying the faces and wounds of those shot and suffering, and walked down a street in Saigon in 1963, spraying an accelerant on Thich Quang Duc, the self-immolating Buddhist monk. Those images that transport pepper-spray cop into a recognizable fictive or historical moment engage the reason-based comparative thinking that Aristotle deems one hallmark of the creation and reception of *mimēsis*. Operations of reckoning and recognizing underlie enthymemes and metaphors as well, according to Aristotle. Humans like to work a little to figure out what parts and connections are missing and to supply them.[82] As it is with making metaphors and analogies, so it is with memes: One must think through points of juncture or disjuncture and whether those points are poignant, parodic, or satirical. Meme-makers overtly rely and play upon resources and reservoirs of communal thinking, enacting the sort of participatory persuasion that Aristotle thinks we find the most captivating.

Replicating Res Publica

> *Plus ça change, plus c'est la même chose.*
> (The more it changes, the more it's the same thing.)

Like enthymemes, memes arise from presumptions about shared assumptions; like metaphors and *mimēseis*, memes arise from observations of similitudes.[83] In a larger sense, memes commonly emerge from moments of rupture, when members of a *dēmos* or other community cleave to or cleave from constitutive principles or practices. Such moments demand a renewed commitment to the old or an adjustment to the new, occasioning a communicative *copia*. Robert Asen names generativity a core quality of a performance of citizenship that is not dependent upon one's official inclusion in the *dēmos*: Civic activities invite further participation in some aspect of shared life. He leads with this image: "If we contextualize a multiple public sphere through metaphors of constellations or networks, then we can examine the ways in which citizenship engagement increases the nodal points in these configurations."[84] Asen does not apply that idiom of networked rhetoric to digitally based examples, but memes generate and constitute countless points of participatory entry into public life.

Our less selfish genealogy of memes divests memes of sociobiologically tinted and, we have argued, tainted agency without discounting their creative and circulatory energy. Indeed, we accounted for that energy by means of *poiēsis*, *genesis*, and *mimēsis*, emphasizing that memes are made and done, generative, and mimetic. In the mode of the meme, the core consideration for rhetoric is not persuasion but rather creation with variation. The move underway in the theoretical humanities to diffuse agency to things should not eclipse humans' agential powers; in that parlance, the meme-making and circulating process is an *assemblage* of the human and nonhuman, a point elided by the idea of a "meme's eye view." Not only publics but also a polity's *dēmos* flux and flex due to democratic iterations such as memes. The reverberant generativity of memes is not theirs alone.[85]

Notes

1. Richard Dawkins, *The Selfish Gene* (New York: Oxford University Press, 1976), 214.

2. See Takis Poulakos and David Depew, eds., *Isocrates and Civic Education* (Austin: University of Texas Press, 2009), and Jeffrey Walker, *The Genuine Teachers of This Art: Rhetorical Education in Antiquity* (Columbia: University of South Carolina Press, 2011), chapters 2–3.

3. Kathleen Welch, *Electric Rhetoric: Classical Rhetoric, Oralism, and a New Literacy* (Cambridge, MA: MIT Press, 1999); David D. Perlmutter, *Blogwars* (New York: Oxford University Press, 2008), 4–5 and 210. See also Rosa A. Eberly and Jeremy D. Johnson's chapter in this volume.

4. See http://rhetcompryangosling.tumblr.com and Danielle Henderson, *Feminist Ryan Gosling: Feminist Theory (as Imagined) from Your Favorite Sensitive Movie Dude* (Philadelphia: Running Press, 2012).

5. Linzi Williamson, Sarah Sangster, and Karen Lawson, "The Effect of Ryan Gosling Feminist Memes on Feminist Identification and Endorsement of Feminist Beliefs." Paper presented at Canadian Psychology Association, Vancouver, British Columbia, 2015.

6. Davi Johnson, "Mapping the Meme: A Geographical Approach to Materialist Rhetorical Criticism," *Communication and Critical/Cultural Studies* 4, no. 1 (2007): 39.

7. Dawkins, *The Selfish Gene*, 206.

8. Ibid., 211–12.

9. Ibid., 211.

10. "Gene talk" originates from Henry Howe and John Lyne, "Gene Talk in Sociobiology," *Social Epistemology* 6, no. 2 (1993): 112.

11. Johnson, "Mapping the Meme," 42–43.

12. Leslie A. Hahner, "The Riot Kiss: Framing Memes as Visual Argument," *Argumentation and Advocacy* 49, no. 3 (2013): 155 [quoting Susan Blackmore, *The Meme Machine* (Oxford: Oxford University Press, 1999), 6], 156.

13. Dawkins also licenses a connection to memory, which has obvious rhetorical potential but is not one we engage.

14. See Quintilian, *Institutio Oratoria Vol. IV*, trans. H. E. Butler (Cambridge, MA: Harvard University Press, 1958), 10.2.

15. Aristotelis, *Ars Rhetorica* (New York: Oxford University Press, 1959), 1394a25ff.

The etymologies of enthymeme and *mimēsis*/meme are unrelated, though the fact that most digital memes function enthymematically invites playing with the similarities between enthymeme and meme.

16. Hillel Schwartz, *The Culture of the Copy: Striking Likenesses, Unreasonable Facsimiles*, updated ed. (New York: Zone Books, 2014).

17. Richard Dawkins, 23rd Saatchi and Saatchi New Directors' Showcase at the Cannes Film Festival, June 20, 2013, https://www.youtube.com/watch?t=305&v=GFn-ixX9edg.

18. See Mari Lee Mifsud, *Rhetoric and the Gift: Ancient Rhetorical Theory and Contemporary Communication* (Pittsburgh: Duquesne University Press, 2015).

19. Such an understanding undermines the strong distinction between memes and virals posited by Limor Shifman in *Memes and Digital Culture* (Cambridge, MA: MIT Press, 2014), 55–63.

20. Seyla Benhabib, "Democratic Iterations: The Local, the National, and the Global," in *Another Cosmopolitanism*, ed. Robert Post (New York: Oxford University Press, 2006), 70 and 48.

21. Ibid., 68.

22. Sasha Weiss, "The Power of #YesAllWomen," *New Yorker*, May 26, 2014, http://www.newyorker.com/online/blogs/culture/2014/05/the-power-of-yesallwomen.html.

23. See Keith Houston's investigation of the complicated provenance of the octothorpe in his chapter "The Octothorpe" in *Shady Characters: The Secret Life of Punctuation, Symbols, and Other Typographic Marks* (New York: W. W. Norton, 2013). Despite disputes over the origin of the name for "#," we find the cartographic consonance inevitable since "thorp" means village in Old English and Old Norse.

24. Consider the affirmation that "black lives matter" (which we treat in more detail below) by several candidates for the 2016 Democratic presidential nomination during the first debate, on October 13, 2015. For a predigital example, consider President Johnson's use of "we shall overcome" in his exhortation to Congress on March 15, 1965, to pass voting rights legislation.

25. Axel Bruns, *Blogs, Wikipedia, Second Life, and Beyond: From Production to Produsage* (New York: Peter Lang, 2008).

26. Eric S. Jenkins, "The Modes of Visual Rhetoric: Circulating Memes as Expressions," *Quarterly Journal of Speech* 100, no. 4 (2014): 445.

27. Rosa A. Eberly, "Public Making and Public Doing: Rhetoric's Productive and Practical Powers," *Rhetoric & Public Affairs* 4, no. 3 (2001): 532–34, 534.

28. Jenkins, "Modes of Visual Rhetoric," 443.

29. Ibid., 454.

30. Walker, *Genuine Teachers*, 1–8.

31. Isocrates, *Against the Sophists*, trans. George Norlin (Cambridge, MA: Harvard University Press, 1980), section 12. All translations by Kennerly.

32. Ibid., section 19.

33. Ibid., section 12.

34. Ibid., section 13.

35. See, e.g., Aristotelis, *Ars Rhetorica* (New York: Oxford University Press, 1959),

1355b25–7, wherein he uses *dunamis* in his definition of rhetoric and *technē* in the very next sentence.

36. See, e.g., Aristotle, *Ethica Nicomachea* (Oxford: Clarendon Press, 1894), I (1094a2) and VI.

37. Aristotelis, *Ars Rhetorica*, 1355b25 and 1377b20–1.

38. Rebecca Radcliffe, "Nobel Scientist Tim Hunt: Female Scientists Cause Trouble for Men in Lab," *Guardian*, June 10, 2015, http://www.theguardian.com/uk-news/2015/jun/10/nobel-scientist-tim-hunt-female-scientists-cause-trouble-for-men-in-labs.

39. Connie St Louis, "Furor Over Tim Hunt Must Lead to Systemic Change," *Scientific American*, June 15, 2015, http://blogs.scientificamerican.com/voices/furor-over-tim-hunt-must-lead-to-systemic-change/.

40. St Louis, "Furor." For a treatment of the ethics of digital shaming and retaliation, see Scott Stroud's chapter in this volume.

41. See, for instance, STEMinist, accessed August 1, 2015, http://steminist.com/about/.

42. For a recommendation that scientific discovery expand to include "external acts of making visible," see Michelle G. Gibbons, "Reassessing Discovery: Rosalind Franklin, Scientific Visualization, and the Structure of DNA," *Philosophy of Science* 79, no. 1 (2012): 63–80.

43. Aristotle, *Ethica Nicomachea*, VI.1140a4.

44. E.g., Jane Bennett, *Vibrant Matter: A Political Ecology of Things* (Durham, NC: Duke University Press, 2010).

45. Plato, *Symposium*, trans. Harold N. Fowler (Cambridge, MA: Harvard University Press, 1925), 206b. Kennerly has adapted Fowler's translations.

46. Ibid., 206c.

47. Ibid., 209a. The adjective *heuretikoi* comes from the Greek noun for rhetorical invention (*heuresis*).

48. Ibid., 209d.

49. Ibid., 209a.

50. Ibid., 209d.

51. Ibid., 209e.

52. Ibid., 208d–e. See also our introduction in this volume (page 7), where we quote Goldstein on *kleos* in the digital age.

53. Pearl S. Buck, "Nobel Lecture—The Chinese Novel," *Nobelprize.org*, Nobel Media AB, 2014, accessed October 15, 2015, http://www.nobelprize.org/nobel_prizes/literature/laureates/1938/buck-lecture.html.

54. Clearly this compulsion to create has a mimetic dimension, too.

55. Elaine Scarry, *On Beauty and Being Just* (Princeton, NJ: Princeton University Press, [1999] 2001), 58. She mentions Diotima explicitly on pages 4 and 8.

56. Ibid., 62, 95.

57. Ibid., 3.

58. Lloyd Bitzer, "The Rhetorical Situation," *Philosophy & Rhetoric* 1, no. 1 (1968): 1–14.

59. Audre Lorde, "The Uses of Anger: Women Responding to Racism," *Women's Studies Quarterly* 25, no. 1/2 (1981): 282.

60. Lorde, "Uses of Anger," 280.

61. Elizabeth Day, "#BlackLivesMatter: The Birth of a New Civil Rights Movement," *Guardian*, July 19, 2015, http://www.theguardian.com/world/2015/jul/19/blacklivesmatter-birth-civil-rights-movement?CMP=share_btn_tw.

62. Ibid.

63. Ibid.

64. Alicia Garza, "A Herstory of the #BlackLivesMatter Movement by Alicia Garza," October 4, 2014, http://www.thefeministwire.com/2014/10/blacklivesmatter-2/.

65. Dara Lind, "Why Martin O'Malley Had to Apologize for Saying 'All Lives Matter,'" July 20, 2015, http://www.vox.com/2015/7/20/9003233/all-lives-matter-black.

66. Amanda Marcotte, "Ben Carson Attacks Reproductive Health Care Access for Women of Color and Rape Victims," *Slate*, August 17, 2015, http://www.slate.com/blogs/xx_factor/2015/08/17/ben_carson_fetal_tissue_gop_presidential_candidate_is_also_willing_to_attack.html.

67. Beenish Ahmed, "Twitter Uses #AllLionsMatter to Mock Race Baiting Over Black Lives Matter," July 31, 2015, http://thinkprogress.org/justice/2015/07/31/3686721/all-lions-matter/.

68. Johnson, "Mapping the Meme," 28.

69. Robert Terrill, "Mimesis, Duality, and Rhetorical Education," *Rhetoric Society Quarterly* 41, no. 4 (2011): 302.

70. See Stephen Halliwell, *The Aesthetics of Mimesis: Ancient Texts and Modern Problems* (Princeton, NJ: Princeton University Press, 2002).

71. For more on philosophical furniture, see Henry Walters, "Consider the Table: Of Socrates, Etymology, and Poetic Logic," *Eidolon*. August 15, 2016, https://eidolon.pub/consider-the-table-b26fc8c940d9#.mxil4gb2f.

72. See Jessica Moss, "What Is Imitative Poetry, and Why Is It Bad?" in *The Cambridge Companion to Plato's* Republic, ed. G. R. F. Ferrari (New York: Cambridge University Press, 2007), 417, and Alexander Nehamas, "Plato's Pop Culture Problem, and Ours," *New York Times' Opinionator Blog*, August 29, 2010, http://opinionator.blogs.nytimes.com/2010/08/29/platos-pop-culture-problem-and-ours/?_r=2.

73. *Republic*, 604e. Moss's translation.

74. Paul Rogat Loeb, *Soul of A Citizen: Living with Conviction in a Cynical Time* (New York: St. Martin's Griffin, 1999), 38.

75. Robert Hariman and John Louis Lucaites, *No Caption Needed: Iconic Photographs, Public Culture, and Liberal Democracy* (Chicago: University of Chicago Press, 2007), 37–39.

76. The photograph was taken by graphic designer and photographer, Adam Anderson, who was on a morning stroll with his camera.

77. See a selection at "The Best Bree Newsome Tributes," June 29, 2015, http://other98.com/best-bree-newsome-tributes/.

78. See her full remarks at "EXCLUSIVE: Bree Newsome Speaks for the First Time

After Courageous Act of Civil Disobedience," June 29, 2015, http://bluenationreview.com/exclusive-bree-newsome-speaks-for-the-first-time-after-courageous-act-of-civil-disobedience.

79. Aristotelis, *De Arte Poetica Liber* (New York: Oxford University Press, 1965), 1448b12–15.

80. Ibid., 1448b16.

81. Ibid., 1450a16–17.

82. Aristotelis, *Ars Rhetorica*, 1410b6–36 and 1356bff.

83. The memes to emerge from the refusal of Rowan County, Kentucky, clerk Kim Davis to issue marriage licenses to same-sex couples are a particularly pert example. See Dawn Ennis, "She'd Better Work: The Best 'Do Your Job, Kim Davis' Memes." September 11, 2015, http://www.advocate.com/media/2015/9/11/shed-better-work-best-do-your-job-kim-davis-memes.

84. Robert Asen, "A Discourse Theory of Citizenship," *Quarterly Journal of Speech* 90, no. 2 (2004): 199.

85. We thank Kaitlyn Patia for her support.

10

Remix, *Śūnyatā*, and *Prosōpopoeia*

Projecting Voice in the Digital Age

Scott Haden Church

Introduction

> Remix generally refers to the practice of recombining preexisting media content—popular songs, films, television programs, texts, web data—to fabricate a new work. Whatever would be called remix is completely contained and exhaustively covered by the existing term collage. Collage is not just an accidental or obvious arrangement of related elements as one might find thrown together; it is a calculated and deliberately subversive strategy that seeks, by way of the juxtaposition of diverse elements, to produce conceptual frictions and illegitimate fusions that are simultaneously unexpected and insightful. Sampling is the key element that makes the act of remixing possible. It is a new way of doing something that's been with us for a long time: creating with found objects. By employing technologically advanced equipment and digital-compression techniques, they create sounds that are more sonically engaging and powerful than even the most skilled instrumentalists can produce. Sampling is not just a musical practice; it is a complex cultural practice. Today's culture hackers meld everything from audio to automobiles, armed with an arsenal of increasingly sophisticated tools that have trickled down from the geek elite into the hands of consumers and triggered a mash-up boom. Every act can be imagined as a process of sampling and mixing that involves the selection, juxtaposition, and manipulation of the world's information as script. In a large sense, one would say there is no pure originality. All minds quote. Old and new make the warp and woof of every moment. There is no thread that is not a twist of these two strands.[1]

In the early twenty-first century, remix artists took advantage of sophisticated studio techniques to help individuals deliver unlikely performances. In 2004, John Lennon (who at that point had been dead for almost two and a half decades) debuted a postmodern iteration of his vocals from the tender ballad "Julia" to accompany Jay Z's "Dirt Off Your Shoulder." In 2010, the voice of the deceased father of golfer Tiger Woods appeared in a Nike commercial, providing voiceover chastising of his son's extramarital affairs. These performances, which I detail more below, have a precedent in ancient cultures. Millennia before the prevailing sampling practices of remix culture, ancient rhetoricians, mindful of Horace's observation that "the voice once sent forth cannot return," discussed the rhetorical strategy of making an abstract or absent essence concrete or present by providing it with a voice.[2] In the first decade of the first century BCE, the anonymous Roman author of the *Rhetorica ad Herennium* explained that "*conformatio* [literally, *giving form to*] consists in representing an absent person as present, or in making a mute thing or one lacking form articulate, and attributing to it a definite form and a language or a certain behavior appropriate to its character" (4.53.66).[3]

Because there is no English cognate of the term *conformatio*, this rhetorical practice of *prosōpopoeia* is often conflated with "personification." Despite its common usage in ancient Roman literature, the term is usefully reconceptualized to analyze contemporary digital culture.[4] As the (re)presentation of a once-absent voice, the ancient rhetorical figure of *prosōpopoeia*—literally, "face-making"—can usefully inform the contemporary popular practice of appropriating and digitally reanimating absent voices. *Prosōpopoeia*, I will argue, is an essential trope of remix.

Similarly, the act of remixing is much richer than merely personifying. When the trope is envisioned as an ancient antecedent for contemporary remix, *prosōpopoeia* is linked to a cluster of additional terms like *voice making*, *voice restoration*, and *voice projection*.[5] Cicero and Quintilian hint at these dimensions by explaining that *prosōpopoiiai* are "bolder" than speech and require "stronger lungs" (85; 9.2.29).[6] As part of its powerful potential for giving form, though, *prosōpopoeia* can give voice only to a person or a thing that has no voice. Historically, the term entails more than simply assuming the mannerisms of a person or thing; it must restore to a thing an essence that was previously lacking.[7] What, then, do we make of *prosōpopoeia* after postmodern critiques of essentialism? The argument of this essay is that to best understand this projection of voice and its relationship to the lack of textual essence, another cultural tradition is needed to explicitly address the essencelessness of remix in a contemporary, digital context. To find overlays between disparate philosophical traditions, this chapter mashes up ancient Greek and Roman rhetorical theories of *prosōpopoeia* with the ancient Mahayana Buddhist philosophy of *śūnyatā*. *Śūnyatā*, the ancient philosophical construct that communicates a simultaneous lack of essence and fullness of possibility, can be a valuable resource for understanding the complexities of how remix plays with form. Like any good remix, the juxtaposition of the different intellectual "tracks" represented by *prosōpopoeia* and *śūnyatā* draws out heretofore-unexplored components of each.

Remix, a chiefly musical phenomenon aided by digital technology, shares essential characteristics with *prosōpopoeia*. Remixes spawn *mash-ups*, which are "the intermingling of different source material that involves seemingly incompatible musical styles, production values, and traditions in popular culture. [The objective of the mash-up] is to produce a new, third term that is arguably greater than the sum of its parts."[8] By its very definition, remix uses *prosōpopoeia* every time a remix artist spins vinyl records on a turntable or copies and pastes samples of music into a digital mash-up. Although it is the rhetorical counterpart of remix, *prosōpopoeia* does not layer multiple voices on top of each other in the same manner as a musical remix. When *prosōpopoeia* projects voice into a new appropriate context, however, the appropriation of the voice combined with the new situation creates a new, recombinant voice in place of the original constitutive voice.

This chapter aims to reconsider remix as a practice more concerned with essence and voice than with "music" narrowly conceived. When conceptualized in this way, remix not only "speaks" the absent recording artists' songs, it opens new possibilities by distilling extracorporeal voices and influences into the body of the speaker. Put another way, the reason that *prosōpopoeia* works with remix is because of its philosophical debt to *śūnyatā*: the essencelessness of the spoken voice is what allows it to be filled by the projection of other voices.

Remix *prosōpoetically* mashes up various voices according to their appropriateness and familiarity. When the DJ or remix artist creates a mash-up, he or she is responsible for choosing the appropriate voices to be sampled and the parts of the composition to be added upon or subtracted from.[9] This is a tricky maneuver because remix artists must allow the original essence of the song to remain intact and yet transform it sufficiently to be considered something novel.[10] This is a rhetorical process, in that the remixer chooses certain songs to sample while deflecting others. Remixers will consider the final product as successful if they, as rhetors, are able to sufficiently sustain the audience's attention as well as help persuade it that the aesthetic choices were appropriate. There is a level of prestige gleaned by DJs when they find an obscure sample and reintroduce it into the limelight in the form of a great mash-up. DJs call it "crate digging," a way to demonstrate to the audience musical proficiency, that the DJ has a vast knowledge of not only songs but also specific (and perhaps forgotten) parts.[11] Because vinyl-record culture is unwieldy and often highly esoteric, remix pervades every aspect of it.

Remix is also at the core of academic scholarship. In fact, the act of research is composed of equal parts *prosōpopoeia* and crate digging. This point is demonstrated by this very chapter, in which overviews of remix, sampling, *prosōpopoeia*, and *śūnyatā* are mashed up from an aggregation of quoted material from dozens of sources. The words included in these sections are not mine; however, I am responsible for their compilation, organization, and recontextualization.[12] To compile them, I needed to scour the literature on remix, digging through the crates of disparate arguments and determining how they could in some way relate to remix. The point to be made here is that academic scholarship is thoroughly *prosōpopoetic*—citation is merely the animation of an absent voice. Because all scholarship is remix, all scholarship is *prosōpopoetic*. Because scholarship aims to build upon past arguments and findings, scholarship considers itself empty of completed essence and full of potential, ready to be used again in the service of epistemology.[13]

Interlude One

> Remix discovers what is there in the emptiness. It is the emptiness of all phenomena, their lack of essence, which makes copying possible; but more important, this emptiness is what makes it possible for anything to appear at all. Similarly, if the essence of a thing were truly fixed, it could not be transported to the copy, and imitation, even as a degradation of the original would not be possible. Buddhist emptiness in its very etymology holds a hint of fullness that is lost in its English translation. Emptiness turns out to be full; that is, full of the immense possibilities that the mind has not yet grasped. When emptiness is possible, everything is remix; were emptiness impossible, nothing would be remix.[14]

Remix highlights an important rhetorical function of contemporary digital artifacts: The ease with which the absent are reanimated, repurposed, and remixed demonstrates how artifacts lack essential properties. This absence of essence recalls the words of the second-century philosopher Nāgārjuna. Nāgārjuna theorized *śūnyatā*, Sanskrit for "emptiness," as a Buddhist approach that sees existence as all-encompassing and hollow at the same time.[15] Nāgārjuna's invitation to understand *śūnyatā* demonstrates the tantalizing potential of the idea: "For whom emptiness is possible/For him all is possible."[16] To the remix artist, the possibilities for creative production are innumerable for the same reason.

Śūnyatā defies the traditional concept of emptiness and the value ascribed to essence. Rather than embodying a lack of essence, *śūnyatā* should be considered a stripping away of all external complications, including the complications posed by meaning, thus freeing the text to become the starting point for something new. *Śūnyatā* is more closely associated with a swelling or hollowness than with an absolute emptiness. It is neither a being nor a nonbeing: "Buddhist emptiness in its very etymology holds a hint of fullness that is lost in its English translation."[17] That swelling is addressed by numerous explications of *śūnyatā* through the metaphor of an expanding seed. Nāgārjuna explained the emptiness of perceived reality in this way: "All phenomena are devoid of inherent existence . . . because the inherent existence of all phenomena is not to be found in causes, conditions, aggregations or individualities."[18] Unpacking these provocative stanzas, David Ross Komito explains that every phenomenon is dependent upon being caused, having a condition, being joined together, or being taken separately. The joining together of one phenomenon with its essential conditions is the *aggregation* of which Nāgārjuna wrote and makes their discrete *individualities* impossible to examine. With a sprout, for example, a seed is the cause and the water and earth are contributing conditions that aggregate with the seed to allow the sprout to take shape.[19] Key to understanding the connection is acknowledging the interdependence of each of the factors—not simply the nature of their individualities but, rather, their essences in totality. As Michael G. Barnhart explains, "*Śūnyata* was dependence; that all things were empty meant that all things were mutually (and thoroughly) dependent."[20] Nothing can emerge in front of the rest if everything is interdependent. Thus, all phenomena are empty of essence, traditionally understood.[21] By contrast, as philosopher Gay Watson argues, Western intellectual traditions have a difficult time comprehending the nature of *śūnyatā* because of their cultural fetishization of meaning and the master narratives of Authority and Absolutes, inspired by Platonic thought.[22]

Remix artists also consider all potential samples as interdependent. When a song is recorded and released to the public, like the seed of *śūnyatā*, it swells into fruit that is available for any perceptive passerby to pick, plant, and cultivate. Because of the relatively affordable and intuitive digital composition technologies available, remixers consider recorded songs to be raw materials just waiting to be mashed up. This perception is emblematic of remix culture's appropriation of the credo embodied by the open-source technological movement over the last few decades: If something has been created, take it, use it, and make it better. Like Nāgārjuna's insistence on the lack of "inherent existence," remix culture flattens *authorities* and multiplies *absolutes* until the seed of new forms of creation can swell unhindered. In this way, remix culture directly challenges norms of authorship enshrined in the codified language of copyright law.

Both remix and *śūnyatā* have a radical potential for resistance or subversion. For Fernando Tola and Carmen Dragonetti, the concept of *śūnyatā* was void of any sacred or religious element whatsoever and so was antagonistic to Buddhist institutions: "[The *śūnyatā*] is not a religious or mystic experience. Every religious or mystic experience supposes a sacred object to which it aims, and this does not happen in the case of the *Madhyamaka* school. The experience of the *śūnyatā* is only an extra-ordinary experience of the true reality. . . . [Nāgārjuna's] abolishing analysis of the empirical reality does not limit itself to the common beings and things of the world; it attacks also, with the same severity, the most valuable and respectable beliefs and doctrines of the Buddhist Church, to which he and his school belong."[23] This subversive potential of *śūnyatā*, though certainly contested by others, does provide a fruitful comparison with remix. Remix attempts to resist hegemonic aesthetic conventions by subverting them, even turning the critical eye upon itself when appropriate. In 2013, Daft Punk, a duo that has thrived by means of its unashamedly derivative music, challenged the myriad digital capabilities of recording technology by being explicitly analog.[24] In their album *Random Access Memories (RAM)*, the duo subverted the trend of sampling influential artists and digitally mashing them together by inviting a few of those actual artists into the studio to collaborate and create new analog music. Ironically, Daft Punk "de-black-boxed Remix" by creating a real-life remix without digital samples.[25] Although they used no samples, they still facilitated a variety of *prosōpopoeia* by inviting other recording artists to provide vocals. Fundamentally, then, the emptiness inherent in both *śūnyatā* and remix, unfettered by anchoring essences, enhances the subversive potential of *prosōpopoeia*. This subversive potential is sympathetic to the mask-wearing purpose of *prosōpopoeia*; the mask will fit any face if the face's features are ever changing.

Interlude Two

> *Prosōpopoeia* can be used to represent the inner thoughts of opponents, recreate conversations, create characters who voice words of pity or reproach, or give voice to the dead, gods, cities, and nations. The goal is to produce a new, third term that is arguably greater than the sum of its parts. A key aspect of any attributed speech in the ancient world is its appropriateness about the character to whom it is attributed and the situation in which it is delivered. She or he is always and already responding to the scene or culture in which one already finds oneself and is, for this reason, responsible only for the manner, method, and means of that particular response.[26]

Prosōpopoeia is a powerful, yet peculiar, trope. For literary critic Paul de Man, *prosōpopoeia* is "the master trope of poetic discourse," as it describes the very act of reading.[27] The functions of the trope have been the subject of controversy. Exceeding the contemporary parameters of the symbolic act, the poetic usage of *prosōpopoeia* in the nineteenth century initiated a microgenre called the "corpse poem."[28] Diana Fuss claims that the act of "giving voice to the voiceless cadaver" allows the poet to use "literary fictions that seek to revivify and reauthorize the dead."[29] Giorgio Agamben acknowledges that those endowed with voice have the capability to act as pseudowitnesses for those who no longer have it. He quickly adds, however, that this *lending* of voice is no different than the fictive *imposition* of voice.[30] Despite its potentially controversial dimensions, the ability of *prosōpopoeia* to reanimate the dead is only one aspect of its rhetorical function. Fundamentally, *prosōpopoeia* is about projecting voice.

The ability of *prosōpopoeia* to project voice to another person or thing is a critical component of its definition. In the bundle of ancient Greek handbooks that preserve the *progymnasmata* (preliminary rhetorical exercises), several teachers address the exercise in similar ways. Some of the accounts distinguish between variations of the term depending on the identity of the orator and the audience. While one of the progymnasmatists, Theon, broadly defines the term as the act of attributing appropriate speech to another, other ancient pedagogues consider *prosōpopoeia* as the attribution of voice to more than absent speakers. While addressing the rhetorical figure of personification, the author of *Rhetorica Ad Herennium* explained that it could "be applied to a variety of things, mute and inanimate" (4.53.66).[31] Save Theon, the progymnasmatists distinguish variants of speech for the absent, like *eidōlopoeia* (image, idol, or ghost making; that is, a speech for a god or dead person) and *ēthopoeia* (character making; that is, a speech for a living person).

Although this chapter generally focuses on *prosōpopoeia*, the ancient variant *ēthopoeia* offers additional insights into the various voice makings that prevail in remix culture. *Ēthopoeia*, or "character making," is concerned with the persuasive possibilities of character portrayal, rather than its mere fidelity.[32] Performing *ēthopoeia* entails representing one individual with another for the purpose of dramatic characterization. This act can exhibit a discourse that is deeply rhetorical because it impersonates the individual to exploit his or her *ēthos*.

Cicero's *Pro Caelio* is one such example. In this defense speech, recognized since its delivery as a "first-rate piece of acting,"[33] Cicero attempts to absolve Marcus Caelius of several charges. Among these charges is the attempted poisoning of Caelius's lover, the infamous Clodia. Cicero, relishing the opportunity to embarrass the sister of one of his own enemies, chastises her while using the persona of the long-deceased Appius Claudius Caecus, a dignified ancestor.[34] Prior to this mask-wearing discourse, Cicero acknowledges that he is choosing to let the dead speak so that he can invoke a harsher tone appropriate to the grievousness of Clodia's immoral act: "I will first inquire of [Clodia], whether she prefers me to deal with her severely, solemnly, and in an old-fashioned manner, or mildly, gently, and in a modern way. If in the old grim mode and method, then I must call up from the dead one of those full-bearded men of old—not with a trim modern beardlet that she delights in, but a rough one, like those we see on old statues and busts—to rebuke the woman and speak instead of me" (14.33).[35] Turning the knife, he then shifts to impersonate Clodia's absent brother to continue the rebuke.[36] In prefacing his rhetorical onslaught, Cicero tells Clodia that her brother is "a perfect man of the world; who loves you most dearly." He then invites her to "imagine him saying . . . , 'Sister, why are you making such a to-do? Why have you lost your senses?'" (15.36).[37] Cicero's *prosōpopoetic* prosecution was so devastating that Quintilian later celebrated Cicero's stylistic use of "force for rousing emotion" (12.10.61).[38] Cicero's masterful technique was so successful in achieving its rhetorical objective that it prompted Clodia to be subsequently "laughed out of court."[39] Cicero's in-court mash-up of his own ideas with the simulated mannerisms of Appius illustrate the potency of the recombinant voice.

Interlude Three

No beings, not even humans, speak on their own, but always through something or someone else. Effective remix, according to Dionysius of Halicarnassus, is not only persuasive but also aesthetically pleasing, as indicated by his emphasis on lifelike, charming composition. The remix when extended as a cultural practice, as a form of discourse, is a second mix of something pre-existent. Remix demands that people recognize some trace of history. Part of the interest in sampling is in taking a bit of music that the listener will recognize, who will in turn most likely become excited when she recognizes the sample. And yet, despite all this, the author does not rest in peace. He is continually reanimated and resuscitated. The more perfect the reproduction, the more it becomes virtual. Who is the real speaker? Who's to say? They've even felt the need to reintroduce certain mannerisms to give it a natural effect, or an effect of the hyper-simulacrum of the natural.[40]

Contemporary remix deploys *prosōpopoeia* for rhetorical purposes. The familial voice in Cicero's *prosōpopoetic* defense offers a point of consonance with a 2010 television commercial by Nike about one of the company's most famous spokespersons, professional golfer Tiger Woods. At the time the company created the ad—informally called "Did You Learn Anything?"—news coverage of Woods was ubiquitous due to his myriad marital infidelities. Inspired by the milieu, the ad begins with a camera slowly closing in on Woods's expressionless face, while a recording of his father's voice apparently urges Woods to take responsibility for his actions: "Tiger? I am more prone to be inquisitive to promote discussion. I want to find out what your thinking was. I want to find out what your feelings are. And did you learn anything?"[41] The remixed recording is especially poignant because Woods's father, Earl, had passed away four years previously.[42] The *prosōpopoetic* familial pleading of the ad is reminiscent of Cicero's invocations more than two millennia before.

By reanimating the elder Woods's fatherly evocations, Nike uses *prosōpopoeia* and thus exploits Earl Woods's paternal propriety. The advertisement implies that Woods's actions gravely disappointed his father, yet, rather than offering a Ciceronian shaming for his behavior, it becomes a vehicle for recuperating his image on the eve of his return to competitive golf. The first phrase of the recording indicates that Woods's actions warranted a clear explanation, though it simultaneously invites a dialogue—an invitation for a cathartic learning moment. The remixed voice of the deceased speaks to Woods in a familial accord that affords permission for the viewing public to voyeuristically participate in expressing its own feelings vicariously for the betrayal of sorts. The aesthetic representation of Earl Woods's voice, offering a fidelity to the speaker that Cicero never could have accomplished in the ancient Roman court, created a *prosōpopoetic* remix that could both chastise and rehabilitate simultaneously. The aesthetic arrangement of the samples of Earl Woods's voice, coupled with the fidelity of portrayal made possible by the digital technology, amplifies the rhetorical potential of *prosōpopoeia* beyond its analog constraints. A contrast of this type of remix with an *ēthopoetic* remix can help make this point clearer.

The "character making" *ēthopoetic* text is analogous to *prosōpopoeia*, but a close reading of the ancient term reveals a crucial point of divergence. In the ancient text *Lysias*, Greek literary critic Dionysius of Halicarnassus focused on the persuasive compositional element of *ēthopoeia* as opposed to the progymnasmatic emphasis on suitable characterization. Dionysius's *ēthopoeia* was unique because it emphasized the creation of a persuasive *ēthos*, one that was stylistically concise and compellingly performed. According to Kristine S. Bruss, most noteworthy of Dionysius's emphases was the importance of "lifelike, charming composition," a manner of aesthetic arrangement that was ostensibly improvised.[43] Despite its clearly aestheticized structure, this *ēthopoetic* manner of composition is called "artless," because the arrangement of the text should not be "contrived or formed by any conscious art . . . [but rather give] the impression that this arrangement has not been deliberately and artistically devised, [that it] is somehow spontaneous and fortuitous."[44] This emergent distinction between *prosōpopoeia* and *ēthopoeia* offers clarification into the structural variations of remix and their associated appeals.

Some remixed texts tend to be *prosōpopoetic* while others are *ēthopoetic*—the distinction emerges from how the text is arranged. An *ēthopoetic* remix is the equivalent of a musical cover; it retains the basic sonic properties of the original song but aims to create a "lifelike, charming" representation rather than a portrayal that is completely accurate. Thus, the *ēthos* of the original artist(s) continues through a straightforward reimagining of the song.

Although it is emphasized in Dionysian *ēthopoeia*, this "art of artlessness" is not a necessary component of *prosōpopoeia*.[45] Appropriateness is lauded as the most important dimension of *prosōpopoeia*. In *Pro Caelio*, Cicero would change his tone of voice and nonverbal delivery to faithfully impersonate the mannerisms of the character he wished to evoke.[46] Theon and his fellow pedagogues in the *Progymnasmata* spent the majority of their lectures on the "propriety of portrayal," in taking on the form of voice of characters or things.[47] Indeed, their discussions of the trope, as well as Quintilian's, focus on the minutiae of the task, including what age-appropriate colors of apparel the orator should wear when impersonating other individuals.[48] Details of such an endeavor were meticulously outlined so that the character being impersonated could retain a semblance of its *ēthos* in the perception of the audience.

Remix also exhibits a strong concern for appropriateness. In selecting samples of songs, the best remixers appropriate appropriately. These standards of appropriateness can be witnessed in the musical articulation of a remix. Just as language has a grammar, music has elements crucial to its aesthetic representation. Among them are tempo (the pace of the music), chords (the sonic interplay of the notes to create harmonies), pitch (the position of the aural output on a scale from low to high), beat (the instrumentation and intensity with which the musician enforces the rhythm and tempo), and phrasing (the musical segments and the transitions that link them together). Just as Cicero altered the tone of his voice when he assumed the *persona* of Appius Claudius, remix artists must alter the pitch of the samples appropriately, slow down the tempo or speed it up, and impose a coalescing beat on top. Remix artists perform a similar dialectical task each time they sample materials and mash them up into a distinctive mosaic.

To meet the standards of appropriateness, remixers rely on *śūnyatā* to see the potential interdependence of the texts, a contingent collection of song snippets that can function as the starting ground for further artistic creation, while conveying the extant social and aesthetic associations inherent to each of the songs. For example, a sampled song that included recorded sound by the Beatles might contain any number of the following musical and/or cultural associations: the British invasion, the political and social tumult of the late 1960s, and the assassination of John Lennon. In the case of remix, the allegorical potential of the samples may be amplified and simultaneously effaced as these musical fragments are accompanied by charged associations even while they are reframed with new juxtapositions.[49] In addition to having a sense of the interdependence of all these factors, the remixer must have an astute ear to find aesthetic connections where others would likely never look.

The dialectical logic of remixes—oscillating between the familiarity of the sampled material and the novelty of the mash-up—is illustrated in the case of "the most popular and historically important mashup up to date,"[50] Danger Mouse's *The Grey Album*. More importantly, the album illustrates a tension between *prosōpopoetic* and *ēthopoetic* types of remix. In this 2004 album, a mash-up of rapper Jay Z's a cappella tracks from his *Black Album* and selections from the Beatles' *The Beatles* (aka the White Album), remix artist Danger Mouse uses aesthetic appropriateness to explore the gray area between allegory and clean slate. Although his experiment has greater success in some tracks than in others, Danger Mouse appropriates musical samples from the iconic album and transforms them so convincingly that they appear to be appropriate backing tracks to Jay Z.

Despite its abundance of samples, *The Grey Album* is an example of an *ēthopoetic* remix. An illustrative song from the album is "Dirt Off Your Shoulder," which samples John Lennon's "Julia." Lennon's original track, perhaps the most tender and haunting track on the White Album, is an homage to both his deceased mother (Julia Lennon) and his partner (Yoko Ono). Setting aside its original authorial meaning and quiet, haunting chords, Danger Mouse adds a stuttering and frenetic drumbeat behind the original samples while savagely splicing them into new configurations. The final effect is stunning, as the intensity of the chords and beat coupled with Jay Z's rhymes work in unexpected ways. Additionally, unsettling elements of the mash-up emerge through the unlikely pairing; on top of the original song's fingerpicked chords, Lennon's repeated vocal utterances are disembodied and incomplete in the mash-up, evincing a haunting quality that becomes more amplified given the premature deaths of Lennon and his mother. Further, the incongruous pairing of the two songs makes some perhaps uncomfortable associations for Beatles fans who hold Lennon's legacy to be sacrosanct: not only is Lennon's song transformed into a nearly unrecognizable sample, Lennon's voice is drowned out by Jay Z's "Now tuned into the [expletive] greatest. Turn the music up in the headphones." The relationship of the voices is ambiguous enough that it can be dually interpreted as valorizing Lennon's original song and *ēthos* as the greatest or as "Now turning to the [expletive] greatest," essentially proclaiming the artistic superiority of Jay Z over Lennon.[51] The literal layering of Jay Z's voice over Lennon's resuscitates Agamben's fears that any sort of *prosōpopoeia* is an imposition of the living's voice over the dead. However, the concerns of Beatles and Lennon fans from their reading of the remix may be rooted in expectations relating to the type of remix from Danger Mouse. This might occur when fans want a *prosōpopoetic* remix but get an *ēthopoetic* remix instead. The remix artist exploited Lennon's *ēthos* in the mash-up. The rendition vivified Lennon's voice by transforming it into a virtually unrecognizable sample. The arrangement of the sample did not strive for a replication of his voice but reanimated it through *ēthopoeia*.

As evidenced in the case of *The Grey Album*, the appropriateness of the samples is paramount to the text's aesthetic and rhetorical success. Yet the conflation of competing *ēthē* that derives from the congruity of musical form and incongruity of extramusical associations further enhances the rhetorical value of the mash-up. For example, one entertaining aspect of remix is its nuanced assemblage of music samples from competing genres.[52] Hip-hop and classic rock mashed together can be surprising; blends of smooth jazz and heavy metal can create a sense of the carnivalesque.[53] Functioning in more profound ways than its mere entertainment value, the congruent mashing up of samples from traditionally incongruent musical genres demonstrates more of the rhetorical power of the practice. Although this species of layering does not directly parallel the *prosōpopoeia* of ancient times, its musical-rhetorical use in music throughout history illustrates how incongruity and *prosōpopoeia* converge in remix.[54] Ultimately, the appropriateness of the character's representation (or the violation thereof) has been one of the most compelling dimensions of *prosōpopoeia* in both ancient and networked eras.

The congruity of ostensibly clashing genres in remix cues us to another dimension of *śūnyatā*: an emphasis on *process* rather than *product*.[55] Nāgārjuna taught that once one could grasp the true meaning of emptiness, one could transcend conventional ways of thinking about reality and perceive the true reality enveloped within the external reality dictated by one's mortal (and thus incomplete) understandings.[56] *Śūnyatā*, the middle path between permanence and nihilism, was to be achieved through a rigorous process resembling contemporary argumentation: "[Nāgārjuna] developed a process of dialectical reasoning which exposed contradictions in ordinary thought and which, by reducing all claims to pairs of negations, sought to dislodge thinking from such extremes, thereby freeing the mind to achieve enlightenment. According to this procedure, when it is recognized that opposing poles of thought may be negated by reasoning, the mind is able to acknowledge that reality is neither of them, and is able to experience *sunyata*."[57] This process of intense interrogation was also designed to penetrate through what the Mādhyamaka School called the "concealment reality" of empirical experience that obscured true reality. This true reality could be discovered only through this rigorous process of intellectual reasoning and reflexive negation. Nevertheless, this discursive process of "abolishing analysis," and the discovery of the true reality "through the [destruction of] the empirical one," was no easy task, nor was the overall objective immediately evident.[58] The monolithic structures of empirical reality were difficult to overcome because they set the parameters for the discussion, after all. Therefore, the school believed in employing aggressive argumentation tactics to challenge the contrary theses. These included using the reductio ad absurdum fallacy or addressing the rival schools' claims by continuously negating rather than engaging them.[59] The individual's phenomenal perception of reality needed to be challenged through a "destructive dialectic" so that the "ultimate truth" could emerge from the obstructions surrounding it.[60] Ultimately, the process of liberating the self from the illusion of empirical reality is the same as the goal of obtaining a true understanding of *śūnyatā*.

Remix also focuses more on the process than on the final product. Like the open-source culture that enabled its growth, remix is situated within a discourse that privileges the process of production and circulation even more than the content itself. A key component of the practice is the DJ's demanding challenge of digging in crates, finding creative ways to mash up the subsequent samples, and *prosōpopoetically* assessing the allegorical aura attached to each sample to please the audience. An appreciation of a remix is closely tied to an understanding of how many hours of work it took the remixer(s) to put it together. For example, former NBC anchor Brian Williams's broadcasts have been remixed by Jimmy Fallon's *The Tonight Show* staff numerous times to make Williams appear as if he is rapping famous hip-hop songs.[61] These remixes are not only entertaining but also a testament to the tedious process of discovering the appropriate vocal snippets from hundreds of hours of film and compiling them into a two-minute rap. Further complicating the process, in the "Rapper's Delight" video at least,

Williams needs to appear to rap colloquialisms that certainly would never appear on an evening news broadcast.[62] The remixers need to edit the clips with such precision that the words would sound appropriate for their new musical context. Additionally, the rap includes other NBC news anchors and commentators in supporting roles, sometimes rapping in unison with Williams. Part of the delight of the clip certainly is the incongruity of seeing Brian Williams rap the iconic song; another part of the delight stems from knowing that a person (or team of people) would go through such effort to make the clip. Although a "real" clip of Williams rapping the song would certainly entertain as well, the remix is more memorable because of the levels of *prosōpopoeia* at work; the remixers have imposed their own voices by selecting the Sugarhill Gang song, the clips of news broadcasts, and Williams himself as the unwitting rapper. The projection of voices and *ēthē* in the remix are imbued with more aesthetic value (hence more *prosōpopoetic* power) because of the painstaking process of finding and remixing the appropriate samples. Hence, like *śūnyatā*, the process of getting there is even more impressive than the remix understood on its own merits.

Outro

> That something is empty of essential lasting identity allows it to be changeful. A remix is not the product of mere promiscuity and simple infidelities. An interesting and compelling remix, for example, does not just fool around with different sounds. It is instead a highly calculated practice that is exceedingly attentive to its source material. That all things were empty meant that all things were mutually (and thoroughly) dependent. *Prosōpopoeia* is predicated on the act of selecting and evaluating the sample, but it is equally dependent on the recontextualization of the sample. Matching is the modus operandi for most sampling, remix, and mash-up productions. Sampling practices put the voice into conversation with its new speaker, creating a synthesis that exceeds the sum of its parts.[63]

Remix is one of the most common practices of the networked era.[64] Its pervasive culture did not arise ex nihilo, however. Rather, remix represents an amalgamation of persuasive tropes, writings, and philosophies from secular and religious societies spanning both hemispheres and thousands of years. Linking these disparate elements is belief in the never-ending potential of the text, empty of essence, ready to receive a voice. Poet Emily Dickinson once ruminated about this vivacity of language, declaring that "A word is dead when it is said/some say/I say it just begins to live/that day."[65] Sonic artist Christian Marclay applies the same sentiment to remix: "recorded sound is dead sound, in the sense that it's not 'live' anymore. . . . I'm trying to bring it back to life through my art."[66] Following Marclay's articulation of *śūnyatā*, remix culture considers a text to be lacking essence when it is not in flux. Through the digital facilitation of *prosōpopoeia* made possible by remix, the once-dead text is transformed, thereby filling its emptiness, or lack of vivacity, with new life. That new essence, in turn, fends off further "death" in favor of an encompassing and theoretically never-ending cycle of textual recirculation.

Using the elements of music as its repertory of persuasive resources, the task for the remix artist is to create a coherent text out of many that, by definition, are tremendously fragmented. Therefore, the remix artist will select samples of songs, videos, or images that provide a starting place for a transformative articulation of the available material. To attempt this creative task, the remix artist must envision the texts' potential for transformation and, thus, sees the raw materials as a palimpsest; the materials being sampled already contain the essence of the soon-to-be text but also absent of the social (and legal) imposition of completeness.

Still, the vocalization that imbues absent or inanimate form with essence, *prosōpopoeia*, only partially describes the rhetorical function of remix. If a rhetor is giving voice to another, it is because that other is not present or cannot speak. This practice of assuming the voice of the absent is heuristically valuable for contemporary discourse. It can be considered a social practice intended to grant voice to marginalized, incapacitated, or deceased populations; this rhetorical maneuver can be empowering or forbidding.[67] But certain conditions must exist for it to take place. Anciently, Cicero could only grant voice to Clodius because there was no essential voice of Clodius to hear in the trial. In the ancient case's networked counterpart, Nike gave Earl Woods a voice because his opportunity to share it had passed. Put differently, *prosōpopoeia* is only plausible for an audience if there is no essence of the voice being projected.

Prosōpopoeia, however, is not theoretically sufficient to account for the essencelessness of texts—a crucial dimension of mash-up logic. For this reason, a turn to the ancient Buddhist tradition of *śūnyatā* is needed to complement the discussion. For a text to be remixed, it must first be simultaneously considered as lacking essence and yet as entirely interdependent with all other texts. Remix projects voices onto texts that are perceived as lacking their own. These essenceless texts are considered discrete and complete. Much as *prosōpopoeia* projects form onto an absent individual by giving him or her a voice, remix reanimates its samples by mashing them up in novel ways. In other words, digital *prosōpopoeia* is possible because the text is philosophically bracketed by *śūnyatā*.

It is *śūnyatā* that considers the finished text empty of essence, because its constituent hooks and melodies have not yet been sampled. In other words, a completed text is dead because it is brought to life only through its artistic appropriation. In so doing, remix injects the text again with a contingent essence. Like a seed's relationship with its nourishing surroundings, the song's constitutive musical elements are all interdependent, as are the entire population of human creative texts. Therein lies another useful element of *śūnyatā* heretofore unexplored in this discussion: its utility in understanding the tension between copyright regulation and remix culture.

Through envisioning all texts as interdependent, *śūnyatā* flattens authority and multiplies absolutes. If everything is contingent and thus devoid of "inherent existence," then authority and absolute values become meaningless. The never-ending potential of the text, illuminated by remix, diverges from cultural norms once initiated by print. Print culture attempted to restore essence by implementing the notion of ownership and intellectual property;[68] remix culture subverts print culture by drawing attention to the essencelessness of the completed text, refiguring *prosōpopoetic* appropriations to amplify the inventional practices of the remix artist.

Analogous to the Platonic view of substance, printing-press culture placed a higher value upon a canon of texts that embody some form of originality and, thus, authority.[69] Print culture initiated the very ideals of substance, being, and finality that *śūnyatā* seeks to subvert. Like the seed caused to grow through its aggregation with the conditions of the sun, water, and soil, *śūnyatā* would consider any text part of an interdependent aggregation with all other texts. Because this intertextuality brings to the fore the metaessence of all texts, they are also therefore essenceless. Thus, a song absent essence or authority cannot lose its essence or authority if it is digitally sampled for use elsewhere.[70] If the presuppositions of *śūnyatā* are indeed the case with digitally remixed music, then the contemporary intellectual property debate spawned by print culture becomes effectively nullified. Remix depends on the appropriation of samples and their subsequent *prosōpopoetic* deployment. A prior understanding of *śūnyatā* can elucidate this process and offer insight into the competing positions in the debate.

Notes

1. This paragraph is a discursive remix. For the most part, the sources quoted are included verbatim here and throughout the interludes in this chapter. In some

cases, I have changed some of the transition words or names to emphasize the interchangeability of the conceptual terms. This paragraph is compiled from the following sources: David J. Gunkel, *Of Remixology: Ethics and Aesthetics After Remix* (Cambridge, MA: MIT Press, 2016); Eduardo Navas, *Remix Theory: The Aesthetics of Sampling* (New York: SpringerWien NewYork, 2012); Richard L. Edwards, "Remixing with Rules: Constraint and Potential in Restrictive Remixes," in *Sampling Media*, ed. David Laderman and Laurel Westrup (New York: Oxford University Press, 2014), 31–42; John Seabrook, *The Song Machine: Inside the Hit Factory* (New York: W. W. Norton, 2015); Jesse Stewart, "Articulating a Hip-Hop Sampling Aesthetic through Film: Music and Memory in Three Hip-Hop Documentaries," in *Sampling Media*, ed. David Laderman and Laurel Westrup (New York: Oxford University Press, 2014), 143–54; Barry Mauer, "Rigorous Infidelity: Whole Text Sampling in the Curatorial Work of Henri Langlois, Dewey Phillips, and Jean-Francois Lyotard," in *Sampling Media*, ed. David Laderman and Laurel Westrup (New York: Oxford University Press, 2014), 60–72.

2. Horace, *The Art of Poetry*, trans. James Hynd and Burton Raffel (Albany: SUNY Press, 1974), 59. In writing about the ephemerality of voice, Horace also pronounced: "Dead words shall live/and live words shall die/and only the mouths of men can decide/only what's said is said/and therefore alive. And therefore correct." (12).

3. *Rhetorica ad Herennium*, trans. Harry Caplan (Cambridge, MA: Harvard University Press, 1954), 399.

4. Quintilian's works offer examples of Roman literature that conflate both terms. See Quintilian, *The Orator's Education, Books 9–10*, trans. Donald A. Russell (Cambridge, MA: Harvard University Press, 2001), 51.

5. For example, Megan Foley addresses *prosōpopoeia*, "the trope of giving voice to a voiceless body," in a democratic context by arguing that it represents the citizens by projecting their voice. More specifically, it described how the trope of "voice" was materially projected to bodies incapable of speaking, like the incapacitated Terri Schiavo. See Foley, "Voicing Terri Schiavo: Prosopopeic Citizenship in the Democratic Aporia between Soverignty and Biopower," *Communication and Critical/Cultural Studies* 7, no. 4 (2010): 381–400, doi:10.1080/14791420.2010.523433.

6. Cicero, *Orator*, in Cicero, *Brutus. Orator*, trans. G. L. Hendrickson and H. M. Hubbell (Cambridge, MA: Harvard University Press, 1939), 369. Quintilian, *Orator's Education, Books 9–10*, 51. The cited numbers refer to the book, chapter, and line of the text, respectively.

7. According to the writings of the progymnasmatists, there are other types of speech for the absent as well, like *eidōlopoeia* (image, idol, or ghost making) and *ēthopoeia* (character making). See *Progymnasmata: Greek Textbooks of Prose Composition and Rhetoric*, trans. George Kennedy (Atlanta, GA: Society of Biblical Literature, 2003).

8. Gunkel, *Of Remixology*, 12–14.

9. Navas, *Remix Theory*, 66.

10. According to Stefan Sonvilla-Weiss, this is at the core of how mash-ups work: "Mashups . . . put together different information, media, or objects without changing

their original source of information, i.e., the original format remains the same and can be retraced as the original form and content, although recombined in different new designs and contexts." Stefan Sonvilla-Weiss, "Introduction: Mashups, Remix Practices and the Recombination of Existing Digital Content," in *Mashup Culture*, ed. Stefan Sonvilla-Weiss (New York: Springer Wien New York, 2010), 9.

11. Kembrew McLeod and Peter DiCola, *Creative License: The Law and Culture of Digital Sampling* (Durham, NC: Duke University Press, 2011), 66. This process here, however, is not exactly congruent to strategies of argumentation. While it is similar in the way that argumentation uses portions of arguments that directly complement those being posed by the rhetor, the relative obscurity of crate digging seems counterintuitive to the *ethos* that needs to be transmitted in argumentation, namely, that it is beneficial to one's credibility if he or she can successfully draw from arguments from reputable sources. Regarding our discussion here about the allegorical properties of remix, DJs generally sample from well-known popular hits to stoke the excitement of the audience. But "crate digging" is different, in that the rare or obscure songs are being rediscovered and deployed. Still, this creates a perverse sense of *ethos* that permeates internet culture. To be deemed worthy to be circulated, internet memes need to be sufficiently funny, fascinating, or shocking—despite their source. In the same way, according to the doctrine of crate digging, if a song is too good to leave untouched, despite its obscurity, it can provide the sampling DJ with a measure of respect for finding it in the dusty confines of retail bins.

12. Indeed, even this idea is sampled. Other examples of discursive remixes may be found in Jonathan Lethem, "The Ecstasy of Influence: A Plagiarism Mosaic," in *Sound Unbound: Sampling Digital Music and Culture*, ed. Paul D. Miller (Cambridge, MA: MIT Press, 2008), 51; and Scott H. Church, "All Living Things Are DJs: Rhetoric, Aesthetics, and Remix Culture" (PhD diss., University of Nebraska-Lincoln, 2013), 1.

13. For a book linking a similar argument to the history of the footnote, see Anthony Grafton, *The Footnote: A Curious History* (Cambridge, MA: Harvard University Press, 1999).

14. This paragraph is compiled from the following sources: Gay Watson, *A Philosophy of Emptiness* (London: Reaktion Books, 2014); Marcus Boon, *In Praise of Copying* (Cambridge, MA: Harvard University Press, 2010); Sor-Ching Low, "Romancing Emptiness," *Contemporary Buddhism* 7, no. 2 (2006), doi:10.1080/14639940601025098.

15. See Boon, *In Praise of Copying*, 27.

16. Jeffrey Hopkins, *Meditation on Emptiness* (Boston, MA: Wisdom Publications, 1996), 404.

17. Watson, *Philosophy of Emptiness*, 15.

18. David Ross Komito, *Nāgārjuna's* Seventy Stanzas*: A Buddhist Psychology of Emptiness* (Ithaca, NY: Snow Lion, 1987), 68.

19. "If we examine causes, conditions, and their aggregations we discover that their individual existence is unfindable because they can not be separated and still retain their own natures in dependence on each other": Komito, *Nāgārjuna*, 102.

20. Michael G. Barnhart, "*Śūnyatā*, Textualism, and Incommensurability," *Philosophy East and West* 44, no. 4 (1994): 649, doi:10.2307/1399756.

21. Komito, *Nāgārjuna*, 102.

22. Watson, *Philosophy of Emptiness*, 10, 14.

23. Fernando Tola and Carmen Dragonetti, "Nagarjuna's Conception of 'Voidness' (Sunyata)," *Journal of Indian Philosophy* 9, no. 3 (1981): 278, 279. This claim appears to be contradicted by Sor-Ching Low, who considers experimental composer John Cage's appropriation of Zen Buddhism in his art offensive because it failed to take "the full religious import of sunyata" into account. See Sor-Ching Low, "Romancing Emptiness," 139. Meanwhile, Barnhart clearly addresses the accusation that Nāgārjuna attacked Buddhist doctrines with *śūnyatā*: "Nāgārjuna devotes an entire chapter to rebuffing the claim that *śūnyatā* disables . . . such truths, Nāgārjuna argues quite to the contrary that only in the equation of *śūnyatā* and *pratītyasamutpāda* can we hope to give these truths any alethic status." See his "*Śūnyatā*, Textualism," 651.

24. Martin Irvine, "Remix and the Dialogic Engine of Culture," in *The Routledge Companion to Remix Studies*, ed. Eduardo Navas, Owen Gallagher, and xtine burrough (New York: Routledge, 2015), 30.

25. Ibid., 30.

26. This paragraph is compiled from the following sources: Kristine S. Bruss, "Persuasive *Ethopoeia* in Dionysius's Lysias," *Rhetorica: A Journal of the History of Rhetoric* 31, no. 1 (2013): 34–57, doi:10.1525/RH.2013.31.1.34; Gunkel, *Of Remixology*; Alicia D. Myers, "Prosopopoetics and Conflict: Speech and Expectations in John 8," *Biblica* 92, no. 4 (2011): 580–96.

27. Colin Davis, "Can the Dead Speak to Us? De Man, Levinas and Agamben," *Culture, Theory & Critique* 45, no. 1 (2004): 79, doi:10.1080/1473578041000l686469.

28. Diana Fuss, "Corpse Poem," *Critical Inquiry* 30, no. 1 (2003): 1–30.

29. Ibid., 1.

30. Giorgio Agamben, *Remnants of Auschwitz: The Witness and the Archive* (New York: Zone Books, 1999), 34. For further discussion on the controversy of imposing one's voice over another's, see Linda Alcoff, "The Problem of Speaking for Others," *Cultural Critique* 20 (1991–1992): 5–32, doi: 10.2307/1354221.

31. *Rhetorica Ad Herennium*, 401.

32. Bruss, "Persuasive *Ethopoeia*," 35. In the writers of the *Progymnasmata*, Hermogenes is explicit about the difference between the variants: "The difference is clear: in [*ēthopoeia*] we imagine words for a real person, in *prosopopoeia* we imagine a non-existing person." *Progymnasmata*, trans. Kennedy, 84. George Kennedy, however, opts to follow Theon's path and not distinguish between both, nor do other recent expert commentaries. The parlance of this chapter will also follow this well-trod path, despite the potential taxonomic differences introduced by these ancient writers. Although clear differences between these terms will emerge in this discussion, the definitional distinction offered by Hermogenes is not so distinct when applied to remix; remix not only "speaks" the absent recording artists' songs, including artists who have long passed on, but also breathes life into previously closed songs or

videos, thereby vivifying them. See also *Progymnasmata*, trans. Kennedy, 47; Myers, "Prosopopoetics," 580–96.

33. R. G. Austin, trans., M. Tulli Ciceronis' *Pro M. Caelio Oratio*, 3rd ed. (London: Oxford Press, 1960), 90.

34. So distinguished was Appius that the Appian Way in Rome would be named for him.

35. Cicero, *Pro Caelio*, in *Cicero XIII*, trans. R. Gardner (Cambridge, MA: Harvard University Press, 1970), 445, 447.

36. See Cicero, *Pro Marco Caelio*, ed. Andrew R. Dyck (New York: Cambridge Press, 2013), 20–21. Also Ciceronis, *Pro M. Caelio*. Later, Quintilian approves of Cicero's stratagem to change his tone to fit the character: "Publius Clodius and Appius Caecus are imagined very differently" (11.1.39). Quintilian, *The Orator's Education, Books 11–12*, trans. Donald A. Russell (Cambridge, MA: Harvard University Press, 2001), 29.

37. Cicero, *Pro Caelio*, trans. Gardner, 451. Considering the historical context, this statement is particularly cutting because it was doubly defamatory toward Clodia and her brother. Cicero certainly did not consider Clodius, his political enemy, to be perfect. Nor did Cicero mean that Clodius loved his sister in the familial sense; rather, this statement was scandalous, a euphemism for the rumored incestuous relationship between the two siblings. Indeed, Cicero wittily hinted at the gross impropriety earlier: "Were I not hindered by my personal enmity to that woman's husband—I meant to say brother; I always make that slip" (13.32). Cicero, *Pro Caelio*, 445. If even her brother, the "petty gangster," considered Clodia's actions inappropriate, as Cicero's masked statement makes clear, her transgressions must have been quite egregious. W. M. F. Rundell, "Cicero and Clodius: The Question of Credibility," *Historia* 28, no. 3 (1979): 301, 318.

38. Quintilian, *The Orator's Education*, *Books 11–12*, 313.

39. Austin, *Pro M. Caelio*, 91.

40. This paragraph is compiled from the following sources: Foley, "Voicing Terry Schiavo"; Bruss, "Persuasive *Ethopoeia*"; Navas, *Remix Theory*; Gunkel, *Of Remixology*.

41. "Tiger Woods Commercial: Earl and Tiger," accessed August 24, 2016, https://www.youtube.com/watch?v=5NTRvlrP2NU.

42. Portions of the audio were taken from a 2004 interview with Earl Woods, as displayed in "Tiger: The Authorized DVD Collection." Russell Goldman, "Tiger Woods Ad Ripped Audio from 2004 Documentary," *ABC News*, April 8, 2010. http://abcnews.go.com/Business/tiger-woods-ad-ripped-audio-2004-documentary/story?id=10323878.

43. Bruss, "Persuasive *Ethopoeia*," 38.

44. Ibid., 50.

45. Ibid., 51.

46. Austin, *Pro M. Caelio*, 90; Quintilian, *Orator's Education, Books 11–12*, 29–31: "The same differentiation must be observed. This is because we use imaginary per-

sons and speak as it were with other men's lips, and so have to provide the appropriate personalities for those to whom we lend our voice. . . . In Prosopopoeia we simulate the emotions of children, women, nations, and even things which cannot speak, and they are all entitled to their appropriate character" (11.1.39–41).

47. Bruss, "Persuasive *Ethopoeia*," 44.

48. "Different styles of eloquence too are appropriate to different people. . . . It is the same too with clothes: old age does not go well with the brilliance of purple and scarlet" (11.1.31); Quintilian, *Orator's Education, Books 11–12*, 25.

49. Eduardo Navas argues that remixes are necessarily allegorical because "their effectiveness depends on the recognition of pre-existing recordings." Navas, *Remix Theory*, 96. This allegorical potential of remixes offers dual readings characterized by the tension between the sampled materials and the mash-up. Navas, *Remix Theory*, 43.

50. *Ibid.*, 96.

51. Regardless of its reading, this is an example of a rhetorical phenomenon that occurs in remix that I call elsewhere "possession by misnomer." As a terminological mash-up of Kenneth Burke's "exorcism by misnomer," this term refers to the "conversion upward" that takes place when one orator's text is mashed up with another orator's text, thus earning the latter's text an extra measure of attention by virtue of its association with the former's text. See Church, "All Living Things Are DJs," 220.

52. John Shiga, "Copy-and-Persist: The Logic of Mash-up Culture," *Critical Studies in Media Communication* 24, no. 2 (2007): 96, doi:10.1080/07393180701262685.

53. For one of the most musically talented remix artists on the internet, please see Andy Rehfeldt's YouTube channel, which showcases such videos as "Judy Garland Sings Death Metal" and "Metallica-Enter Sandman (Smooth Jazz version)." Although he does not collect and mash up samples of recorded music, like other remix artists do, he dedicates his considerable musical talent to performing the incongruous generic musical background while retaining the original vocals. In other videos, he retains only the video and title, while recording a new death metal backing track and employing the guttural death metal growl as the new vocal.

54. The ancient rhetorical texts strongly caution against incongruous representations with *prosōpopoeia*. However, as the trope's parameters of decorum expanded from its transepochal applications, other methods for its use materialized. As would be the case with digital remix composition centuries later, musicians and composers would use incongruous representations as *prosōpopoeia*. Funeral music during the Baroque period, for example, would personify the dead through their placement in the church during the service or by the timbre of their voice. However, Gregory S. Johnston has documented how the seventeenth-century German composer Gottfried Ernst Brechthold personified a deceased adult man through a solo performance. Interestingly, rather than asking a tenor or bass with a lower voice to personify the deceased, a much higher soprano or alto voice would be chosen to represent him. Johnston speculates that the incongruity of the representation may have provided a more ethereal or angelic personification of the deceased to the au-

dience. Gregory S. Johnston, "Musical-Rhetorical *Prosopopoeia* and the Animation of the Dead in Seventeenth-Century Funeral Music," *Canadian University Music Review* 10, no. 1 (1990): 28.

55. Watson, *Philosophy of Emptiness*, 170.

56. Tola and Dragonetti, "Nagarjuna's Conception."

57. Diané Collinson, Kathryn Plant, and Robert Wilkinson, *Fifty Eastern Thinkers: Routledge Key Guides* (New York: Routledge, 2000), 95.

58. Tola and Dragonetti, "Nagarjuna's Conception," 276–77.

59. Ibid., 275; Hopkins, *Meditation*, 494.

60. Tola and Dragonetti, "Nagarjuna's Conception," 276; Hopkins, *Meditation*, 401. This notion of a "destructive dialectic" contains provocative parallels with a classical sophistic practice called *dissoi logoi*. Although both the Sophists and the Mādhyamaka School addressed a similar deliberative phenomenon, they conflicted in how they conceptualized the purpose of countervailing arguments. An ancient Greek fragment, *dissoi logoi* presupposes that argumentation is made more robust when the rhetor addresses both sides of the debate, often with the pedagogical objective of making the weaker argument appear stronger. However, the Mādhyamaka School considers opposition to be less ontologically defined. To the followers of Nāgārjuna, the opposition, rather, is the perspective of "concealment reality" held by the interlocutor instead of one that can be discursively articulated. Although each school holds a similar objective, the goal of *dissoi logoi* is perhaps more easily obtainable because the task is more discursive than ontological. Hence, *dissoi logoi* aims for persuasion through perspective-taking engagement; the more difficult perspective-shattering process that leads to *śūnyatā* can be fructified only through a sustained process of dialectical "abolishment" and "destruction." For more on *dissoi logoi*, see Rosamond Kent Sprague, trans., "Dissoi Logoi or Dialexeis," in *The Older Sophists: A Complete Translation by Several Hands of the Fragments in Die Fragmente Der Vorsokratiker*, ed. Rosamond Kent Sprague (Indianapolis, IN: Hackett, 1972). I offer my thanks to Damien Smith Pfister for calling my attention to these parallels.

61. These clips have showcased Williams "rapping" to songs such as Sir Mix-a-Lot's "Baby Got Back," Snoop Dogg's "Gin and Juice," and Sugarhill Gang's "Rappers Delight."

62. "Brian Williams Raps 'Rapper's Delight,'" accessed June 2, 2015, https://www.youtube.com/watch?v=7CYJ73pVpVc.

63. This paragraph is compiled from the following sources: Watson, *Philosophy of Emptiness*; Gunkel, *Of Remixology*; Barnhart, "*Śūnyatā*, Textualism"; Laurel Westrup, "Thinking through Sampling, Literally," in *Sampling Media*, ed. David Laderman and Laurel Westrup (New York: Oxford University Press, 2014), 228–46.

64. This claim refers to remix in all its iterations. Science fiction writer William Gibson explains, "The remix is the very nature of the digital. . . . The recombinant (the bootleg, the remix, the mash-up) has become the characteristic pivot at the turn of our two centuries." Quoted in Gunkel, *Of Remixology*, xvii.

65. Quoted in Fuss, "Corpse Poem," 28.

66. Christian Marclay, quoted in *Audio Culture: Readings in Modern Music*, ed. Christoph Cox and Daniel Warner (New York: Continuum, 2004), 327.

67. For more on the trope's potential for empowerment, see Foley, "Voicing Terry Schiavo." For more on the trope's potential to preclude others from speaking, see Johanna Hartelius, "Face-ing Immigration: *Prosopopeia* and the 'Muslim-Arab-Middle Eastern' Other," *Rhetoric Society Quarterly* 43, no. 4 (2013): 311–34, doi:10.1080/02773945.2013.819990.

68. Thomas Pettitt, "Bracketing the Gutenberg Parenthesis," *Explorations in Media Ecology* 11, no. 2 (2012): 102, doi:10.1386/eme.11.2.95_1.

69. Ibid., 102.

70. Boon, *In Praise of Copying*, 27.

11

The Jaina Rhetoric of Nonviolence and the Culture of Online Shaming

Scott R. Stroud

"Dongle." This neologism from the technology world refers to a software key that is physically inserted into a computer or to labeling an add-on device to create network connectivity. Heard with a certain mindset, it can induce giggling and sexualized imagery. In some contexts, it could get you fired. That is what happened to one of two software engineers who, at the programming convention PyCon in 2013, began bantering in a sexualized way about "big dongles" and "forking a guy's repo [software repository]."[1] "Forking" is programmer-speak for a division or a "forking of the paths" in the use of software code or ideas. Adria Richards, a "developer evangelist" for a software company, was sitting in front of these two men. Richards had experienced firsthand the challenges that women faced in the programming world, knew of the emphasis that PyCon was putting on including and encouraging women, and concluded that this use of sexualized humor worked at cross-purposes with these goals. So, she did what many do now in a world infused with smartphones: she turned to Twitter, the popular microblogging application, to try to digitally right offline wrongs. She quickly shot out a tweet—properly hashtagged so as to alert PyCon organizers and other attendees—that said: "Not cool. Jokes about forking repo's in a sexual way and 'big' dongles. Right behind me." She also included a picture of the smiling men. A second tweet asked the public audience attending to this message in a technological bottle: "Can someone talk to these guys about their conduct?" She followed this with a description of her seating location in the auditorium. Eventually, she found a representative of the conference and identified the two men, who were subsequently asked to leave the event.[2]

Like so many moments in a world increasingly structured by digital communication networks, however, nothing ended up where one might expect. Eventually, a boss of one of the joking men saw this picture and fired the programmer because of the negative publicity from the Twittersphere toward this instance of putative misogyny. The fired programmer posted—under an assumed name—on "Hacker News" that he was sorry for his utterance, but criticized Richards's very damaging way of handling the situation. Instead of directly addressing the matter with the offenders, Richards had turned to a very public way of shaming identifiable individuals to correct what they thought was a minor infraction. This perspective inflamed segments of the internet and programming worlds, leading the anarchic online community "4chan" to organize efforts to get Richards fired from her job. Such efforts involved large masses of passionate, anonymized individuals pursuing a project they saw as aligned with the public good. Richards was eventually fired, and an online posting supposedly authored by her contained expressions of remorse about how the situation was handled.[3]

Whether one agrees more with the group publicly shaming an instance of misogynistic joking or with the group systematically complaining to and boycotting Richard's company, the issue of mobilizing intricate and damaging campaigns via online communicative means presents an opportunity to question the ethical worth of emerging digital habits of using approbation for putatively good causes. While Richards and the photographed men were identifiable individuals, many of those involved in shaming companies and unethical programmers were anonymous or pseudonymous. Our online identity is inherently changeable and thereby can occlude our offline selves.[4] In the most extreme cases, we can be something radically different online than how we and others identify our offline selves. This is a double-edged sword: Some have noted the benefits *and* drawbacks of pseudo-anonymity for web forums, tweeting, and blogs.[5] It allows us to explore aspects to our present or future selves that may be repressed or threatened by dominant societal forces, such as sexual preferences that large parts of a given society may not condone or approve of. On the other hand, anonymity may encourage us to be vicious and hurtful toward others. What this present chapter hopes to bring out are the ethics of the rhetorical practices implicated when online selves attempt to cast a shameful light on others and thereby harm them for what they judge to be a good cause, whether it is in regard to a tasteless Halloween costume, the killing of a beloved lion on a trophy hunt, the public naming of supposed adulterers, the shaming of bullies, the outing of supposed pedophiles, or the many other modern examples of crowdsourced condemnation.[6]

Public shaming is nothing new. Whether it is in the form of stockades or pillories or visible identifying labels, it exists to alert and involve others in harming the individual in an attempt to get them to change their supposedly evil ways. I will argue that online shaming seems to be ethically suspect because of the forces it marshals. Given the harm it brings to an accused individual, one could see it as a sort of violence wrought with symbols. Building on this concept of symbolic violence, my investigation into online shaming is grounded in an engagement with a non-Western tradition that has explored the meaning of violence and nonviolence for well over two thousand years: the Jainas. The Jainas are an important intellectual group in the Indian subcontinent that have negotiated elaborate belief systems devised by various strands of Hinduism and the reactionary movement of Buddhism. The Jainas offer a case study of a minority group that must survive through pluralistic tolerance mixed with an assertive individuation of its doctrinal claims from majority systems of thought. More important than its example are its doctrines. Jainism so extensively elaborated and practiced the concept of nonviolence (*ahimsā*) that it was a major influence on one of the twentieth century's icons of nonviolence Mahatma Gandhi.[7] This chapter will examine the Jaina rhetoric of nonviolence and analyze online discourse in light of a derived notion of symbolic violence. Using the Jaina theory of passions as inherently self-harming, digital discourse used to marshal public shaming is worrisome because of the self-focused passions and ego-bolstering that it entails. From the Jaina perspective, violence is not primarily worrisome because it results in actual harm to others. It is undesirable because it involves intense, unenlightened, self-focused passions.[8] Violence is thus the harming of one's self through self-focused, distorting desires that may or may not culminate in overt action. One can apply this account of violence to symbolic actions to certain online interactions, as the extreme thoughts and communicative acts that comprise online shaming represent instances of violence and a problem we must address in creating rhetorically competent and ethical agents. Put simply, Jainism gives us a radically novel way to diagnose what is wrong with so much of the online discourse we find ourselves hating or loving.

Shaming in Digital Forums

Let us start this exploration by noting a few typical categories of using shaming—or the casting of some negatively judged agent in a socially undesirable light—to motivate online courses of action. The creation of negative consequences for this other agent is considered an integral part of public campaigns of shaming. They deserved what happened to them, or so the shaming agents think. Often this is couched in the verbiage of "justice," but often it is simply noted as revenge for perceived wrongs or character flaws. In many cases, online revenge is very harmful. Take, for instance, the modern issue of "revenge pornography": the nonconsensual posting of identifiable nude images of a former relational partner. These are sometimes, but not always, accompanied by accusations of relational transgression or negative character judgments. Such embarrassing or negative publicity sometimes results in the loss of one's job, of future relational partners, and of one's dignity.[9] These are often motivated by the pursuit of revenge for perceived relational wrongs.[10] Few would say it is ethical to harm a former romantic partner in this manner. But there are tougher cases in which one may find more widespread support for the idea that the acts of a certain agent merit a type of harmful consequence. In other words, ethical duty or moral considerations are said to justify a certain sort of harm being attached to an infringing action, even if the state and its laws will not do so. This is the case in instances of public shaming, in which otherwise legally allowable speech or action is connected to negative consequences through the speech of others out of a sense of unrequited justice. Such public shaming becomes online shaming when it uses the modalities of digital communication networks to gather interested parties, spread information, and harm the targeted individual. The evocation of shame through digital networks is an ethical concern, as it travels where the law allows one to tread rhetorically but where conflicting ethical reasoning often stimulates debate over the rightness of such actions. To put it simply, while we are *legally* allowed to use our online rhetorical activity to bring shame and disrepute on those whom we think deserve it, are we *ethically* justified in doing so?

Even though there is a range of cases of online shaming that could be brought forth, one contemporary example is particularly noteworthy.[11] The Tumblr blog *Racists Getting Fired* is an exemplar of the ways that contemporary activists and online agents meet to pursue their own views of justice and ethical duty.[12] This blog exists to expose those who post racist comments and content on social media outlets such as Facebook or Twitter. The blog took off after the shooting of Michael Brown in Ferguson, Missouri, and the consequent protests in which the protestors accused the police and government of being racist. Many individuals heatedly expressed sentiments for and against the protestors on social media. Not all these utterances were civil, and many of them could be judged as racist in content. This Tumblr blog found a role in public discourse: It serves as a clearinghouse for individuals to post "captured" images of racist comments and content for others to see. Beyond this publicity was the additional impact of the racist labeling and shaming. As Soraya Nadia MacDonald puts it, "Racists Getting Fired doesn't just publicly shame—it adds consequences by rounding up those willing to call a business to say they don't want to patronize a place with an employee who says things like '#Ferguson one less n– – on #foodstamps.'"[13] The blog allows a grouping of others—mostly anonymous individuals—to contact a putative racist's employer and request that an accused individual be fired. Thus, this blog goes beyond the tactics of the "Yes, You're Racist" Twitter account, which simply relies on shaming racists through retweeting offending material. *Racists Getting Fired* marshals a crowd of enthusiastic others to deprive the racist individual of his or her job as a consequence for their racist utterance.

Submitters to this blog are asked to provide links to the offensive posts and to provide a "screen capture" of the original post. The content must be negative in regard to race and will be rejected by the administrators of the *Racists Getting Fired* blog if the content is "not explicitly hateful (e.g., talks down on Ferguson protestors but doesn't refer to race)."[14] Browsing through the content on the site, one sees a range of materials from various sites on the internet, ranging from posts involving the explicit use of racial slurs to comments about all members of a specific minority group, distasteful jokes, or inferences that can be taken as motivated by racism. For instance, one posting "calls out" a tweet by an individual who said of the Baltimore rioters, "I'm sure all those pieces of shit went right to the pharmacy and stole every drug in that CVS."[15] The inference that is made by the submitters is that "pieces of shit" refers to the rioters and that their behavior is being caused by essential racial characteristics. Other examples are clearer, with pictures and words targeting individuals and groups *as* racial minorities. Connected to many posts are responses from other Tumblr users. Some of these report on the efforts to get a person fired or expelled from a postsecondary educational institution. For instance, one comment reads: "So this woman think she can not lose her job. It's crazy in Baltimore right now and She saying racists things. Please call her job. . . . Please REBLOG! We came together before, let's put a end to ignorance! I will be calling the corporate office Monday morning!"[16] After this comment, another Tumblr user adds "DESTROY HER INCOME."[17] In another example, a role-playing game company and writer are posted, along with a picture of their description of a character type modeled after indigenous peoples. Commenters react negatively to this posting: "The level of willful ignorance and desire to profit off the genocide of indigenous people is astounding. The amount of stereotypical amalgamating of such diverse cultures into one RIDICULOUS group that is both plains native and pacific northwest is vomit inducing. I need, NEED, to know. . . . What the HELL are these white people thinking?!"[18] In a more high-profile instance, the University of Oklahoma fraternity members caught on video singing a racist song were quickly outed by name on *Racists Getting Fired*, along with comments such as "The expelled OU racists' names are [. . .]. Spread their names. Make them exist in infamy. Make sure they can never get jobs beyond (maybe) working for their parents."[19] Contact information for individuals and workplaces typically accompanies screen captures of Facebook or Twitter profile pages.

Racists Getting Fired is geared toward mobilizing an acting crowd of persuasive activists. The *Racists Getting Fired* blog gives advice on how to contact employers in an effort to get the accused racist fired. One mentions that he or she follows this protocol: "i personally just ask 'why do you choose to employ such a blatant racist? does your company condone racism? do you only hire racists? do you allow (x-variable poc [person of color]) in your establishment? do you not hire (x-variable poc)?'"[20] Another tip on the *Racists Getting Fired* blog provides a detailed script to use when calling a business:

1. "Hello can I speak to a manager on duty? I have a complaint to make about one of your employees"
2. "Yes hello, my name is ______ and I've been seeing some very disturbing and disgusting racist comments coming from one of your works [*sic*]. Their name is ____"
3. "I know that your business is not one that would condone such behaviour. Considering that this person has your company listed as their workplace, they represent it and the ideals your company was built on. I would like to think that your company isn't one that supports these vile comments and actions, and that's the purpose of my call today"
4. "I was also wondering if there is an email address I can send the proof of these comments too, just to solidify these claims, if you find it necessary"
5. IF THEY SAY THEIR "Hands are tied" say this: "Well if you are not willing to take care of this issue at your level I will be forced to take it to your supervisor and go from there" that usually gets them. And if not ask for their supervisor's number and call them.[21]

This advice is followed up with practical tips for insuring that callers get the desired outcome: termination. One is advised to be calm but insistent: "It is absolutely vital that you CALL BACK to check on the status of the issue. Make sure you hold them to their word and that something is being done. Stay calm and cordial. Channel your inner slightly annoyed PTA soccer mom and you'll be fine."[22]

This path of persuasion appears to be effective. Pages showcase past postings that have been "gotten," the site's parlance for fired or expelled. Other Tumblr users post the responses they have received after contacting places of employment, such as one company's statement that "we thank you for bringing this post to our attention. [. . .] is a zero tolerance corporation and have since released this employee." Another reports, "UPDATE: I just got off of the phone with her school and the dean of admissions has informed me that this [. . .] has been kicked out of college for GOOD; I'm not the only complainer. They received many calls about this girl and they've handled the situation. Nobody's gonna tweet Happy N[. . .] Day and get away with it."[23] Such a blog appears to be justified primarily by creating more tangible consequences for free speech. One is free to say and believe what they want, but others are under no obligation to agree with these statements or sentiments. Nor are others obligated by law or ethics to remain quiet. In this way, one might judge that the tactic used by this blog emphasizes *counterspeech*, or the idea that the solution to offensive speech is more speech. This approach to dealing with hated or unpopular opinions has its roots in John Stuart Mill's arguments against forceful government censorship.[24] The speech used by the blog, however, does not follow the line of truth-searching discourse that Mill would advocate in regard to hated viewpoints. Instead of Mill's desired tactic of using more speech (and not government force) to show how these views are false, a person animated by this blog simply attempts to cast targeted comments in a shameful light (viz., as belonging to the negative category of "racist") and to thereby connect negative consequences to the speaker in question (e.g., to get them fired). Instead of firming up one's grasp on the truth through highlighting the falsity of the opposing viewpoint, the strategy seems to be grounded in the certainty one feels in one's own views and judgments. One knows what racist views are; the task remains to hurt those who utter such thoughts. As one commenter confidently puts it about the opposing viewpoints, "LOL white men like to act tough as shit until they realize that for the first time in their life there might actually be consequences for their actions."[25] The anonymous crowd reaction to one's racist tweet is legally and ethically justifiable, according to this line of reasoning, because it attaches a consequence (losing one's job) that is judged as merited by acting like a racist (through one's public utterances on Facebook, Twitter, and so forth).

If the predominant justification for this site and the activities it encourages is consequentialist—namely, matching a social punishment to the offense of being a racist—what is the nature of ethical objections to this practice? Primarily, the objections are consequentialist as well, pointing out (1) the chance for error in this practice of shaming and (2) the incommensurability between punishment and "crime." Regarding the first objection, one could point to the well-known case wherein a jealous ex-boyfriend faked a social media profile in the name of his ex-girlfriend ("Brianna"), posted racist drivel, and then submitted this forgery to *Racists Getting Fired*. By the time the blog realized it had been fooled, the "smear racked up tens of thousands of reblogs and notes and prompted readers to bombard the real Brianna's workplace with phone calls and tweets."[26] Yet, as the "retraction" notice published by *Racists Getting Fired* makes clear, the fake information and screen captures (with her face and name) remained up on the blog filled with racists: "This post was faked to frame Brianna [. . .]. Do not signal boost this. Do not contact any workplace or school. This was a serious issue and we are issuing a call to tumblr to help heal the harm that Brianna was subjected to and may still be subjected to. We cannot remove the images as we are not the original posters. Please help us report this post to Tumblr to get it removed and out of circulation. We edited the following text to censor the phone number of her employer since this issue has been brought to light to her community." Some would worry about a blog that has the hard-to-undo power to ruin the lives and livelihoods of racists *and* those falsely accused of being a racist. Due process and rights for the accused are protections that defendants have in court, but such jurisprudential safeguards are lacking in "internet court." Second, one could worry that the harm to the accused—even if they are the utterer of a deeply problematic phrase—is out of balance with the offense. Why should one lose their job (and potentially future jobs, given the tenacious followers of this blog) because of one bad joke or one racial slur? Harking back to the opening example, why should the programmers lose their jobs because of that joke and not just be ejected from the conference venue where Richards felt threatened by them? The arguments for or against this practice all seem to be consequentialist in a specific way: they all fixate on the sort of harm we might cause *to others*, justifiably or not, through our rhetorical actions. Is there another way of analyzing this new issue in digital rhetoric?

Jaina Views on Karma and the Solitary War on the Passions

An interesting feature of Jaina thought is its self-focused consequentialism. In the West, violence is typically seen as bad because it represents a harming of the interests of *others*; for the Jainas, violence is bad because it primarily represents *self-harm*. Instead of thinking about sites like *Racists Getting Fired* in terms of the dichotomy of justified or unjustified violence toward others, what insights could we gain if we instead focus on how such sites develop or stunt one's own self-development and cultivation? I will use the ancient system of Jaina thought to sidestep approaches to this question that blindly assume that some people deserve to be harmed and to highlight other wrong-making features of these online shaming practices. To make this point, I first explain a bit about Jainism and then apply parts of its approach to the very modern issue of online mobs and shaming campaigns such as those enabled by *Racists Getting Fired*. At issue here will not be the *legal* concerns with speech but the *ethical* concerns with rhetoric and persuasion: When and how should we persuade others to do things that might harm another person we dislike or with whom we disagree? In pursuit of a response, I will first provide a general sketch of Jaina thought and then propose an account of Jaina rhetorical ethics for our actions in the digital networks in which we so often find ourselves.

Jainism arose, or at least gained its contemporary contours, next to Buddhism as a critique of Vedic thought in India. Instead of privileging rituals and sacrifice (sometimes of animals) as means of procuring favors from the Gods, Jainism and Buddhism focused on rejecting violence and on cultivating one's self. Hinduism was influenced by these critiques, taking on vegetarianism as an expression of nonviolence and changing to the inner focus exemplified by the *Upanishads*. While this is a simplification of the long story of Indian philosophical thought, it provides some historical context to support a sketch of Jainism as a philosophical system. Jainism is said to be at least 2,500 years old, explicated through a line of twenty-four teachers known as *Jinas* (translated as "conqueror" or "victor"). The last in this line of enlightened beings is Vardhamāna Mahāvīra (599–527 BCE), a contemporary of the Buddha.[27] Mahāvīra, like those who followed him in the Jaina tradition, emphasized nonviolence, truth-telling, and self-discipline as vital parts of the path to enlightenment.[28]

What Jaina thought provides is a novel way to critique the violence or harm that occurs in these campaigns of public shaming. Most of us in the West see violence as intentional harm rendered upon another agent. This violence is unjustified when that other person, through his acts or character, does not deserve that harm to their person or interests. In the case of an unanswered wrong against us, violence assumes a balancing role by punishing the other. Such restorative justice can occur within the confines of the law or through extralegal pursuits. Notice that the issue of force and acceptable harm is primarily an ethical issue, and only in some cases is it approached from the legal realm. The primary problem with unjustified violence in the West, however, is the unjustified harm to an individual agent or that agent's interests that it entails; the individual is sacred and must be allowed to flourish in her own life and on her own terms. Happiness occurs when this individual quest proceeds without setbacks.

Jainism, however, gives us a different way to analyze the relation of individuals, harm, and violence. In Jaina thought, the concept of *ahimsā* (nonviolence) is integrally tied to a theory of karma, or the causal conditioning of future experiences by past experiences. For the Jainas, individual souls or selves (*jīvas*) exist. These are coupled with nonsoul (*ajīva*) or material particles called *pudgala*. Karma denotes physical matter that accrues to the *jīva* or self, causing various states of pleasure or pain for the *jīva* and rebirth in future earthly forms. Karmic matter does not lead us to enlightenment; it promises only more temporary pleasures and pains. True enlightenment on the Jaina view, like so many other Indian views, consists in escape from the karma-driven cycle of birth, death, and rebirth. What causes karmic matter to "stick" to a pure soul (*jīva*)? The standard Jaina answer is *himsā* (violence), which can manifest as violence in intent, in execution, or against a large variety of organisms, from the microscopic to the visible (such as other humans).[29] In contrast to the dominant Western conception of violence, the Jain conception of violence orients our attention to harm done to the self of the agent in question.

Given the complexity of Jaina thought and tradition, let us ground our expansion of the harm of violence in word and deed in a determinate text: the *Purusārtha Siddyupāya* by the tenth-century Jaina author Amṛtacandra Sūri. This text extends traditional Jaina thought, but with reference to its employment by householders in everyday life.[30] The *Purusārtha Siddyupāya* is centrally concerned with *ahimsā* or nonviolence. It begins with a foundational question: What is the goal of moral cultivation? The soul must realize that it (the *jīva*) is not that which it sees in the body (*ajīva*). The root of harm is the false conflation of self with empirical world. Thinking that *I* am this specific *body* binds me to this commitment: "The soul, due to its ignorance, since eternity and incessantly, gets conditioned by passions like attachment and is thus the creator of those conditions and also the enjoyer of the fruits thereof."[31] Passions are at the root of the bondage denoted by the karmic attachment of *jīva* and *pudgala* (karmic matter). According to the *Purusārtha Siddyupāya*, the self creates those conditions through illusions fostered by the passions. Thus, the self has the power to undo those illusions. Karma accrues not simply through harming other beings but through the passions present in the self: "As a consequence of the rise of passions like attachment in the soul, physical matter gets bonded with the soul, transforming itself into karmic matter like knowledge-obscuring karma."[32] The power of the soul is not removed from the karmic process; instead, the individual soul or self begins to "buy into" or be persuaded by the illusion that self is the empirical, physically embodied karmic self: "Thus, the soul, although distinct from its modifications due to the influence of the karmic matter including the quasi-karmic matter (particles of matter fit for the three kinds of bodies and the six kinds of completion and development), it appears to be one with these modifications to the ignorant. And this misapprehension is truly the seed of the worldly cycle of births and deaths (*samsāra*)."[33] Thus, the cycle of rebirth driven by karmic material is primarily a cycle of illusion that the self, driven by passions, creates. We believe that we are *this* physical, suffering, and striving individual organism or being. We are thereby bound by this passionate commitment. In a real sense, the self creates its own prison and must find the resolve to undo this subjugation. No one else can do this, since it is a matter of seeing the self and world in a specific way. The Jaina paths to purification (the *vratas*) are then said by the *Purusārtha Siddyupāya* to fixate on the removal of *himsā*. This is not the *himsā* involved in harming other beings that have a right not to be harmed; *himsā* (and immoral acts in general) are violence to one's self through the arising of passions that bind one's self: "All these subdivisions (injury, falsehood, stealing, unchastity, and attachment) are '*himsā*' as indulgence in these sullies the pure nature of the soul. Falsehood etc. have been mentioned separately only to make the disciple understand through illustrations. . . . Acting under the influence of passions, an injury caused to physical or psychical vitalities is deliberate *himsā*."[34] Doesn't *ahimsā* mean not harming or doing violence to other things? This understanding misplaces the focus on harm onto others instead of onto self via the presence of passions. It is passion in the self that causes karmic bondage, not the pain caused to animals or organisms: "Therefore, a person following right conduct, due to the absence of passions like attachment, does not commit slightest of injury (*himsā*) although he may have occasioned severance of vitalities."[35] Harm of others is optional; it is the *illusion* that is reified in actions that harm (or could harm) others that is important, and it is passion that achieves this effect: "Infatuated by passions like attachment, when a person acts carelessly, he commits *himsā*, whether a living being is killed or not."[36] One does not actually have to harm another being for the karmic effects of *himsā* to occur: "In reality, one may be responsible for the sinful act of *himsā* without actually causing injury; the other, while having caused injury, may not be responsible for the act."[37]

It is the passion associated with privileging the self over other beings that accrues bad karma. Thus, we arrive at the core of Jaina *ahimsā*: the avoidance of violence or harm to the self due to that self's passions. Harm to others is optional and always occurs with self-harm: "It is due to the fact that under the influence of passions, the self first injures himself, later on, he may or may not cause injury to other living beings."[38] The focus on violence to others is only an indirect way to focus on self-violence. It is the presence of passion—in intentionally harming others, or intentionally not caring about the harm we may cause—that harms one's self. The self causes harm to itself, and only the self can stop such a process through an individual's war on the passions, the root of *himsā* and karmic bondage.

Passions and the Illusions of Online Shaming

Jaina thought gives us a new way to see these instances of public shaming on the internet as morally undesirable. Something that is undesirable in a moral sense denotes something we want to avoid; the "unethical" points to an action, character, or type of consequence that we see reason to avoid. Thus, the *ethics* of using one's speech to encourage or participate in group shaming of individuals through online means is separate from the *legality* of such counterreactions. Even though it is perfectly legal for me to point out a racist's awful tweets to their employer and voice my opinions about a business that would employ such an individual, *should* I do this? For most of those participating or commenting on the above noted instances, the criterion for the good life seems to be an individually directed sense of flourishing. Racism prevents this for certain groups and the individuals that comprise them, so calling out and perhaps financially destroying individuals who perpetuate systems of oppression is good. It is coordinated action that attempts to achieve the state of affairs—individual flourishing for all the members of some group or community—identified as the guiding ethical or moral end to behavior. Shaming thereby is ethical *if* it contributes to reaching this end. This is the consequentialism inherent in the arguments supporting blogs such as *Racists Getting Fired* and the crowd activities they spawn. Violence done with words is fine, as long as it pushes us toward the right end point.

Jaina thought gives us a new way to think about our rhetorical powers and persuasive activities. Instead of thinking of the consequences to others and whether they deserve the harm or benefit that results, Jaina philosophy emphasizes the results that accrue to ourselves through the state we create in our activities. As noted above, violence (*himsā*) should be avoided but not simply because it hurts a sovereign being that does not deserve to be harmed. For the Jainas, that would leave a class of beings that we could harm: those guilty of wrongdoing and deserving of harm. Jaina thought sees the same problematic feature behind me hurting someone I hate or helping someone I love at the cost of someone else's happiness. Both cases of helping and hurting foreground my passions or desires, which are always self-focused. When I love someone intensely, it is because of their relation to *me*; I fear their loss or their absence, which I will do anything to prevent. In the cases that concern us most—violence to others—what is worrisome is that we value ourselves so much in these actions. In harming, or even thinking about harming, another person, we safeguard our self, we feel better and stronger than the one we have the capacity to harm. Their loss will satisfy us in some mental or physical way. Our violence, in words, thoughts, or physical deeds, increases our passions, which in turn increase our belief that this physical body and its passing interests *are* who I really am. For Jainism, who I am is a *jīva*, something beyond my clothes, car, job, bank account, or family relations for the years comprising this life. All these considerations are important, but they are not cosmically important. And these worldly things can easily derail us on our path to cultivating the right attitude toward our passions and our self, if we let them.

In this Jaina view, rhetoric and persuasion would emphasize the use of symbols to diminish the distance between self and other, since the only other alternative is some sort of self-violence through increased passions. This is the critique of online shaming that Jaina rhetoric proffers: the naming and shaming of unknown others is pursued because of a powerful feeling of self-worth, mission, and value concomitant with the act. These passions can remain—and perhaps be intensified—even if the shaming agent intends to effect significant social change with their patterns of critique. The very act of aggressively and publically shaming others creates strong self-focused passions in the one doing the shaming. Jon Ronson notes this phenomenon in his study of online crowd shaming of individuals.[39] Some group defines those who need to be shamed, and the feedback loops within that group reaffirm the group's virtue and efficacy in pursuing these actions. Those doing the shaming on blogs such as *Racists Getting Fired* are not inclined to radically change course and admit they are wrong or went too far. Instead, they push on with their mission, morally valenced and righteous as the like-minded group defines it. Like the crowds in the 1800s who happened to wander by a criminal being shamed in stocks or being otherwise pilloried, "we have the power to determine the severity of some punishments."[40] Blogs like *Racists Getting Fired* only provide a more effective, more insulated means (due to online anonymity) by which to determine how harshly we will punish the offending other, and they provide access to a network of actors willing to assist in this matter. As Danielle Keats Citron notes, online groups or publics transcend the usual problems in cost and spatial location associated with rallying others to punish a transgressor.[41] Online, collections of passionate individuals can form out of web denizens from around the world and coordinate their efforts on a target in a destructive fashion.

Online shaming is clearly destructive, but are there constructive aspects to this collective activity? Besides harming lives and livelihood, we can see this activity as an exercise in passionate self-building or ego construction. This is a feature of protest rhetoric noted by Richard Gregg, who connects the persuasive force of rhetoric to its "*ego-function*." Trying to discern the logic behind extreme protest rhetoric that denied room for effective change or compromise between opposing sides, Gregg postulates that such rhetoric focuses on "*constituting* self-hood through expression" or on the process of "establishing, defining, and affirming one's self-hood as one engages in a rhetorical act."[42] Thus, even though it is directed at opposing others, such rhetoric is primarily aimed at convincing one's self and allies that one's group and way of being matters. It constructs a self that is respected and valued, and it tends to do this by noting how one's ego (self) is ignored by opposing others and is morally valuable and how the "foreign ego" of those who oppose them is morally inferior.[43] In other words, much protest rhetoric achieves the laudable and useful goal of constituting and bolstering one's self vis-à-vis the asserted immoral selves of the opposition.

Gregg's account explains one function of much protest rhetoric. This account can be applied to the communities being formed by online sites such as *Racists Getting Fired* in their collation of like-minded individuals in reporting and punishing transgressive behavior. The Jaina view discussed here, however, would characterize the self-bolstering that occurs in a moralized quest against wrongdoers (e.g., racists) as problematic. Through public shaming, do we get the sort of self and the sort of relationship with others that we desire? The Jaina perspective would hold that in demonizing the racist other, we are strengthening a false idea of our selves as the ultimate moral judge and punisher; our thoughts and desires are the ultimate arbiter of such matters, and we must find a way to push them on others. We all desire to live in respectful and equal relations among others, in real life and online. Do we achieve this through passionate, violent online rhetoric such as that involved in "getting racists fired?" Most likely not. As Gregg notes about protest rhetoric, "The struggle for a resurrected self seems to be aided by locating other selves, establishing personality typologies among them, and using these as targets for arrows of scorn, ridicule, condemnation, and charges of character defect. . . . The rhetoric of attack becomes at the same time a rhetoric of ego-building."[44] By focusing so much on our self and by amplifying this concern in our rhetorical uplifting of "our team" and our denigration of "their team," we create conditions for more symbolic violence and more passionate selves not coexisting respectfully. We depersonalize or dehumanize our enemies, because our thoughts, judgments, and views all have a passionate, self-focused importance: *I* am right, but *they* are evil or animated by a pernicious ideology. They are not reasonable like me, or they are not mistaken; they are simply a "racist." Such simple, uniform, and inescapable acts of labeling do violence to the other person's humanity, complexity, and capability to reform. There is no attempt to get them to change or to see things with the same justification that we possess. There is simply the urge to ruin them and their livelihood through communicative reactions and to make sure future employers see the discursive horrors that we see when we search the web for their names. All the passions and signs of importance point to my dear self.

Others are selves too, and the Jaina view I have described in this chapter would also worry about what our online shaming practices might cultivate in them. Most likely, those shamed as "racists" will react to the crowd's efforts even more forcefully, or at least withdraw, having learned nothing more than the lesson that their evil opponents can wield force effectively. How we act online in dealing with shameful displays such as a racist tweet not only forms our character but also pushes others to act and think in certain ways. Given the high stakes and the dehumanizing rhetoric employed by those organizing crowd-based shaming, it is likely that little true attitude change will occur in the targeted individual, who next time may simply think, rather than tweet, those awful thoughts. Perhaps they will only become more entrenched in their racism, given the aggressive antics of the "mindless" mob they sense as oppressing them. What we never see on *Racists Getting Fired* is evidence of accused racists repenting or apologizing. Most of the time, the responses we see are defenses of those accused or targeted by this blog. For instance, "angrydaughter1897" rises to defend her mother, who was posted on the site, with the following message: "I ask that you remove a post about my mother from this site. Someone did get her fired and now I have to pay the price. We live in a very small town and this is affecting my employment opportunities and we do not deserve to have to live in deep poverty because of a moment of anger. My mother nor myself are truly racist, despite what she said out of anger. Our lives don't need to be ruined because someone got offended."[45] Of course, this is not an apology; it is merely a request to remove the damaging post from the blog that specializes in marshaling publics to undertake damaging action through publicizing such posts. To those convinced of the guilt residing in the original poster's actions, this reads as merely an empty denial. Thus, the blog responds as follows: "I don't believe that for a second, and I don't care. You're both racist, and I'm glad you have to deal with this because you never gave a shit until now. But People of Color continuously deal with your people's violence against us. We can't escape it; it can't be simply deleted. You want me to remove the post? I'll do so as soon as you end institutional racism. Go back to Europe."[46] I doubt there is any hope for compromise or mutual understanding here, unless it is on the terms of the accusing party. But the inner change that we may be used to connecting with shaming behaviors is not necessary here; those using the blog to shame these supposedly racist individuals may not ultimately "care" about racists coming around to see the error of their ways, since what they stand for and enable harms others. Striking a blow against them—regardless of whether they acknowledge the blow—weakens this system, and that promise is what the activist using such methods feels empowered by. Self-change in accused racists would only be icing on the cake; it is not necessary for those naming and shaming racists. They already are convinced of the truth and power of their judgments, and acquiescence is a secondary point. Such public shaming aims primarily at effecting consequences that the dear self of the activist desires, and only secondarily at rectifying the root cause of the behavior in the transgressor.

What if we *really* wanted to confront the signs of racist attitudes we see in others on social media? The Jaina philosophy would encourage us to pursue this with an eye to making ourselves—and others—less prone to the problems that come with a false identification and valuing of our physical or worldly characteristics. *I* am not my class, race, occupation, nationality, and so forth, so I should not overly value or fixate on these features in my thoughts or in my critiques of others. Yet here I see another person overly valuing their racial identity through a racist utterance. What am I to do? At this point, we can see how the debate between "call out" and "call in" culture connects with the Jaina view of nonviolence in language. Instead of "calling out" someone publically—perhaps through alerting their employer to exact a financial punishment for the offender's putatively racist views—one could "call in" and alert the offending individual to why you find that utterance offensive.[47] As Asam Ahmad puts it, "calling in means speaking privately with an individual who has done some wrong, in order to address the behaviour without making a spectacle of the address itself."[48] Ahmad notes that calling others out immediately creates "in groups" and "out groups" and does nothing to set the stage for cooperation and mutual self-growth. Instead, others become incentivized to remove themselves from you or to fight you more strenuously. In the Jaina sense, violence begets more violence, and strengthening one's sense of self only encourages others to further define their selves as opposed to you. As Ahmad argues, "No matter the wrong we are naming . . . there are ways of calling people out that are compassionate and creative, and that recognize the whole individual instead of viewing them simply as representations of the systems from which they benefit."[49] The Jaina view that I have been elaborating here would largely agree with this use of more private modalities of persuading and reforming an offending speaker.

There are at least three objections to employing Jaina thought to argue for the problematic nature of online shaming. First, one could argue that "calling in" privately does nothing to challenge the power dynamics involved. Perhaps it will change the transgressor, the argument goes, but it will not set back larger systems of racism. A second, and related, objection relinquishes a focus on changing the transgressor and merely wants to use publicity to change elements of the larger culture that might be observing this interaction. Thus, this objection worries that "calling in" will be ineffective in changing the thing such online shamers want to change—discourses of and on racism, sexism, animal rights, and so forth. Notice how both responses see communication between and among humans as consequential; it matters for shaping larger systems of interaction. It is unclear, however, that such objections are decisive in affecting change at such a level. What we know for a fact are the effects on the target's economic status, the reaction their offending utterance elicited in a certain group, and the orientation toward others that they portray. One would be hard-pressed to empirically substantiate the claim that being loud and public in a specific act of criticism changes public discourse, especially when placed alongside other ways of influencing public discourse (e.g., protests, adoption of rhetorical labeling by mainstream politicians, etc.). The person worrying about what the Jainas worry about—the dear self and its quest to always be right—would be concerned with the reactance such strategies sow in the humans targeted for change, as well as the likely change it makes in the habits of one's self. The more one thinks about "systems" or "races," the greater the chance of losing track of the concrete individuality of interaction, conflict, and agency. We detect this habit of harshness when the antiracist snarls at a concerned daughter, "Go back to Europe." Such a rhetorical strategy does little to change the targeted person, and it doesn't seem to highlight a truly egalitarian mindset in the utterer. An overbearing concern with and faith in abstract (and self-empowering) concepts such as those systemic critiques proffer should not eclipse the humanness of ourselves, our passions, and the persons of other communicators.

The third objection follows the previous lines of argument. We may demur to this whole enterprise, saying Jaina thought is a religious orientation that is not required or necessary in how we cut up the world. This may be the reaction of those who imbibe Marxist and critical thought and who frequent such causes as those advanced by online publics like the one that drives *Racists Getting Fired*. Jainism is clearly metaphysical insofar as it pushes claims that are more than empirical, as any rich statements about the nature of the self are bound to be. But I offer two lines of response to those who want to dismiss the Jaina critique. The first wonders about the level of faith and contingent belief that underlies a subscription to a multitude of other ways of taking and talking about the world. It bespeaks an extreme positivism to hold the position that there is a set of views or intellectual orientations that are "given," and a set of others that are only adopted contingently and on faith. Reading Marx, Foucault, or Fanon does not give one more access to the real, according to this response, than do the sacred texts of India. The former texts and traditions may be more accepted or taken for granted in certain interpretive communities and publics, but it does not follow that this popularity necessarily correlates to veracity. Thus, being swayed by the Jaina critique is no more a leap of faith than talking a lot like what one reads in Marx or Mill. A second response acquiesces to the incipient positivism in the objection and breaks Jainism down into a "scientific" view. Such a view would deemphasize soul and instead foreground the creation of a rich empirical self through the habits we develop in our critiques and treatment of others. Such a "naturalized" version of this critique would worry about self-focus less for its metaphysical effects on karma and more for its karmic-like effects on the self it creates here and now. With a bit of synthetic imagination and charity, we can see just such a view peeking through in William James's ruminations on the ethical importance of tending to our habits:

> The hell to be endured hereafter, of which theology tells, is no worse than the hell we make for ourselves in this world by habitually fashioning our characters in the wrong way. . . . We are spinning our own fates, good or evil, and never to be undone. Every smallest stroke of virtue or of vice leaves its never so little scar. The drunken Rip Van Winkle, in Jefferson's play, excuses himself for every fresh dereliction by saying, 'I won't count this time!' Well! he may not count it, and a kind Heaven may not count it; but it is being counted none the less. Down among his nerve-cells and fibres the molecules are counting it, registering and storing it up to be used against him when the next temptation comes. Nothing we ever do is, in strict scientific literalness, wiped out. Of course, this has its good side as well as its bad one. As we become permanent drunkards by so many separate drinks, so we become saints in the moral, and authorities and experts in the practical and scientific spheres, by so many separate acts and hours of work.[50]

One can see here a similar point to the Jaina reading of self and effect. We should worry about our habits of self-aggrandizing and zero-sum critique of others because they seem poised to create the conditions for further conflict in the communities we wish to rectify. We embolden our opponents to fight for their discursive lives all the harder, and we fuel the fires of the dear self in the form of the seemingly always right critic. This is merely the recipe for more and harsher conflict, not for mutual understanding and change that is acknowledged as the road to improved community interactions. On such a naturalized reading, the Jaina has much to offer us in warning about the violence we do to self and others at the hands of the passions we indiscriminately feed in present rhetorical activities of shaming. Should we not consider better, more productive ways to effect change in others *and* in ourselves through our acts of social critique? The Jaina would submit that "calling in" might be the best option to meet these goals.

What Jainism adds to this debate about the use of internet publics to correct and punish disliked activity is the imperative to find a way of respecting and reforming self and other that does not bolster the root cause of oppression or harm: that is, the illusory sense or habits of self that work against moral cultivation or enlightened treatment of others. Compassion for others is merely another way of indicating the practice of nonviolent rhetoric or persuasion. It is a practice that does not elevate one's self to the harmful levels of worth that may be responsible for one's "enemy" behaving in a problematic way. It is a way of persuading others that does not simply push one expression of another's self down, only to have it gush forward in a new geyser of self valuing that is socially undesirable elsewhere. "Calling in" would be an attempt at persuasion that does not overvalue our agency or self and that adjusts social pressures such that we have a reasonable chance of inducing self-change in our supposed opponent or enemy. In other words, it follows through on the Jaina philosophy of communication: to exist and communicatively assert our views in as nonviolent a way as possible and to always guard against those passions that fixate our attention on the temporary features of our social and natural existence. The phenomenon of online shaming does the precise opposite of these, no matter how righteous the cause feels to us and how evil our enemies seem.

Notes

1. Kris Holt, "How a 'Big Dongle' Joke brought out the worst of the Internet," *Daily Dot*, March 20, 2013, http://www.dailydot.com/society/pycon-dongle-joke-misogyny-sexism-adria-richards/.

2. Ibid.

3. Kris Holt, "Inside 4chan's Plot to get Adria Richards Fired," *Daily Dot*, March 21, 2013, http://www.dailydot.com/society/adria-richards-4chan-raid-sendgrid-donglegate/.

4. Nancy K. Baym, *Personal Connections in the Digital Age* (Malden: Polity, 2010).

5. Judith S. Donath, "Identity and Deception in the Virtual Community," in *Communities in Cyberspace*, ed. Marc Smith and Peter Kollock (New York: Routledge, 1999): 27–56; Scott R. Stroud and Danee Pye, "Kant on Unsocial Sociability and the Ethics of Social Blogging," in *New Agendas in Communication: Ethics in Communication Professions*, ed. Minette E. Drumwright (New York: Routledge, 2013), 41–64.

6. See "Boston Marathon Bombing Victim Halloween Costume Prompts Online Fury," *Huffington Post*, November 3, 2013, http://www.huffingtonpost.com/2013/11/03/boston-marathon-victim-costume_n_4208720.html; Christina Cappechi and Katie Rogers, "Killer of Cecil the Lion Finds Out That He Is a Target Now, of Internet Vigilantism," *New York Times*, July 29, 2015, www.nytimes.com/2015/07/30/us/cecil-the-lion-walter-palmer.html; "Cheating Website Ashley Madison Hacked, Personal Info Posted," *US News and World Report*, July 20, 2015, https://www.usnews.com/news/business/articles/2015/07/20/cheating-website-ashley-madison-hacked-personal-info-posted; Scott R. Stroud, "'Be a Bully to Beat a Bully': Twitter Ethics, Online Identity, and the Culture of Quick Revenge," in *Controversies in Digital Ethics*, ed. Amber Davisson and Paul Booth (New York: Bloomsbury Press, 2016), 264–78.

7. Jeffrey D. Long, *Jainism* (London: I. B. Tauris, 2011). For work on Gandhi's rhetoric, see Arvind Singhal, "The Mahatma's Message: Gandhi's Contributions to the Art and Science of Communication," *China Media Research* 6, no. 3 (2010): 103–6;

William J. Starosta and Lili Shi, "Alternate Perspectives on Gandhian Communication Ethics," *China Media Research* 3, no. 4 (2007): 7–14.

8. The importance of resisting the "dear self" and its self-focused passions is not foreign to Western theories of ethics. One can find versions of it in ancient schools of thought dating back to the Epicureans and Stoics, as well as in more well-known normative orientations such as that of Kant's ethics.

9. Danielle Keats Citron, *Hate Crimes in Cyberspace* (Cambridge, MA: Harvard University Press, 2014).

10. Scott R. Stroud, "The Dark Side of the Online Self: A Pragmatist Critique of the Growing Plague of Revenge Porn," *Journal of Mass Media Ethics* 29, no. 3 (2014): 168–83.

11. Jon Ronson, *So You've Been Publicly Shamed* (New York: Riverhead Books, 2015).

12. This Tumblr blog can be found at http://racistsgettingfired.tumblr.com.

13. Soraya Nadia McDonald, "'Racists Getting Fired' Exposes Weaknesses of Internet Vigilantism, No Matter How Well-Intentioned," *Washington Post*, December 2, 2014, http://www.washingtonpost.com/news/morning-mix/wp/2014/12/02/racists-getting-fired-exposes-weaknesses-of-internet-vigilantism-no-matter-how-well-intentioned/.

14. Archived at http://web.archive.org/web/20150206044123/http://racistsgettingfired.tumblr.com/submit.

15. Originally at http://racistsgettingfired.tumblr.com/tagged/GETTIN; screen capture on file with the author.

16. I will redact identifiable names of individuals and companies, given that this information may only perpetuate justified or unjustified harm to specific individuals. I will also redact racial slurs so as not to reproduce the harm such utterances may have originally caused. Given the changeable and revisable nature of the internet, I also captured and saved screenshots of all the Tumblr postings analyzed in this chapter.

17. http://racistsgettingfired.tumblr.com/tagged/GETTIN.

18. http://racistsgettingfired.tumblr.com/tagged/GETTIN%27/page/3.

19. Ibid.

20. http://racistsgettingfired.tumblr.com/tagged/TIPS/page/2.

21. Ibid.

22. Ibid.

23. http://racistsgettingfired.tumblr.com/tagged/GOTTEN.

24. See John Stuart Mill, *On Liberty* (Indianapolis, IN: Hackett, 1978), especially chapter 2.

25. http://racistsgettingfired.tumblr.com/tagged/GOTTEN.

26. Sam Biddle, "Psycho Frames Ex with Fake Racist Facebook Posts," Gawker, December 1, 2014, http://gawker.com/psycho-frames-ex-with-fake-racist-facebook-posts-1665418226.

27. For more on the history of Jainism, see Kendall W. Folkert, *Scripture and Community: Selected Essays on the Jains* (Atlanta: Scholars Press, 1993).

28. Much more can be said on the Jaina tradition and rhetoric. For a detailed ac-

count of the evolution of Jaina thought, see Paul Dundas, *The Jains*, 2nd ed. (London: Routledge, 2002). For accounts of the specific contours of Jaina rhetorical practices, see Peter Flügel, "Power and Insight in Jaina Discourse," in *Logic and Belief in Indian Philosophy*, ed. Piotr Balcerowicz (Delhi: Motilal Banarsidas, 2009), 85–217; Bimal Krishna Matilal, *The Central Philosophy of Jainism (Anekānta-Vāda)* (Ahmedabad: L.D. Institute of Indology, 1981); Scott R. Stroud, "Anekāntavāda and Engaged Rhetorical Pluralism: Explicating Jaina Views on Perspectivism, Violence, and Rhetoric," *Advances in the History of Rhetoric* 17, no. 2 (2014): 131–56.

29. Padmanabh S. Jaini, *The Jaina Path of Purification* (Delhi: Motilal Banarsidass Publishers, 1998).

30. The translation used is *Purusārtha Siddyupāya: Realization of the True Self*, trans. Vijay K. Jain (Dehradun, India: Vikalp Printers, 2012).

31. Ibid., 10.

32. Ibid., 12.

33. Ibid., 14.

34. Ibid., 42–43.

35. Ibid., 45.

36. Ibid., 46.

37. Ibid., 51.

38. Ibid., 47.

39. Jon Ronson, *So You've Been Publicly Shamed*, 280–81.

40. Ibid., 275.

41. Citron, *Hate Crimes*, 61.

42. Richard B. Gregg, "The Ego-Function of the Rhetoric of Protest," *Philosophy & Rhetoric* 4, no. 2 (1971): 74.

43. Ibid., 76.

44. Ibid., 81–82.

45. http://racistsgettingfired.tumblr.com/tagged/GETTIN.

46. http://racistsgettingfired.tumblr.com/tagged/GETTIN.

47. One could see predecessors to "calling in" strategies in the characterization of invitational rhetoric or in the notion of true persuaders as helpful "lovers." See Sonja K. Foss and Cindy L. Griffin, "Beyond Persuasion: A Proposal for an Invitational Rhetoric," *Communication Monographs* 62, no. 1(1995): 2–18; Wayne Brockriede, "Arguers as Lovers," *Philosophy & Rhetoric* 5, no. 1 (1972): 1–11.

48. Asam Ahmad, "A Note on Call-Out Culture," *Briarpatch Magazine*, March 2, 2015, http://briarpatchmagazine.com/articles/view/a-note-on-call-out-culture.

49. Ibid.

50. William James, *Principles of Psychology* (Cambridge, MA: Harvard University Press, [1890] 1981), 130–31.

Bibliography

Adams, Carol J. *The Sexual Politics of Meat: A Feminist-Vegetarian Critical Theory.* New York: Continuum, 1990.

Aeschylus. *Agamemnon, Libation Bearers, and Eumenides*, Vol. 2. Translated by Herbert Weir Smyth. Cambridge, MA: Harvard University Press, 1971.

——. *Suppliant Maidens, Persians, Prometheus, Seven Against Thebes*, Vol. 1. Edited by Hugh Lloyd-Jones and translated by Herbert Weir Smyth. Cambridge, MA: Harvard University Press, 2006.

Agamben, Giorgio. *Remnants of Auschwitz: The Witness and the Archive.* New York: Zone Books, 1999.

Ahmed, Sara. "Collective Feelings Or, the Impressions Left by Others." *Theory, Culture and Society* 21, no. 2 (2004): 25–42.

——. *The Cultural Politics of Emotion.* New York: Routledge, 2004.

Alcoff, Linda. "The Problem of Speaking for Others." *Cultural Critique* 20 (1991–1992): 5–32.

Anderson Jr., R. Dean. *Glossary of Greek Rhetorical Terms Connected to Methods of Argumentation, Figures and Tropes from Anaximenes to Quintilian.* Leuven, Belgium: Peeters, 2000.

Angles, Stephen C. *Contemporary Confucian Political Philosophy.* Boston: Polity, 2012.

Apollonius Rhodius. *Argonautica.* Edited by George W. Mooney. London: Longmans, Green, 1912.

Arendt, Hannah. *The Human Condition*, 2nd ed. Chicago: University of Chicago Press, 1998 [1958].

Aristotelis. *Ars Rhetorica.* New York: Oxford University Press, 1959.

——. *De Arte Poetica Liber.* New York: Oxford University Press, 1965.

Aristotle. *The "Art" of Rhetoric.* Translated by John Henry Freese. Cambridge, MA: Harvard University Press, 1982.

——. *Ethica Nicomachea.* Oxford: Clarendon Press, 1894.

——. *History of Animals, Volume I: Books 1–3.* Translated by A. L. Peck. Cambridge, MA: Harvard University Press, 1965.

——. *History of Animals, Volume III: Books 7–10.* Translated by D. M. Balme. Cambridge, MA: Harvard University Press, 1991.

———. *Metaphysics, Books 1–9*. Translated by Hugh Tredennick. Cambridge, MA: Harvard University Press, 1980.

———. *Parts of Animals*. Translated by A. L. Peck. Cambridge, MA: Harvard University Press, 1961 [1937].

———. *On Rhetoric*, 2nd ed. Translated by George Kennedy. New York: Oxford University Press, 2007.

———. *Poetics*. Translated by W. Hamilton Fyfe. Cambridge, MA: Harvard University Press, 1932 [1927].

———. *Politics*. Translated by H. Rackam. Cambridge, MA: Harvard University Press, 1977.

———. *Rhetoric*. Translated by Rhys Roberts. New York: Modern Library, 1954.

———. *Topica*. Translated by E. S. Forster. Cambridge, MA: Harvard University Press, 1960.

Arroyo, Sarah. *Participatory Composition: Video Culture, Writing, and Electracy*. Carbondale: Southern Illinois Press, 2013.

Asen, Robert. "A Discourse Theory of Citizenship." *Quarterly Journal of Speech* 90, no. 2 (2004): 189–211.

Atwill, Janet M. *Rhetoric Reclaimed: Aristotle and the Liberal Arts Tradition*. Ithaca, NY: Cornell University Press, 1998.

Aune, James Arnt. "Coping with Modernity: Strategies of 20th-Century Rhetorical Theory." In *The SAGE Handbook of Rhetorical Studies*, edited by Andrea A. Lunsford, Kirt H. Wilson, and Rosa A. Eberly, 85–109. Thousand Oaks, CA: SAGE, 2009.

Ayers, Gael. "The Evolutionary Nature of Genre: An Investigation of the Short Texts Accompanying Research Articles in the Scientific Journal *Nature*." *English for Specific Purposes* 27, no. 1 (2008): 22–41.

Balme, D. M. "Aristotle's Biology Was Not Essentialist." *Archiv für Geschichte der Philosophie* 62, no. 1 (1980): 1–12.

———. "*Γένος* and *Εἶδος* in Aristotle's Biology." *Classical Quarterly* 12, no. 1 (1962): 81–98.

Barber, E. J. W. "The Peplos of Athena." In *Goddess and Polis: The Panathenaic Festival in Ancient Athens*, edited by Jennifer Niels, 103–17. Princeton, NJ: Princeton University Press, 1992.

———. *Prehistoric Textiles: The Development of Cloth in the Neolithic and Bronze Ages with Special Reference to the Aegean*. Princeton, NJ: Princeton University Press, 1991.

Bardini, Thierry. *Junkware*. Minneapolis: University of Minnesota Press, 2011.

Barnhart, Michael G. "*Śūnyatā*, Textualism, and Incommensurability." *Philosophy East and West* 44, no. 4 (1994): 647–58.

Barnouw, Jeffrey. *Odysseus, Hero of Practical Intelligence: Deliberation and Signs in Homer's* Odyssey. Lanham, MD: University Press of America, 2004.

Bataille, Georges. *Visions of Excess*. Edited and translated by Allan Stoekl with Carl R. Lovitt and Donald M. Leslie Jr. Minneapolis: University of Minnesota Press, 1985.

Bauerlein, Mark. *The Dumbest Generation: How the Digital Age Stupefies Young Americans and Jeopardizes Our Future (Or, Don't Trust Anyone Under 30)*. New York: Penguin, 2008.

Bawarshi, Anis S., and Mary Jo Reiff. *Genre: An Introduction to History, Theory, Research, and Pedagogy*. West Lafayette, IN: Parlor Press, 2010.

Baym, Nancy K. *Personal Connections in the Digital Age*. Malden, MA: Polity, 2010.

Bazerman, Charles. "Systems of Genres and the Enactment of Social Intentions." In *Genre and the New Rhetoric*, edited by Aviva Freedman and Peter Medway, 79–101. London: Taylor and Francis, 1994.

———. *Shaping Written Knowledge: The Genre and Activity of the Experimental Article in Science*. Madison: University of Wisconsin Press, 1988.

Beacock, Ian. "Humanist among Machines." *Aeon* (June 25, 2015). http://aeon.co/magazine/society/why-we-need-arnold-toynbees-muscular-humanism/.

Benhabib, Seyla. "Democratic Iterations: The Local, the National, and the Global." In *Another Cosmopolitanism*, edited by Seyla Benhabib and Robert Post, 45–82. New York: Oxford University Press, 2006.

Benitez, Eugenio. "Argument, Rhetoric, and Philosophic Method: Plato's *Protagoras*." *Philosophy & Rhetoric* 25, no. 3 (1992): 222–52.

Bennett, Jane. *Vibrant Matter: A Political Ecology of Things*. Durham, NC: Duke University Press, 2010.

Bentley, Russell. "On Plato's *Phaedrus*: Politics Beyond the City Walls." *Polis: The Journal of the Society for Greek Political Thought* 22, no. 2 (2005): 230–49.

Benveniste, Émile. "Gift and Exchange in Indo-European Vocabulary." In *The Logic of the Gift: Toward an Ethic of Generosity*, edited by Alan Schrift, 33–43. New York: Routledge, 1997.

Beresford, Adam. "Fangs, Feathers, and Fairness: Protagoras on the Origins of Right and Wrong." In *Protagoras of Abdera: The Man, His Measure*, edited by Johannes M. van Ophuijsen, Marlein Van Raalte, and Peter Stork Leiden, 139–62. The Netherlands: Brill, 2013.

Berger, Jonah, and Katherine L. Milkman. "What Makes Online Content Viral?" *Journal of Marketing Research* 49, no. 2 (2012): 192–205.

Berkenkotter, Carol, and Thomas N. Huckin. "Rethinking Genre from a Sociocognitive Perspective." *Written Communication* 10, no. 4 (1993): 475–509.

———. *Genre Knowledge in Disciplinary Communication: Cognition/Culture/Power*. Hillsdale, NJ: Lawrence Erlbaum Associates, 1995.

Berry, David M. *Critical Theory and the Digital*. New York: Bloomsbury, 2014.

Biesecker, Barbara A., and John Louis Lucaites, eds. *Rhetoric, Materiality, and Politics*. New York: Peter Lang, 2009.

Birkets, Sven. *The Gutenberg Elegies: The Fate of Reading in an Electronic Age*. New York: Fawcett Columbine, 1994.

Bitzer, Lloyd F., and Edwin Black, eds. *The Prospect of Rhetoric*. Englewood Cliffs, NJ: Prentice-Hall, 1971.

Black, Edwin. *Rhetorical Criticism: A Study in Method*. Madison: University of Wisconsin Press, 1978 [1965].

Blackmore, Susan. *The Meme Machine*. Oxford: Oxford University Press, 1999.

Blum, Lawrence. "Compassion." In *Explaining Emotions*, edited by Amélie Oksenberg Rorty, 507–18. Berkeley: University of California Press, 1980.

Blumenberg, Hans. "An Anthropological Approach to the Contemporary Significance of Rhetoric." In *After Philosophy: End or Transformation?*, edited by Kenneth Baynes, James Bohman, and Thomas McCarthy, 429–58. Cambridge, MA: MIT Press, 1987.

Boellstorff, Tom. *Coming of Age in Second Life: An Anthropologist Explores the Virtually Human*. Princeton, NJ: University of Princeton Press, 2008.

Bogost, Ian, Simon Ferrari, and Bobby Schweizer. *Newsgames: Journalism at Play*. Cambridge, MA: MIT Press, 2010.

Boler, Megan. "Introduction." In *Digital Media and the Democracy: Tactics in Hard Times*, edited by Megan Boler, 1–50. Cambridge, MA: MIT Press, 2008.

Bollmer, Grant. *Inhuman Networks: Social Media and the Archaeology of Connection*. New York: Bloomsbury, 2016.

Bolt, Richard A. *The Human Interface: Where People and Computers Meet*. Belmont, CA: Wadsworth, 1984.

———. "Spatial Data-Management." In *Multimedia: From Wagner to Virtual Reality*, edited by Randall Packer and Ken Jordan, 185–97. New York: Norton, 2001.

Bolter, Jay David, and Richard Grusin. *Remediation: Understanding New Media*. Cambridge, MA: MIT Press, 1998.

Bolter, Jay David. *Writing Space: Computers, Hypertext, and the Remediation of Print*. Mahwah, NJ: Lawrence Erlbaum Associates, 1991.

Bond, Sarah. "Facebook Before Facebook: Tagging in Antiquity." *medievalbooks* (November 21, 2014). http://medievalbooks.nl/2014/11/21/facebook-before-facebook-tagging-in-antiquity/.

Bone, Jennifer Emerling, Cindy L. M. Griffin, and T. M. Linda Scholz. "Beyond Traditional Conceptions of Rhetoric: Invitational Rhetoric and a Move Toward Civility." *Western Journal of Communication* 72, no. 4 (2008): 434–62.

Boon, Marcus. *In Praise of Copying*. Cambridge, MA: Harvard University Press, 2010.

Brand, Stewart. *The Media Lab: Inventing the Future at MIT*. New York: Viking, 1987.

Brennan, Teresa. *The Transmission of Affect*. Ithaca, NY: Cornell University Press, 2004.

Brock, Roger. *Greek Political Imagery: From Homer to Aristotle*. New York: Bloomsbury, 2013.

Brockriede, Wayne. "Arguers as Lovers." *Philosophy & Rhetoric* 5, no. 1 (1972): 1–11.

Brooke, Collin G. *Lingua Fracta: Toward a Rhetoric of New Media*. New York: Hampton Press, 2009.

Brown, Emily, and Paul Cairns. "A Grounded Investigation of Game Immersion." Paper presented at the Annual Meeting of the Computer-Human Interaction Conference, Vienna, Austria, 2004.

Bromwich, David. "Trapped in the Virtual Classroom." *New York Review of Books* (July 9, 2015). http://www.nybooks.com/articles/2015/07/09/trapped-virtual-classroom/.

Brummett, Barry. *A Rhetoric of Style*. Carbondale: Southern Illinois University Press, 2008.

Bruns, Axel. *Blogs, Wikipedia, Second Life, and Beyond: From Production to Produsage*. New York: Peter Lang, 2008.

Bruss, Kristine S. "Persuasive *Ethopoeia* in Dionysius's *Lysias*." *Rhetorica: A Journal of the History of Rhetoric* 31, no. 1 (2013): 34–57.

Bryant, Donald. "Rhetoric: Its Function and Scope." *Quarterly Journal of Speech* 39, no. 4 (1953): 401–24.

Buck, Pearl S. "Nobel Lecture—The Chinese Novel." Nobelprize.org (September 25, 2017). http://www.nobelprize.org/nobel_prizes/literature/laureates/1938/buck-lecture.html.

Burdea, Grigore C. "Keynote Address: Haptic Feedback for Virtual Reality." *Proceedings of International Workshop on Virtual Prototyping*, Laval, France, 1999.

Burke, Kenneth. "Catharsis—Second View." *The Centennial Review* 5, no. 2 (1961): 107–32.

———. *Counter-Statement*. Berkeley: University of California Press, 1968.

———. "On Catharsis, or Resolution." *Kenyon Review* 21, no. 3 (1959): 337–75.

———. *Permanence and Change: An Anatomy of Purpose*. Indianapolis: Bobbs-Merrill, 1965 [1935].

———. *A Rhetoric of Motives*. Berkeley: University of California Press, 1969.

Burnyeat, Myles. "Protagoras and Self-Refutation in Plato's *Theaetetus*." *Philosophical Review* 85, no. 2 (1976): 172–95.

Campbell, Karlyn Kohrs, and Kathleen Hall Jamieson. "Form and Genre in Rhetorical Criticism: An Introduction." In *Form and Genre: Shaping Rhetorical Action*, edited by Karlyn Kohrs Campbell and Kathleen Hall Jamieson, 9–32. Falls Church, VA: Speech Communication Association, 1978.

Capra, Andrea. "Notes and Discussions: Protagoras' Achilles: Homeric Allusion as a Satirical Weapon." *Classical Philology* 100, no. 3 (2005): 274–77.

Carey, James W. *Communication as Culture: Essays on Media and Society*. New York: Routledge, 1992.

———. "Historical Pragmatism and the Internet." *New Media and Society* 7, no. 4 (2005): 443–55.

Carey, James W., and John Quirk. "The Mythos of the Electronic Revolution." In *Communication as Culture: Essays on Media and Society*, edited by James Carey, 113–41. Cambridge, MA: Unwin Hyman, 1988.

Chan, Joseph. *Confucian Perfectionism: A Political Philosophy for Modern Times*. Princeton, NJ: Princeton University Press, 2015.

Church, Scott H. "All Living Things Are DJs: Rhetoric, Aesthetics, and Remix Culture." PhD diss., University of Nebraska-Lincoln, 2013.

———. "A Rhetoric of Remix." In *The Routledge Companion to Remix Studies*, edited by Eduardo Navas, Owen Gallagher, and xtine burrough, 43–53. New York: Routledge, 2014.

Cicero. *De Inventione*. Translated by H. M. Hubbell. Cambridge, MA: Harvard University Press, 1949.

——. *Orator*. Translated by G. L. Hendrickson and H. M. Hubbell. Cambridge, MA: Harvard University Press, 1939.

——. *Pro Caelio*, in *Cicero XIII*. Translated by R. Gardner. Cambridge, MA: Harvard University Press, 1970.

——. *Pro Marco Caelio*. Edited by Andrew R. Dyck. New York: Cambridge Press, 2013.

Ciceronis, M. Tulli. *Pro M. Caelio Oratio*, 3rd ed. Translated by R. G. Austin. London: Oxford Press, 1960.

Citron, Danielle Keats. *Hate Crimes in Cyberspace*. Cambridge, MA: Harvard University Press, 2014.

Clark, Malcolm, Ian Ruthven, and Patrik O'Brian Holt. "The Evolution of Genre in *Wikipedia*." *Journal for Language Technology and Computational Linguistics* 24, no. 1 (2009): 1–22.

Cloud, Dana. "'Civility' as a Threat to Academic Freedom." *First Amendment Studies* 49, no. 1 (2015): 13–17.

Clough, Patricia Ticineto, ed. *The Affective Turn*. Durham, NC: Duke University Press, 2007.

Collinson, Diané, Kathryn Plant, and Robert Wilkinson. *Fifty Eastern Thinkers: Routledge Key Guides*. New York, NY: Routledge, 2000.

Combs, James E., and Dan Nimmo. *The Comedy of Democracy*. Westport: Praeger, 1996.

Confucius. *The Analects*. Translated by D. C. Lau. Hong Kong: Chinese University Press, 1992.

Consigny, Scott. *Gorgias: Sophists and Artist*. Columbia: University of South Carolina Press, 2001.

Cook, Albert. "Dialectic, Irony, and Myth in Plato's *Phaedrus*." *American Journal of Philology* 106, no. 4 (1985): 427–41.

Corbett, Edward P. J. "Rhetoric in Search of a Past, Present, and Future." In *The Prospect of Rhetoric: Report of the National Developmental Project*, edited by Edwin Black and Lloyd Bitzer, 166–78. Englewood Cliffs, NJ: Prentice-Hall, 1971.

——. "The Theory and Practice of Imitation in Classical Rhetoric." *College Composition and Communication* 22, no. 3 (1971): 243–50.

Cox, Christoph, and Daniel Warner, eds. *Audio Culture: Readings in Modern Music*. New York, NY: Continuum, 2004.

Crespi, John. "Beyond Satire: The Pictorial Imagination of Zhang Guangyu's 1945 *Journey to the West in Cartoons*." In *Oxford Handbook of Modern Chinese Literatures*, edited by Carlos Rojas and Andrea Bachner, 215–44. New York: Oxford University Press, 2016.

Crick, Nathan. "The Sophistical Attitude and the Invention of Rhetoric." *Quarterly Journal of Speech* 96, no. 1 (2010): 25–45.

Crosswhite, James. *Deep Rhetoric: Philosophy, Reason, Violence, Justice, Wisdom*. Chicago: University of Chicago Press, 2013.

Crowley, Sharon. *Toward a Civil Discourse: Rhetoric and Fundamentalism*. Pittsburgh, PA: University of Pittsburgh Press, 2006.

Csikszentmihalyi, Mihaly. *Flow: The Psychology of Optimal Experience*. New York: Harper, 1990.

Damiris, Niklas, and Helga Wild. "The Internet: A New Agora?" In *An Ethical Global Information Society: Culture and Democracy Revisited*, edited by Jacques J. Berleur and Diane Whitehouse, 307–17. New York: Chapman and Hall, 1997.

D'Angour, Armand. *The Greeks and the New: Novelty in Ancient Greek Imagination and Experience*. New York: Cambridge, 2011.

Dawkins, Richard. *The Selfish Gene*. New York: Oxford University Press, 1976.

Davies, Malcolm, and Jeyareney Kathirithamby. *Greek Insects*. New York: Oxford University Press, 1986.

Davis, Colin. "Can the Dead Speak to Us? De Man, Levinas and Agamben." *Culture, Theory and Critique* 45, no. 1 (2004): 77–89.

Davis, Cynthia J. "Contagion as Metaphor." *American Literary History* 14, no. 4 (2002): 828–36.

de Bary, Wm. Theodore and Tu Weiming. *Confucianism and Human Rights*. New York: Columbia University Press, 1999.

Dede, Chris. "Immersive Interfaces for Engagement and Learning." *Science* 323 (5910): 66–69.

Del Caro, Adrian. *Grounding the Nietzsche Rhetoric of Earth*. Berlin: Walter de Gruyter, 2004.

DeLuca, Kevin Michael, and Jennifer Peeples. "From Public Sphere to Public Screen: Democracy, Activism, and the 'Violence' of Seattle." *Critical Studies in Media Communication* 19, no. 2 (2002): 125–51.

Dennett, Daniel C. *Darwin's Dangerous Idea: Evolution and the Meanings of Life*. New York: Simon and Schuster, 1995.

Derrida, Jacques. *Dissemination*. London: Continuum, 1981.

Desjardins, Rosemarie. "Why Dialogues? Plato's Serious Play." In *Platonic Writings, Platonic Readings*, edited by Charles Griswold, 110–25. London: Routledge, 1988.

Detienne, Marcel, and Jean-Pierre Vernant. *Cunning Intelligence in Greek Culture and Society*. Translated by Janet Lloyd. Chicago: University of Chicago Press, 1991.

Devitt, Amy. "Re-fusing Form in Genre Study." In *Genres in the Internet: Issues in the Theory of Genre*, edited by Janet Giltrow and Dieter Stein, 27–48. Amsterdam: John Benjamins, 2009.

Diamond, Larry. "Liberation Technology." *Journal of Democracy* 21, no. 3 (July 2010): 70–71.

Donath, Judith S. "Identity and Deception in the Virtual Community." In *Communities in Cyberspace*, edited by Marc Smith and Peter Kollock, 27–56. New York: Routledge, 1999.

Dondlinger, Mary Jo. "Educational Video Game Design: A Review of the Literature." *Journal of Applied Educational Technology*, 4, no. 1 (2007): 21–31.

Douglas, Mary. "No Free Gifts." In *The Gift: Form and Reason for Exchange in Archaic Societies*, edited by Marcel Mauss, translated by W. D. Hall, vii–xviii. New York: W. W. Norton, 2000.

Downey, Sharon D. "The Evolution of the Rhetorical Genre of Apologia." *Western Journal of Communication* 57 (1993): 42–64.

Dubrow, Heather. *Genre*. London: Methuen, 1982.

Dundas, Paul. *The Jains*, 2nd ed. London: Routledge, 2002.

Dyehouse, Jeremiah. "The Cyberspace Incrementum: Technology Development for Communicative Abundance." *Rhetoric Society Quarterly* 39, no. 3 (2009): 281–302.

Eagleton, Terry. *Literary Theory: An Introduction*. Minneapolis: University of Minnesota Press, 1983.

Eberly, Rosa A. "Public Making and Public Doing: Rhetoric's Productive and Practical Powers," *Rhetoric & Public Affairs* 4, no. 3 (2001): 532–34.

Edwards, Richard L. "Remixing with Rules: Constraint and Potential in Restrictive Remixes." In *Sampling Media*, edited by David Laderman and Laurel Westrup, 31–42. Oxford: Oxford University Press, 2014.

Ehninger, Douglas. "On Systems of Rhetoric." *Philosophy & Rhetoric* 1, no. 3 (1968): 131–44.

Enos, Theresa. "A Call for Comity." In *Beyond Postprocess and Postmodernism: Essays on the Spaciousness of Rhetoric*, edited by Theresa Enos, Keith D. Miller, and Jill McCracken, 131–57. New York: Routledge, 2002.

Entralgo, Pedro Laín. *The Therapy of the Word in Classical Antiquity*, edited and translated by L. J. Rather and John M. Sharp. New Haven, CT: Yale University Press, 1970.

Eördögh, Fruzsina. "Gamergate and the New Horde of Digital Saboteurs." *Christian Science Monitor*, November 25, 2014, Accessed October 6, 2017. https://www.csmonitor.com/Technology/Tech-Culture/2014/1125/Gamergate-and-the-new-horde-of-digital-saboteurs.

Eyman, Douglas. *Digital Rhetoric: Theory, Method, Practice*. Ann Arbor: University of Michigan Press, 2015.

Ezell, Margaret. *Social Authorship and the Advent of Print*. Baltimore, MD: Johns Hopkins University Press, 2003.

Faigley, Lester. *Fragments of Rationality: Postmodernity and the Subject of Composition*. Pittsburgh: University of Pittsburgh Press, 1992.

Farrell, Joseph. "Classical Genre in Theory and Practice." *New Literary History* 34, no. 3 (2003): 383–400.

Farrell, Thomas B. *Norms of Rhetorical Culture*. New Haven, CT: Yale University Press, 1993.

———. "Sizing Things Up: Colloquial Reflection as Practical Wisdom." *Argumentation* 12, no. 1 (1998): 1–14.

Ferrari, G. R. F. *Listening to the Cicadas: A Study of Plato's* Phaedrus. Cambridge, UK: Cambridge University Press, 1987.

Fingarette, Herbert. "Following the 'One Thread' of the *Analects*." *Journal of the American Academy of Religion* 47, no. 3 (1979): 373–405.

Finnegan, Cara A. "Recognizing Lincoln: Image Vernaculars in Nineteenth-Century Visual Culture." In *Visual Rhetoric: A Reader in Communication and American*

Culture, edited by Lester Olson, Cara A. Finnegan, and Diane S. Hope, 61–78. New York: Sage, 2008.

Fitzpatrick, David. *The Facebook Effect: The Inside Story of the Company That Is Connecting the World*. New York: Simon and Schuster, 2010.

Flaxman, Gregory. "Networked: Politics in the Age of Information." *Media Res* (Oct. 5, 2012). http://mediacommons.futureofthebook.org/imr/2012/10/05/networked-politics-age-information.

Flügel, Peter. "Power and Insight in Jaina Discourse." In *Logic and Belief in Indian Philosophy*, edited by Piotr Balcerowicz, 85–217. Delhi: Motilal Banarsidas, 2009.

Foley, John Miles. *Oral Tradition and the Internet*. Urbana: University of Illinois Press, 2012.

Foley, Megan. "Voicing Terri Schiavo: Prosopopeic Citizenship in the Democratic Aporia between Sovereignty and Biopower." *Communication and Critical/Cultural Studies* 7, no. 4 (2010): 381–400.

Folkert, Kendall W. *Scripture and Community: Selected Essays on the Jains*. Atlanta: Scholars Press, 1993.

Foss, Sonja K., and Cindy L. Griffin. "Beyond Persuasion: A Proposal for an Invitational Rhetoric." *Communication Monographs* 62, no. 1 (1995): 2–18.

Fowler, Alastair. *Kinds of Literature: An Introduction to the Theory of Genres and Modes*. Cambridge, MA: Harvard University Press, 1982.

Franklin, James. "Aristotle on Species Variation." *Philosophy* 61, no. 236 (1986): 242–52.

Freadman, Anne. "Uptake." In *The Rhetoric and Ideology of Genre: Strategies for Stability and Change*, edited by Richard M. Coe, Lorelei Lingard, and Tatiana Teslenko, 39–53. Cresskill, NJ: Hampton Press, 2002.

Freedman, Aviva. "Beyond the Text: Towards Understanding the Teaching and Learning of Genres." *TESOL Quarterly* 33, no. 4 (1999): 764–67.

Frow, John. *Genre: The New Critical Idiom*, 2nd ed. London: Routledge, 2014.

Fuss, Diana. "Corpse Poem." *Critical Inquiry* 30, no. 1 (2003): 1–30.

Galloway, Alexander R., and Eugene Thacker. *The Exploit: A Theory of Networks*. Minneapolis: University of Minnesota Press, 2007.

Galloway, Alexander R. "Networks." In *Critical Terms for Media Studies*, edited by W. J. T. Mitchell and Mark B. N. Hansen, 280–96. Chicago: University of Chicago Press, 2010.

Garfinkel, Harold. *Studies in Ethnomethodology*. Cambridge: Polity Press, 1967.

Garver, Eugene. "Can Virtue Be Bought?" *Philosophy & Rhetoric* 37, no. 4 (2004): 353–82.

Geisler, Cheryl. "How Ought We to Understand the Concept of Rhetorical Agency? Report from the ARS." *Rhetoric Society Quarterly* 34, no. 3 (Summer 2004): 9–17.

Geitgey, Adam. "The Kaycee Nicole (Swenson) FAQ." In *We've Got Blog: How Weblogs Are Changing Our Culture*, edited by John Rodzvilla, 89–98. Cambridge, MA: Perseus Publishing, 2002.

Gencarella, Stephen Olbrys. "The Myth of Rhetoric: Korax and the Art of Pollution." *Rhetoric Society Quarterly* 37, no. 3 (2007): 251–73.

———. "Purifying Rhetoric: Empedocles and the Myth of Rhetorical Theory." *Quarterly Journal of Speech* 96 (2010): 231–56.

Genette, Gerard. *The Architext: An Introduction*. Translated by Jane E. Lewin. Berkeley: University of California Press, 1992 [1979].

Gibbons, Michelle G. "Reassessing Discovery: Rosalind Franklin, Scientific Visualization, and the Structure of DNA." *Philosophy of Science* 79, no. 1 (2012): 63–80.

Gibson, William. *Neuromancer*. New York: Ace Books, 1984.

Giddens, Anthony. *The Constitution of Society: Outline of the Theory of Structuration*. Berkeley: University of California Press, 1984.

Gilbert, Christopher J. "Playing with Hitler: Downfall and Its Ludic Uptake." *Critical Studies in Media Communication* 30, no. 5 (2013): 407–24.

Gill, Christopher. "Ancient Concepts of Personal Identity." In *The Oxford Handbook of Hellenic Studies*, edited by George Boys-Stones, Barbara Graziosi, and Phiroze Vasunia, 260–70. Oxford: Oxford University Press, 2009.

———. *Personality in Greek Epic, Tragedy, and Philosophy: The Self in Dialogue*. Oxford: Clarendon Press, 1998.

———. *The Structured Self in Hellenistic and Roman Thought*. Oxford: Oxford University Press, 2009.

Giltrow, Janet. "Form Alone: The Supreme Court of Canada Reading Historical Treaties." In *Genre Studies around the Globe: Beyond the Three Traditions*, edited by Natasha Artemeva and Aviva Freedman, 207–24. Edmonton, AB: Inkshed Publications, 2015.

Giltrow, Janet, and Dieter Stein. "Genres in the Internet: Innovation, Evolution, and Genre Theory." In *Genres in the Internet: Issues in the Theory of Genre*, edited by Janet Giltrow and Dieter Stein, 1–26. Amsterdam: John Benjamins, 2009.

Gladwell, Malcolm. *The Tipping Point: How Little Things Can Make a Big Difference*. Boston: Back Bay Press, 2002.

Godin, Seth. *Unleashing the Idea Virus*. New York: Do You Zoom, 2001.

Gohr, Siegfried. *Magritte: Attempting the Impossible*. New York: Distributed Art Producers, 2009.

Goldin, Paul R. "The Theme of the Primacy of the Situation in Classical Chinese Philosophy and Rhetoric." *Asia Major* 18 no. 2 (2005): 1–25.

———. "When Zhōng Does Not Mean 'Loyalty,'" *Dao* 7, no. 2 (2008): 169.

Goldstein, Rebecca Newberger. *Plato at the Googleplex: Why Philosophy Won't Go Away*. New York: Pantheon Books, 2014.

Gonzalez, Francisco, ed. *The Third Way: New Directions in Platonic Studies*. Lanham, MD: Rowman and Littlefield, 1995.

Gordon, Jill. *Plato's Erotic World: From Cosmic Origins to Human Death*. Cambridge, UK: Cambridge University Press, 2012.

Gorgias. "Encomium of Helen." In *The Older Sophists*, edited by Rosamond Kent Sprague, 50–4. Indianapolis, IN: Hackett, 1972.

———. "On Not Being." In *Early Greek Political Thought from Homer to the Sophists*, edited and translated by Michael Gagarin and Paul Woodruff, 190–210. Cambridge, UK: Cambridge University Press, 1995.

Grafton, Anthony. *The Footnote: A Curious History*. Cambridge, MA: Harvard University Press, 1999.

Granger, Herbert. "Aristotle and the Genus-Species Relation." *Southern Journal of Philosophy* 18, no. 1 (1980): 37–50.

Grau, Oliver. *Virtual Art: From Illusion to Immersion*. Cambridge, MA: MIT Press, 2003.

Graves, Robert. *The Greek Myths*, Vol. 1. Baltimore: Penguin Books, 1974.

Gregg, Richard B. "The Criticism of Symbolic Inducement: A Critical-Theoretical Connection." In *Speech Communication in the 20th Century*, edited by Thomas W. Benson, 41–62. Carbondale: Southern Illinois University Press, 1985.

———. "The Ego-Function of the Rhetoric of Protest." *Philosophy & Rhetoric* 4, no. 2 (1971): 71–91.

Gries, Laurie E. *Still Life with Rhetoric: A New Materialist Approach for Visual Rhetorics*. Boulder: University Press of Colorado, 2015.

Griffiths, Allison. *Shivers Down Your Spine: Cinema, Museums, and the Immersive View*. New York: Columbia University Press, 2008.

Gronbeck, Bruce. "Foreword: Visual Rhetorical Studies: Traces of Power Through Time and Space." In *Visual Rhetoric: A Reader in Communication and American Culture*, edited by Lester Olson, Cara A. Finnegan, and Diane S. Hope, xxi–xxvi. New York: Sage, 2008.

Gronnvoll, Marita, and Jamie Landau. "From Viruses to Russian Roulette to Dance: A Rhetorical Critique and Creation of Genetic Metaphors." *Rhetoric Society Quarterly* 40, no. 1 (2010): 46–70.

Gross, Alan G., and Jonathan Buehl, eds. *Science and the Internet: Communicating Knowledge in a Digital Age*. Amityville, NY: Baywood, 2016.

Gruber, David R. "The (Digital) Majesty of All Under Heaven: Affective Constitutive Rhetoric at the Hong Kong Museum of History's Multi-Media Exhibition of Terracotta Warriors," *Rhetoric Society Quarterly* 44, no. 2 (2014): 148–67.

Gunkel, David J. *Of Remixology: Ethics and Aesthetics After Remix*. Cambridge, MA: MIT Press, 2016.

Gunter, Glenda A, Robert F. Kenny, and Erik H. Vick. "Taking Educational Games Seriously: Using the RETAIN Model to Design Endogenous Fantasy into Standalone Educational Games." *Educational Technology Research and Development* 56, no. 5 (2008): 511–37.

Gurak, Laura, and Smiljana Antonijevic. "Digital Rhetoric and Public Discourse." In *The SAGE Handbook of Rhetorical Studies*, edited by Andrea Lunsford, Kirt Wilson, and Rosa Eberly, 497–507. Thousand Oaks, CA: Sage Publications, 2009.

Guthrie, W. K. C. *The Sophists*. Cambridge, UK: Cambridge University Press, 1971.

Hahner, Leslie A. "The Riot Kiss: Framing Memes as Visual Argument." *Argumentation and Advocacy* 49, no. 3 (2013): 151–66.

Håland, Evy Johnanne. "Athena's Peplos: Weaving as a Core Female Activity in Ancient and Modern Greece." *Cosmos* 20 (2004): 155–82.

Hall, David L., and Roger T. Ames. *The Democracy of the Dead: Dewey, Confucius, and the Hope for Democracy in China*. Chicago: Open Court, 1999.

——. *Thinking through Confucius*. Albany: SUNY, 1987.

Hall, Edith. *Introducing the Ancient Greeks: From Bronze Age Seafarers to Navigators of the Western Mind*. New York: W. W. Norton, 2014.

Halliwell, Stephen. *The Aesthetics of Mimesis: Ancient Texts and Modern Problems*. Princeton, NJ: Princeton University Press, 2002.

Haraway, Donna J. *Simians, Cyborgs, and Women: The Reinvention of Nature*. New York: Routledge, 1991.

Hariman, Robert, and John Louis Lucaites. *No Caption Needed: Iconic Photographs, Public Culture, and Liberal Democracy* Chicago: University of Chicago Press, 2007.

Hariman, Robert. *Political Style: The Artistry of Power*. Chicago: University of Chicago Press, 1995.

Harold, Christine, and Kevin Michael DeLuca. "Behold the Corpse: Violent Images and the Case of Emmett Till." In *Visual Rhetoric: A Reader in Communication and American Culture*, edited by Lester Olson, Cara A. Finnegan, and Diane S. Hope, 257–72. New York: Sage, 2008.

Harold, Christine. *OurSpace: Resisting the Corporate Control of Culture*. Minneapolis: University of Minnesota Press, 2007.

Hart, Roderick P., Jay P. Childers, and Colene J. Lind. *Political Tone: How Leaders Talk and Why*. Chicago: University of Chicago Press, 2013.

Hartelius, E. Johanna, and Gordon R. Mitchell. "Big Data and New Metrics of Scholarly Expertise." *Review of Communication* 14, no. 3–4 (2014): 288–313.

Hartelius, E. Johanna. "Face-ing Immigration: *Prosopopeia* and the 'Muslim-Arab-Middle Eastern' Other." *Rhetoric Society Quarterly* 43, no. 4 (2013): 311–34.

——. "'Leave a Message of Hope or Tribute': Digital Memorializing as Public Deliberation." *Argumentation and Advocacy* 47, no. 3 (2010): 67–85.

——. *The Rhetoric of Expertise*. Lanham, MD: Lexington Books, 2011.

Haskins, Ekaterina. "Ephemeral Visibility and the Art of Mourning: Eyes Wide Open Traveling Exhibit." In *Rhetoric, Remembrance and Visual Form: Sighting Memory*, edited by Anne T. Demo and Bradford Vivian, 89–112. London: Routledge, 2012.

——. *Logos and Power in Isocrates and Aristotle*. Columbia: University of South Carolina Press, 2004.

Hauser, Gerard A., and Carole Blair. "Rhetorical Antecedents to the Public." *PRE/TEXT* 3 (1982): 139–67.

Havelock, Eric. *The Muse Learns to Write: Reflections on Orality and Literacy from Antiquity to the Present*. New Haven, CT: Yale University Press, 1986.

Hawhee, Debra. *Bodily Arts: Rhetoric and Athletics in Ancient Greece*. Austin: University of Texas Press, 2004.

——. "Kairotic Encounters." In *Perspectives on Rhetorical Invention*, edited by Janet M. Atwill and Janice M. Lauer, 16–35. Knoxville: University of Tennessee Press, 2002.

——. "Looking into Aristotle's Eyes: Toward a Theory of Rhetorical Vision." *Advances in the History of Rhetoric* 14, no. 2 (2011): 139–65.

———. *Moving Bodies: Kenneth Burke at the Edges of Language*. Columbia: University of South Carolina Press, 2009.

Hays, J. N. *The Burdens of Disease: Epidemics and Human Response in Western History*. New Brunswick: Rutgers University Press, 1998.

Hayles, N. Katherine. *How We Became Posthuman: Virtual Bodies in Cybernetics, Literature, and Informatics*. Chicago: University of Chicago Press, 1999.

Heidegger, Martin. *Of the Sublime: Presence in Question*. Translated by Jeffrey S. Librett. Albany: SUNY Press, 1993.

Heinrichs, Jay. *Thank You for Arguing: What Aristotle, Lincoln, and Homer Simpson Can Teach Us About the Art of Persuasion*. New York: Three Rivers Press, 2007.

Henderson, Danielle. *Feminist Ryan Gosling: Feminist Theory (as Imagined) from Your Favorite Sensitive Movie Dude*. Philadelphia: Running Press, 2012.

Henderson, Jeffrey. *Aristophanes: Birds, Lysistrata, Women at the Thesmophoria*. Cambridge, MA: Harvard University Press, 2000.

Hesford, Wendy S. *Spectacular Rhetorics: Human Rights Visions, Recognitions, Feminisms*. Durham, NC: Duke University Press, 2011.

Hess, Aaron. "'You Don't Play, You Volunteer': Narrative Public Memory Construction in Medal of Honor: Rising Sun," *Critical Studies in Media Communication* 24, no. 4 (2007): 339–56.

Hoffman, David C. "Structural *Logos* in Heraclitus and the Sophists." *Advances in the History of Rhetoric* 9, no. 1 (2006): 1–32.

Holmes, Brooke. *The Symptom and the Subject: The Emergence of the Physical Body in Ancient Greece*. Princeton, NJ: Princeton University Press, 2010.

Homer. *Homeri Opera: Iliadis*, Vols. 1–2. London: Oxford University Press, 1920.

———. *Homeri Opera: Odysseae*, Vols. 3–4. London: Oxford University Press, 1974.

———. *The Odyssey of Homer*. Translated by Richmond Lattimore. New York: Harper Perennial, 1991.

———. *Odyssey*, 2 Vols. Translated by A. T. Murray. Cambridge, MA: Harvard University Press, 1984.

Hopkins, Jeffrey. *Meditation on Emptiness*. Boston, MA: Wisdom Publications, 1996.

Horace. *The Art of Poetry*. Translated by James Hynd and Burton Raffel. Albany: SUNY Press, 1974.

Houston, Keith. "The Octothorpe." In *Shady Characters: The Secret Life of Punctuation, Symbols, and Other Typographic Marks*. New York: W. W. Norton, 2013.

Howe, Henry, and John Lyne. "Gene Talk in Sociobiology." *Social Epistemology* 6, no. 2 (1993): 109–63.

Huang, Chen-chieh. "East Asian Conceptions of the Public and Private Realms." In *Taking Confucian Ethics Seriously: Contemporary Theories and Applications*, edited by Kam-por Yu, Julia Tao, and Philip J. Ivanhoe, 73–98. Albany: SUNY Press, 2010.

Hudson, Hoyt H. "Rhetoric and Poetry." *Quarterly Journal of Speech Education* 10, no. 2 (1924): 143–54.

Hume, David. *The History of England, from the Invasion of Julius Caesar to the Revolution in 1688*. London: W. Bulmer and Co. Cleveland Row, 1803.

Irvine, Martin. "Remix and the Dialogic Engine of Culture." In *The Routledge Companion to Remix Studies*, edited by Eduardo Navas, Owen Gallagher, and xtine burrough, 15–42. New York: Routledge, 2015.

Isocrates. "Against the Sophists." In *Isocrates*, vol. 2. Translated by George Norlin. Cambridge, MA: Harvard University Press, 1929.

———. *Isocrates*. Translated by George Norlin and Larue Van Hook. Cambridge, MA: Harvard University Press, 1980.

———. *Isocrates I*. Translated by David Mirhady and Yun Lee Too. Austin: University of Texas Press, 2004.

Jackson, Maggie. *Distracted: The Erosion of Attention and the Coming Dark Age*. Amherst, NY: Prometheus Books, 2008.

Jacobson, Murrey. "Google Finally Discloses Its Diversity Record, and It's Not Good." *PBS NewsHour*. (May 28, 2014). http://www.pbs.org/newshour/updates/google-discloses-workforce-diversity-data-good/.

Jaini, Padmanabh S. *The Jaina Path of Purification*. Delhi: Motilal Banarsidass Publishers, 1998.

Jamieson, Kathleen Hall. "Antecedent Genre as Rhetorical Constraint." *Quarterly Journal of Speech* 61, no. 4 (1975): 406–15.

Jarratt, Susan. "Recontextualizing Ancient Rhetoric." In "Manifesting a Future for Comparative Rhetoric," special issue, *Rhetoric Review* 34, no. 3 (2015): 250–52.

———. *Rereading the Sophists: Classical Rhetoric Refigured*. Carbondale: Southern Illinois University Press, 1998.

Jasper, James. *The Art of Moral Protest: Culture, Biography, and Creativity in Social Movements*. Chicago: University of Chicago Press, 1997.

Jenkins, Eric S. "The Modes of Visual Rhetoric: Circulating Memes as Expressions." *Quarterly Journal of Speech* 100, no. 4 (2014): 442–66.

Jenkins, Henry. "If It Doesn't Spread, It's Dead (Part One): Media Viruses and Memes." Confessions of an Aca-Fan, February 11, 2009. http://henryjenkins.org/2009/02/if_it_doesnt_spread_its_dead_p.html.

Jenkins, Henry, Sam Ford, and Joshua Green. *Spreadable Media: Creating Value and Meaning in a Networked Culture*. New York: New York University Press, 2013.

Johnson, Davi. "Mapping the Meme: A Geographical Approach to Materialist Rhetorical Criticism." *Communication and Critical/Cultural Studies* 4, no. 1 (2007): 27–50.

Johnson, Matthew. "Rekindling Rhetoric: Oratory and Marketplace Culture in *Guild Wars*." In *Writing and the Digital Generation: Essays on New Media Rhetoric*, edited by Heather Urbanski, 162–73. Jefferson, NC: McFarland, 2010.

Johnstone, Christopher Lyle. "Sophistical Wisdom: *Politikē Aretē* and 'Logosophia.'" *Philosophy & Rhetoric* 39, no. 4 (2006): 265–89.

Johnston, Gregory S. "Musical-Rhetorical *Prosopopoeia* and the Animation of the Dead in Seventeenth-Century Funeral Music." *Canadian University Music Review* 10, no. 1 (1990): 12–39.

Jouanna, Jacques. *Greek Medicine from Hippocrates to Galen: Selected Papers*. The Netherlands: Brill, 2012.

Kalimtzis, Kostas. *Aristotle on Political Enmity and Disease: An Inquiry into* Stasis. Albany: SUNY Press, 2000.

Kastely, James. "Plato's Protagoras: Revisionary History as Sophisticated Comedy." *Rhetoric Review* 15, no. 1 (1996): 26–43.

Keen, Andrew. *Cult of the Amateur: How Blogs, MySpace, YouTube, and the Rest of Today's User-generated Media are Destroying Our Economy, Our Culture, and Our Values*. New York: Doubleday, 2007.

Kennedy, George A. *The Art of Persuasion in Ancient Greece*. Princeton, NJ: Princeton University Press, 1963.

———. *Classical Rhetoric and Its Christian and Secular Tradition from Ancient to Modern Times*, 2nd ed. Chapel Hill: University of North Carolina Press, 1999.

Kennedy, George, trans. *Progymnasmata: Greek Textbooks of Prose Composition and Rhetoric*. Atlanta, GA: Society of Biblical Literature, 2003.

Kennerly, Michele. "Getting Carried Away: How Rhetorical Transport Gets Judgment Going." *Rhetoric Society Quarterly*, 40, no. 3 (2010): 269–91.

———. "Micrologia." Paper presented at the annual meeting of the National Communication Association, November 19–23, 2014, Chicago.

———. "The Stoic Nature of Early Dramatistic Theory." *Advances in the History of Rhetoric* 14, no. 1 (2011): 33–48.

Kibby, Marjorie D. "Email Forwardables: Folklore in the Age of the Internet." *New Media and Society* 7, no. 6 (2005): 770–90.

Kirby, John T. "The 'Great Triangle' in Early Greek Rhetoric and Poetics." *Rhetorica* 8, no. 3 (1992): 213–28.

Kirk, Rita, and Dan Schill. "A Digital Agora: Citizen Participation in the 2008 Presidential Debates." *American Behavioral Scientist* 55, no. 3 (2011): 325–47.

Kirkpatrick, Robert. "A Conversation with Robert Kirkpatrick, Director of United Nations Global Pulse." *SAIS Review* 34, no. 1 (2014): 3–8.

Komito, David Ross. *Nagarjuna's* Seventy Stanzas*: A Buddhist Psychology of Emptiness*. Ithaca, NY: Snow Lion, 1987.

Landow, George. *Hypertext: The Convergence of Contemporary Critical Theory and Technology*. Baltimore, MD: Johns Hopkins University Press, 1992.

Lanham, Richard A. *Analyzing Prose*. New York: Scribner's, 1983.

———. *The Economics of Attention: Style and Substance in the Information Age*. Chicago: University of Chicago Press, 2006.

———. *The Electronic Word: Democracy, Technology, and the Arts*. Chicago: University of Chicago Press, 1993.

Lantz, Frank. "Something Called the Immersive Fallacy." Presentation at the annual meeting for the Game Developers Conference, San Francisco, California, 2006, http://www.gdcvault.com/play/1013249/Burn-Baby-Burn-Game-Developers.

Latour, Bruno. *Reassembling the Social: An Introduction to Actor-Network-Theory*. Oxford: Oxford University Press, 2005.

Lavin, Chad. "Factory Farms in a Consumer Society." *American Studies* 50, no. ½ (2009): 71–92.

Lebduska, Lisa. "Emoji, Emoji, What for Art Thou?" *Harlot* 12 (2014). http://harlotofthearts.org/index.php/harlot/article/view/186/157.

Leff, Michael. "Things Made by Words: Reflections on Textual Criticism." *Quarterly Journal of Speech* 78, no. 2 (1992): 223–31.

Leiss, William, and Stephen Kline, and Sut Jhally. *Social Communication in Advertising*, 2nd ed. New York: Routledge, 1997.

Lennox, James G. "Aristotle on Genera, Species, and 'the More and the Less.'" *Journal of the History of Biology* 13, no. 2 (1980): 321–46.

Leroi, Armand Marie. *The Lagoon: How Aristotle Invented Science*. New York: Viking, 2014.

Lethem, Jonathan. "The Ecstasy of Influence: A Plagiarism Mosaic." In *Sound Unbound: Sampling Digital Music and Culture*, edited by Paul D. Miller, 25–52. Cambridge, MA: MIT Press, 2008.

Liddell, Henry George, Robert Scott, Henry Stuart Jones, and Roderick McKenzie, eds. *A Greek English Lexicon*. Oxford: Clarendon Press, 1996.

Liu, Alan. *The Laws of Cool: Knowledge Work and the Culture of Information*. Chicago: University of Chicago Press, 2004.

Lloyd, G. E. R. "The Development of Aristotle's Theory of the Classification of Animals." *Phronesis* 6, no. 1 (1961): 59–81.

———. *In the Grip of Disease: Studies in the Greek Imagination*. New York: Oxford University Press, 2003.

———. *Polarity and Analogy: Two Types of Argumentation in Early Greek Thought*. Cambridge, UK: Cambridge University Press, 1966.

Loeb, Paul Rogat. *Soul of A Citizen: Living with Conviction in a Cynical Time*. New York: St. Martin's Griffin, 1999.

Long, Jeffrey D. *Jainism*. London: I. B. Tauris, 2011.

Low, Sor-Ching. "Romancing Emptiness." *Contemporary Buddhism* 7, no. 2 (2006): 129–47.

Lozano-Reich, Nina M., and Dana L. Cloud. "The Uncivil Tongue: Invitational Rhetoric and the Problem of Inequality." *Western Journal of Communication* 73, no. 2 (2009): 220–26.

Lu, Xing. *Rhetoric in Ancient China, Fifth to Third Century B.C.E.: A Comparison with Classical Greek Rhetoric*. Columbia: University of South Carolina Press, 1998.

Lucas, Stephen E. "Genre Criticism and Historical Context: The Case of George Washington's First Inaugural Address." *Southern Speech Communication Journal* 51, no. 4 (1986): 354–70.

Lyon, Arabella. *Deliberative Acts: Democracy, Rhetoric, and Rights*. University Park: Pennsylvania State University Press, 2013.

———. "Tricky Words: *Rhetoric* and *Comparative*." In "Manifesting a Future for Comparative Rhetoric," special issue, *Rhetoric Review* 34, no. 3 (2015): 243–46.

Macé, Arnaud. "Privatising Perception: Plato's Protagoreanism (*Theaetetus* 154B-157C)." In *Protagoras of Abdera: The Man, His Measure*, edited by Johannes M. van Ophuijsen, Marlein Van Raalte, and Peter Stork Leiden, 195–216. The Netherlands: Brill, 2013.

Manovich, Lev. *The Language of New Media*. Cambridge, MA: MIT Press, 2001.

Matilal, Bimal Krishna. *The Central Philosophy of Jainism (Anekānta-Vāda)*. Ahmedabad: L. D. Institute of Indology, 1981.

Mattelart, Armand. *Networking the World, 1794–2000*. Translated by Liz Carey-Libbrecht and James A. Cohen. Minneapolis: University of Minnesota Press, 2000.

Mauer, Barry. "Rigorous Infidelity: Whole Text Sampling in the Curatorial Work of Henri Langlois, Dewey Phillips, and Jean-Francois Lyotard." In *Sampling Media*, edited by David Laderman and Laurel Westrup, 60–72. Oxford: Oxford University Press, 2014.

Mauss, Marcel. *The Gift: Form and Reason for Exchange in Archaic Societies*. Translated by W. D. Hall. New York: W. W. Norton, 2000.

Mayr, Ernst. *The Growth of Biological Thought: Diversity, Evolution, and Inheritance*. Cambridge, MA: Belknap Press, 1982.

McChesney, Robert. *Digital Disconnect: How Capitalism Is Turning the Internet Against Democracy*. New York: New Press, 2013.

McComiskey, Bruce. *Gorgias and the New Sophistic Rhetoric*. Carbondale: Southern Illinois University Press, 2002.

McCorkle, Ben. *Rhetorical Delivery as Technological Discourse: A Cross-historical Study*. Carbondale: Southern Illinois University Press, 2012.

McGlew, James F. *Tyranny and Political Culture in Ancient Greece*. Ithaca, NY: Cornell University Press, 1993.

McKeon, Richard. "Literary Criticism and the Concept of Imitation in Antiquity." *Modern Philology* 34, no. 1 (1936): 1–35.

———. "The Uses of Rhetoric in a Technological Age: Architectonic Productive Arts." In *The Prospect of Rhetoric*, edited by Lloyd F. Bitzer, Edwin Black, and Karl Richard Wallace, 44–63. Englewood Cliffs, NJ: Prentice-Hall, 1971.

McLeod, Kembrew, and Peter DiCola. *Creative License: The Law and Culture of Digital Sampling*. Durham, NC: Duke University Press, 2011.

McLuhan, Eric. "Literacy in a New Key." *Proceedings of the Media Ecology Association*, vol. 10, edited by Paul A. Soukup. http://www.media-ecology.org/publications/MEA_proceedings/v10/3_literacies.pdf.

McLuhan, Marshall. *Understanding Media: The Extensions of Man*. Cambridge, MA: MIT Press, 1994.

McMahan, Alison. "Immersion, Engagement, and Presence: A New Method for Analyzing 3-D Video Games." In *The Video Game Theory Reader*, edited by Mark J. P. Wolf and Bernard Perron, 67–86. New York: Routledge, 2003.

Mendelson, Michael. *Many Sides: A Protagorean Approach to the Theory, Practice, and Pedagogy of Argument*. Dordrecht, the Netherlands: Kluwer Academic Publishers, 2002.

Mifsud, Mari Lee. *Rhetoric and the Gift: Ancient Rhetorical Theory and Contemporary Communication*. Pittsburgh, PA: Duquesne University Press, 2015.

Mill, John Stuart. *On Liberty*. Indianapolis, IN: Hackett Publishing Company, 1978.

Miller, Carolyn R., and Dawn Shepard. "Blogging as Social Action: A Genre Analysis of Weblogs." In *Into the Blogosphere: Rhetoric, Community, and the Culture of*

Weblogs, edited by Laura Gurak, Smiljana Antonijevic, Laurie Johnson, Clancy Ratliff, and Jessica Reyman. University of Minnesota Digital Conservancy, 2004. http://hdl.handle.net/11299/172818.

———. "Questions for Genre Theory from the Blogosphere." In *Genres in the Internet: Issues in the Theory of Genre*, edited by Janet Giltrow and Dieter Stein, 263–90. Amsterdam: John Benjamins, 2009.

Miller, Carolyn R. "Genre as Social Action." *Quarterly Journal of Speech* 70, no. 2 (1984): 151–76.

———. "New Genres, Now and Then." In *Literature, Rhetoric, and Values*, edited by Shelley Hulan, Murray McArthur, and Randy Allen Harris, 127–49. Newcastle upon Tyne: Cambridge Scholars Publishing, 2012.

———. "Where Do Genres Come From?" In *Emerging Genres in New Media Environments*, edited by Carolyn R. Miller and Ashley R. Kelly, 1–34. London: Palgrave Macmillan, 2017.

———. "Writing in a Culture of Simulation: *Ethos* Online." In *Towards a Rhetoric of Everyday Life: New Directions in Research on Writing, Text, and Discourse*, edited by Martin Nystrand and John Duffy, 58–83. Madison: University of Wisconsin Press, 2003.

Mitchell, Peta. *Contagious Metaphor*. New York: Bloomsbury, 2012.

Mitchell, W. J. T. *Cloning Terror: The War of Images, 9/11 to the Present*. Chicago: University of Chicago Press, 2011.

Mittell, Jason. *Genre and Television: From Cop Shows to Cartoons in American Culture*. New York: Routledge, 2004.

Montiglio, Silvia. *Wandering in Ancient Greek Culture*. Chicago: University of Chicago Press, 2005.

Moore, Robert E. "From Genericide to Viral Marketing: On 'Brand.'" *Language and Communication* 23, no. 3 (2003): 331–57.

Morozov, Evgeny. *The Net Delusion: The Dark Side of Internet Freedom*. London: Penguin, 2011.

Moss, Jessica. "What Is Imitative Poetry, and Why Is It Bad?" In *The Cambridge Companion to Plato's* Republic, edited by G. R. F. Ferrari, 425–44. New York: Cambridge University Press, 2007.

Moss, Richard. "Haptic Technology: The Next Frontier in Video Games, Wearables, Virtual Reality, and Mobile Electronics." *GizMag*. (January 15, 2015). http://www.gizmag.com/haptic-tech-vr-wearables-games-sightlence/35616/.

Muckelbauer, John. "Imitation and Invention in Antiquity: An Historical-Theoretical Revision." *Rhetorica* 21, no. 2 (2003): 61–88.

Mueller, Melissa. *Objects as Actors: Props and the Poetics of Performance in Greek Tragedy*. Chicago: University of Chicago Press, 2016.

Murphy, David J. "Isocrates and the Dialogue." *Classical World* 106, no. 3 (2013): 311–53.

Murphy, James J., Richard A. Katula, and Michael Hoppmann. *A Synoptic History of Classical Rhetoric*, 4th ed. New York: Routledge, 2014.

Murray, Janet H. *Hamlet on the Holodeck: The Future of Narrative in Cyberspace*. Cambridge, MA: MIT Press, 1998.

Myers, Alicia D. "Prosopopoetics and Conflict: Speech and Expectations in John 8." *Biblica* 92, no. 4 (2011): 580–96.

Nahon, Karine, and Jeff Hemsley. *Going Viral.* Malden: Polity Press, 2013.

Navas, Eduardo. *Remix Theory: The Aesthetics of Sampling.* New York: Springer-Wien NewYork, 2012.

Ndalianis, Angela. *The Horror Sensorium: Media and the Senses.* Jefferson, NC: McFarland, 2012.

Negroponte, Nicholas, and Richard A. Bolt. *Spatial Data Management.* Cambridge, MA: MIT Press, 1979.

Nesselroade, J. R., and P. C. M. Molenaar. "Emphasizing Intraindividual Variability in the Study of Development Over the Lifespan." In *The Handbook of Life-span Development: Cognition, Biology, and Methods across the Lifespan*, edited by W. F. Overton, 30–54. Hoboken, NJ: Wiley, 2010.

Newman, James. "The Myth of the Ergodic Videogame: Some Thoughts on Player-Character Relationships in Videogames." *Game Studies* 2, no. 1 (2002). http://www.gamestudies.org/0102/newman/.

Nightingale, Andrea Wilson. *Genres in Dialogue: Plato and the Construct of Philosophy.* Cambridge, UK: Cambridge University Press, 1995.

———. *Spectacles of Truth in Classical Greek Philosophy: Theory in its Cultural Context.* Cambridge, UK: Cambridge University Press, 2004.

Notomi, Noburu. "A Protagonist of the Sophistic Movement? Protagoras in Historiography." In *Protagoras of Abdera: The Man, His Measure*, edited by Johannes M. van Ophuijsen, Marlein Van Raalte, and Peter Stork Leiden, 11–36. The Netherlands: Brill, 2013.

Nussbaum, Martha. "Golden Rule Arguments: A Missing Thought?" In *The Moral Circle and the Self: Chinese and Western Approaches*, edited by Kim-chong Chong, Sor-hoon Tan, and C. L. Ten, 3–16. Chicago: Open Court, 2003.

Ogden, Susan. *Inklings of Chinese Democracy.* Cambridge, MA: Harvard University Press, 2002.

Olson, Lester C., Cara A. Finnegan, and Diane S. Hope. "Visual Rhetoric in Communication: Continuing Questions and Contemporary Issues." In *Visual Rhetoric: A Reader in Communication and American Culture*, edited by Lester Olson, Cara A. Finnegan, and Diane S. Hope, 1–14. New York: Sage, 2008.

Ong, Walter. *Orality and Literacy: The Technologizing of the Word.* New York: Methuen, 1982.

———, and John Hartley. *Orality and Literacy: 30th Anniversary Edition.* New York: Routledge, 2015.

Ostherr, Kirsten. *Cinematic Prophylaxis: Globalization and Contagion in the Discourse of World Health.* Durham, NC: Duke University Press, 2005.

Packer, Jeremy, and Stephen B. Crofts Wiley, eds. *Communication Matters: Materialist Approaches to Media, Mobility, and Networks.* New York: Routledge, 2012.

Padel, Ruth. *In and Out of the Mind: Greek Images of the Tragic Self.* Princeton, NJ: Princeton University Press, 1992.

Papacharissi, Zizi. "The Unbearable Lightness of Information and the Impossible

Gravitas of Knowledge: Big Data and the Makings of Digital Orality." *Media, Culture, and Society* 37, no. 7 (2015): 1095–1100.

Parikka, Jussi. *Digital Contagions: A Media Archaeology of Computer Viruses*. New York: Peter Lang, 2007.

Pariser, Eli. *The Filter Bubble: How the New Personalized Web Is Changing What We Read and How We Think*. New York: Penguin Books, 2012.

Payne, David. "Rhetoric, Reality, and Knowledge: A Re-Examination of Protagoras' Concept of Rhetoric." *Rhetoric Society Quarterly* 16, no. 3 (1986): 187–97.

Payne, Robert. *The Promiscuity of Network Culture: Queer Theory and Digital Media*. New York: Routledge, 2015.

Perelman, Chaim, and Lucie Olbrechts-Tyteca. *The New Rhetoric: A Treatise on Argumentation*. Notre Dame, IN: University of Notre Dame Press, 1991.

Perlmutter, David D. *Blogwars*. New York: Oxford University Press, 2008.

Peters, F. E. *Greek Philosophical Terms: A Historical Lexicon*. New York: New York University Press, 1967.

Peters, John Durham. "Information: Notes Toward a Critical History." *Journal of Communication Inquiry* 12, no. 2 (1988): 9–23.

———. *The Marvelous Clouds: Toward a Philosophy of Elemental Media*. Chicago: University of Chicago Press, 2015.

———. *Speaking into the Air: A History of the Idea of Communication*. Chicago: University of Chicago Press, 2001.

Petropoulos, J. C. B. *Heat and Lust: Hesiod's Midsummer Festival Scene Revisited*. Lanham, MD: Rowman and Littlefield, 1994.

Pettitt, Thomas. "Bracketing the Gutenberg Parenthesis." *Explorations in Media Ecology* 11, no. 2 (2012): 95–114.

———. "Media Dynamics and the Lessons of History: The 'Gutenberg Parenthesis' as Restoration Topos." In *Companion to New Media Dynamics*, edited by John Hartley, Jean Burgess, and Axel Bruns, 53–72. Somerset, NJ: John Wiley and Sons, 2013.

Pfister, Damien Smith. "The *Logos* of the Blogosphere: Flooding the Zone, Invention, and Attention in the Lott Imbroglio." *Argumentation and Advocacy* 47, no. 3 (2011): 141–67.

———. *Networked Media, Networked Rhetorics: Attention and Deliberation in the Early Blogosphere*. University Park: Pennsylvania State University Press, 2014.

Plato. *Euthyphro, Apology, Crito, Phaedo, Phaedrus*. Translated by H. N. Fowler. Cambridge, MA: Harvard University Press, 1995 [1914].

———. *Republic*. In *The Collected Dialogues*, edited by Edith Hamilton and Huntington Cairns. Princeton, NJ: Princeton University Press, 1961.

———. *The Statesman*. Translated by Harold North Fowler and W. R. M. Lamb. Cambridge, MA: Harvard University Press, 1925.

———. *Symposium*. Translated by Harold N. Fowler. Cambridge, MA: Harvard University Press, 1925.

———. *Timaeus, Critias, Cleitophon, Menexenus, Epistles*. Translated by R. G. Bury. Cambridge, MA: Harvard University Press, 1914.

Porter, James. "Recovering Delivery for Digital Rhetoric and Human-Computer Interaction." *Computers & Composition* 26, no. 4 (2009): 207–24.

Poster, Mark. *What's the Matter with the Internet?* Minneapolis: University of Minnesota Press, 2001.

Poulakos, John. *Sophistical Rhetoric in Classical Greece*. Columbia: University of South Carolina Press, 1995.

———. "Toward a Sophistic Definition of Rhetoric." *Philosophy & Rhetoric* 16, no. 1 (1983): 35–48.

Poulakos, Takis, and David Depew, eds. *Isocrates and Civic Education*. Austin: University of Texas Press, 2009.

Powers, Shawn, and Michael Jablonski. *The Real Cyber War: The Political Economy of Internet Freedom*. Urbana: University of Illinois Press, 2015.

Powers, William. *Hamlet's Blackberry: Building a Good Life in the Digital Age*. New York: HarperCollins, 2010.

Press, Gerald A., ed. *Plato's Dialogues: New Studies and Interpretations*. Lanham, MD: Rowman and Littlefield, 1993.

Pruchnic, Jeff, and Kim Lacey. "The Future of Forgetting: Rhetoric, Memory, Affect." *Rhetoric Society Quarterly* 41, no. 5 (2011): 472–94.

Pruchnic, Jeff. "Rhetoric, Cybernetics, and the Work of the Body in Burke's Body of Work." *Rhetoric Review* 25, no. 3 (2006): 275–96.

———. *Rhetoric and Ethics in the Cybernetic Age: The Transhuman Condition*. New York: Routledge, 2014.

Pucci, Pietro. *Odysseus Polutropos: Intertextual Readings in the Odyssey and the Iliad*. Ithaca, NY: Cornell University Press, 1987.

Purusārtha Siddyupāya: Realization of the True Self. Translated by Vijay K. Jain. Dehradun, India: Vikalp Printers, 2012.

Quintilian. *Institutio Oratoria Vol. IV*. Translated by H. E. Butler. Cambridge, MA: Harvard University Press, 1958.

———. *The Orator's Education*. Translated by Donald A. Russell. Cambridge, MA: Harvard University Press, 2001.

Rademaker, Adriaan. "The Most Correct Account: Protagoras on Language." In *Protagoras of Abdera: The Man, His Measure*, edited by Johannes M. van Ophuijsen, Marlein Van Raalte, and Peter Stork Leiden, 87–112. The Netherlands: Brill, 2013.

Rainie, Lee, and Barry Wellman. *Networked: The New Social Operating System*. Cambridge, MA: MIT Press, 2012.

Raley, Rita. *Tactical Media*. Minneapolis: University of Minnesota Press, 2009.

Rand, Erin J. "Bad Feelings in Public: Rhetoric, Affect, and Emotion." *Rhetoric & Public Affairs* 18, no.1 (2015): 161–75.

Rettberg, Jill. *Blogging*. Cambridge, UK: Polity Press, 2008.

Rhet/Comp Ryan Gosling. November 14, 2011. Image retrieved April 21, 2017. https://web.archive.org/web/20150822134018/http://rhetcompryangosling.tumblr.com/post/12787928449.

Rhetorica ad Herennium. Translated by Harry Caplan. Cambridge, MA: Harvard University Press, 1954.

Ricciardone, C. T. "'We are the Disease': Truth, Health, and Politics from Plato's *Gorgias* to Foucault." *Epoché* 18, no. 2 (2014): 287–310.

Rice, Jeff. *Digital Detroit: Rhetoric and Space in the Age of the Network*. Carbondale: Southern Illinois Press, 2012.

Rickert, Thomas. *Ambient Rhetoric: The Attunements of Rhetorical Being*. Pittsburgh, PA: University of Pittsburgh Press, 2013.

Roetz, Heiner. *Confucian Ethics of the Axial Age: A Reconstruction Under the Aspect of the Breakthrough Toward Postconventional Thinking*. Albany: SUNY Press, 1993.

Ronson, Jon. *So You've Been Publicly Shamed*. New York: Riverhead Books, 2015.

Rorty, Richard. "The Historiography of Philosophy: Four Genres." In *Philosophy in History: Essays in the Historiography of Philosophy*, edited by Richard Rorty, Jerome B. Schneewind, Quentin Skinner, 49–75. New York: Cambridge University Press, 1984.

Rosas, Ricardo, Miguel Nussbaum, Patricio Cumsille, Vladimir Marlanov, Mónica Correa, and Patricia Flores. "Beyond Nintendo: Design and Assessment of Educational Games for First and Second Grade Students." *Computers and Education* 40, no.1 (2003): 71–94.

Rosmarin, Adena. *The Power of Genre*. Minneapolis: University of Minnesota Press, 1985.

Roth, Adam David. "Reciprocal Influences between Rhetoric and Medicine in Ancient Greece." PhD diss., University of Iowa, 2008. http://ir.uiowa.edu/etd/3.

Rubel, Alexander. *Fear and Loathing in Ancient Athens: Religion and Politics During the Peloponnesian War*. New York: Routledge, 2014.

Rundell, W. M. F. "Cicero and Clodius: The Question of Credibility." *Historia* 28, no. 3 (1979): 301–28.

Rushkoff, Douglas. *Media Virus! Hidden Agendas in Popular Culture*. New York: Ballantine Books, 1996.

Ryan, Marie. *Narrative as Virtual Reality: Immersion and Interactivity in Literature and Electronic Media*. Baltimore, MD: Johns Hopkins University Press, 2001.

Ryle, Gilbert. *Dilemmas*. New York: Cambridge University Press, 1954.

Sachs, Jonah. *Winning the Story Wars*. Cambridge, MA: Harvard Business School Publishing, 2012.

Sammons, Benjamin. *The Art and Rhetoric of the Homeric Catalogue*. Oxford: Oxford University Press, 2010.

Scaliger, Julius Caesar. *Poetices libri septem*. Lyon: Joannes Crispinus, 1561.

Scarry, Elaine. *On Beauty and Being Just*. Princeton, NJ: Princeton University Press, 2001 [1999].

Schenker, David J. "The Strangeness of the Phaedrus," *American Journal of Philology* 127, no. 1 (2006): 67–88.

Schiappa, Edward. "Isocrates' *Philosophia* and Contemporary Pragmatism." In *Rhetoric, Sophistry, Pragmatism*, edited by Steven Mailloux, 33–60. Cambridge, UK: Cambridge University Press, 1995.

———. *Protagoras and Logos*, 2nd ed. Columbia: University of South Carolina Press, 2003.

Schryer, Catherine F., and Philippa Spoel. "Genre Theory, Health-Care Discourse, and Professional Identity Formation." *Journal of Business and Technical Communication* 19, no. 3 (2005): 249–78.

Schryer, Catherine. "Records as Genre." *Written Communication* 10, no. 2 (1993): 200–234.

Schulte, Stephanie. *Cached: Decoding the Internet in Global Popular Culture*. New York: New York University Press, 2013.

Schwartz, Hillel. *The Culture of the Copy: Striking Likenesses, Unreasonable Facsimiles*. Updated Edition. New York: Zone Books, 2014.

Seabrook, John. *The Song Machine: Inside the Hit Factory*. New York: W. W. Norton, 2015.

Segal, Judy Z. *Health and the Rhetoric of Medicine*. Carbondale: Southern Illinois University Press, 2005.

Shaffer, Diana. "The Shadow of Helen: The Status of the Visual Image in Gorgias's *Encomium of Helen*." *Rhetorica* 16, no. 3 (1998): 243–57.

Shepherd, Michael, and Carolyn Watters. "The Evolution of Cybergenres." In *31st Hawaii International Conference on System Sciences*, edited by Ralph H. Sprague Jr., 97–109. Maui: IEEE Computer Society Press, 1998.

Shifman, Limor. *Memes and Digital Culture*. Cambridge, MA: MIT Press, 2014.

Shiga, John. "Copy-and-Persist: The Logic of Mash-up Culture." *Critical Studies in Media Communication* 24, no. 2 (2007): 93–114.

Shouse, Eric. "Feeling, Emotion, Affect," *Media/Culture* 8, no. 6 (2005). http://journal.media-culture.org.au/0512/03-shouse.php.

Singhal, Arvind. "The Mahatma's Message: Gandhi's Contributions to the Art and Science of Communication." *China Media Research* 6, no. 3 (2010): 103–6.

Skågeby, Jörgen. "Dismantling the Guitar Hero? A Case of Prodused Parody and Disarmed Subversion." *Convergence: The International Journal of Research into New Media Technologies* 19, no. 1 (2013): 63–76.

Sloane, Thomas O. "Prologue: *The Prospect* As Prospectus." In *Reengaging the Prospects of Rhetoric: Current Conversations and Contemporary Challenges*, edited by Mark J. Porrovecchio, 1–4. New York: Routledge, 2010.

Smith, Craig R. "*Ethos* Dwells Pervasively: A Hermeneutic Reading of Aristotle on Credibility." In *The Ethos of Rhetoric*, edited by Michael J. Hyde, 1–19. Columbia: University of South Carolina Press, 2004.

Smith, Robin. "Aristotle's Logic." In *The Stanford Encyclopedia of Philosophy*, edited by Edward N. Zalta. Palo Alto, CA: Center for the Study of Language and Information, Stanford University, 2015. http://plato.stanford.edu/archives/spr2016/entries/aristotle-logic/.

Sontag, Susan. *Illness as Metaphor and AIDS and Its Metaphors*. New York: Anchor Books, 1988.

Sonvilla-Weiss, Stefan. "Introduction: Mashups, Remix Practices and the Recombination of Existing Digital Content." In *Mashup Culture*, edited by Stefan Sonvilla-Weiss, 8–23. New York: SpringerWien NewYork, 2010.

Sophocles. *Ajax, Electra, Oedipus*, Vol. 1. Edited and translated by Hugh Lloyd-Jones. Cambridge, MA: Harvard University Press, 1994.

Sparks, Colin. "The Internet and the Public Sphere." In *Mediated Politics: Communication in the Future of Democracy*, edited by W. Lance Bennett and Robert M. Entman, 75–95. Cambridge, UK: Cambridge University Press, 2001.

Sprague, Rosamond Kent, trans. "*Dissoi Logoi* or *Dialexeis*." In *The Older Sophists: A Complete Translation by Several Hands of the Fragments of Die Fragmente Der Vorsokratikeri*. Edited by Rosamond Kent Sprague. Indianapolis, IN: Hackett, 1972.

———, ed. *The Older Sophists*. Columbia: University of South Carolina Press, 1972.

Stahl, Roger. "A Clockwork War: Rhetorics of Time in a Time of Terror." *Quarterly Journal of Speech* 94, no. 1 (2008): 73–99.

Standage, Tom. *Writing on the Wall: Social Media-The First 2,000 Years*. New York: Bloomsbury, 2013.

Starosta, William J., and Lili Shi. "Alternate Perspectives on Gandhian Communication Ethics." *China Media Research* 3, no. 4 (2007): 7–14.

Stewart, Jesse. "Articulating a Hip-Hop Sampling Aesthetic through Film: Music and Memory in Three Hip-Hop Documentaries." In *Sampling Media*, edited by David Laderman and Laurel Westrup, 143–54. New York: Oxford University Press, 2014.

Strate, Lance. *Amazing Ourselves to Death: Neil Postman's Brave New World Revisited*. New York: Peter Lang, 2014.

Stroud, Scott R. "Anekāntavāda and Engaged Rhetorical Pluralism: Explicating Jaina Views on Perspectivism, Violence, and Rhetoric." *Advances in the History of Rhetoric* 17, no. 2 (2014): 131–56.

———. "'Be a Bully to Beat a Bully': Twitter Ethics, Online Identity, and the Culture of Quick Revenge." In *Controversies in Digital Ethics*, edited by Amber Davisson and Paul Booth, 264–78. New York: Bloomsbury Press, 2016.

———. "The Dark Side of the Online Self: A Pragmatist Critique of the Growing Plague of Revenge Porn." *Journal of Mass Media Ethics* 29, no. 3 (2014): 168–83.

Stroud, Scott R., and Danee Pye. "Kant on Unsocial Sociability and the Ethics of Social Blogging." In *New Agendas in Communication: Ethics in Communication Professions*, edited by Minette E. Drumwright, 41–64. New York: Routledge, 2013.

Stuckey, Mary E., and Sean Patrick O'Rourke. "Civility, Democracy, and National Politics." *Rhetoric & Public Affairs* 17, no. 4 (2014): 711–36.

Sullivan, Robert G. "*Eidos/idea* in Isocrates." *Philosophy & Rhetoric* 34, no. 1 (2001): 79–92.

Sunstein, Cass. *Republic.com*. Princeton, NJ: Princeton University Press, 2001.

Sutton, Jane S., and Mari Lee Mifsud. *A Revolution in Tropes: Alloiostrophic Rhetoric*. Lanham, MD: Lexington Books, 2015.

Tao, Julia. "Trust Within Democracy: A Reconstructed Confucian Perspective." In *Taking Confucian Ethics Seriously: Contemporary Theories and Applications*, edited by Kam-por Yu, Julia Tao, and Philip J. Ivanhoe, 99–122. Albany: SUNY Press, 2010.

Thalmann, William G. *Apollonius of Rhodes and the Spaces of Hellenism*. Oxford: Oxford University Press, 2011.

Tiwald, Justin. "Is Sympathy Naive?: Dai Zhen on the Use of Shu to Track Well-

Being." In *Taking Confucian Ethics Seriously: Contemporary Theories and Applications*, edited by Kam-por Yu, Julia Tao, and Philip J. Ivanhoe, 145–62. Albany: SUNY Press, 2010.

Todorov, Tzvetan. *The Fantastic: A Structural Approach to a Literary Genre*. Translated by Richard Howard. Ithaca, NY: Cornell University Press, 1975 [1970].

———. "The Origin of Genres." *New Literary History* 8, no. 1 (1976): 159–70.

Tola, Fernando, and Carmen Dragonetti. "Nagarjuna's Conception of 'Voidness' (*Śūnyatā*)." *Journal of Indian Philosophy* 9, no. 3 (1981): 273–82.

Turow, Joseph. *Playing Doctor: Television, Storytelling, and Medical Power*. Ann Arbor: University of Michigan Press, 2010.

Vaidhyanathan, Siva. *The Googlization of Everything (And Why We Should Worry)*. Berkeley: University of California Press, 2011.

van Berkel, Tazuko A. "Made to Measure: Protagoras' *Metron*." In *Protagoras of Abdera: The Man, His Measure*, edited by Johannes M. van Ophuijsen, Marlein Van Raalte, and Peter Stork Leiden, 37–68. The Netherlands: Brill, 2013.

Van Looy, Jan. "Uneasy Lies the Head that Wears a Crown: Interactivity and Signification in *Head Over Heels*." *Game Studies* 3, no. 2 (2003). http://www.gamestudies.org/0302/vanlooy/.

van Ophuijsen, Johannes. "Introduction: Protagoras of Abdera: Amicus Homo Magis Amica Veritas?" In *Protagoras of Abdera: The Man, His Measure*, edited by Johannes M. van Ophuijsen, Marlein Van Raalte, and Peter Stork Leiden, 1–10. The Netherlands: Brill, 2013.

Vivian, Bradford. "Style, Rhetoric, and Postmodern Culture." *Philosophy & Rhetoric* 35, no. 3 (2002): 223–43.

Wald, Priscilla. *Contagious: Cultures, Carriers, and the Outbreak Narratives*. Durham, NC: Duke University Press, 2008.

Walker, Jeffrey. *The Genuine Teachers of This Art: Rhetorical Education in Antiquity*. Columbia: University of South Carolina Press, 2011.

———. *Rhetoric and Poetics in Antiquity*. New York: Oxford University Press, 2000.

Walters, Henry. "Consider the Table: Of Socrates, Etymology, and Poetic Logic." *Eidolon* (August 15, 2016). https://eidolon.pub/consider-the-table-b26fc8c940d9#.mxil4gb2f.

Wang, Bo. "Plural Local Terms, Dialectical Processes, and Co-participants: Re-envisioning Transnational Rhetorical Studies." *JAC* 34 (forthcoming).

Warnick, Barbara. *Rhetoric Online: Persuasion and Politics on the World Wide Web*. New York: Peter Lang, 2007.

Watson, Gay. *A Philosophy of Emptiness*. London: Reaktion Books, 2014.

Welch, Kathleen. *Electric Rhetoric: Classical Rhetoric, Oralism, and a New Literacy*. Cambridge, MA: MIT Press, 1999.

Westrup, Laurel. "Thinking through Sampling, Literally." In *Sampling Media*, edited by David Laderman and Laurel Westrup, 228–46. New York: Oxford University Press, 2014.

Wiener, Norbert. *Cybernetics, or Control and Communication in the Animal and the Machine*, 2nd ed. Cambridge, MA: MIT Press, 1962.

Williamson, Linzi, Sarah Sangster, and Karen Lawson, "The Effect of Ryan Gosling Feminist Memes on Feminist Identification and Endorsement of Feminist Beliefs." Paper presented at Canadian Psychology Association, Vancouver, British Columbia, 2015.

Wing, Fok Chun. Screenshot of Fok Chun Wing's Facebook page. October 25, 2014. Image retrieved January 6, 2015. https://www.facebook.com/photo.php?fbid=10152323544227531&set=a.124481952530.115709.608252530&type=3&theater.

Wohl, Victoria. "Rhetoric of the Athenian Citizen." In *The Cambridge Companion to Ancient Rhetoric*, edited by Erik Gunderson, 162–77. New York: Cambridge University Press, 2009.

Woodhouse, S. C. *English-Greek Dictionary: A Vocabulary of the Attic Language*. London: George Routledge and Sons, 1910.

Xi, Jinping. *The Governance of China*. Beijing: Foreign Language Press, 2014.

Yán Channel. October 23, 2014. Image retrieved on January 6, 2015, https://www.facebook.com/ArmChannelTV/photos/pb.330920327037381.-2207520000.1435245766./578431132286298/?type=3&theater.

Yates, Frances A. *The Art of Memory*. Chicago: University of Chicago Press, 1966.

Yates, JoAnne. "The Emergence of the Memo as a Managerial Genre." *Management Communication Quarterly* 2, no. 4 (1989): 485–510.

Yates, JoAnne, and Wanda Orlikowski. "Genres of Organizational Communication: A Structurational Approach to Studying Communication and Media." *Academy of Management Review* 17, no. 2 (1992): 299–326.

Yeo, I. S. "The Concept of Disease in Galen." *Uisahak* 12, no. 1 (2003): 54–65.

Yoos, George E. *Politics and Rhetoric: Coming to Terms with Terms*. New York: Palgrave MacMillan, 2009.

You, Xiaoye. *The Devil's Tongue: A History of English Composition in China*. Carbondale: Southern Illinois University Press, 2010.

Zilioli, Ugo. *Protagoras and the Challenge of Relativism: Plato's Subtlest Enemy*. Hampshire, England: Ashgate, 2007.

———. "Protagoras through Plato and Aristotle: A Case for the Philosophical Significance of Ancient Relativism." In *Protagoras of Abdera: The Man, His Measure*, edited by Johannes M. van Ophuijsen, Marlein Van Raalte, and Peter Stork Leiden, 233–59. The Netherlands: Brill, 2013.

Contributors

Scott Haden Church is an assistant professor in the School of Communications at Brigham Young University. His research primarily uses rhetoric, aesthetics, and media ecology as analytic lenses for digital media and mediated popular texts. His research has been published in *The Information Society*, *The Journal of Information Technology and Politics*, and *The Routledge Companion to Remix Studies*.

Nathan Crick is a professor of rhetoric at Texas A&M University. His work explores the relationship between and power throughout different periods of political and social change, focusing specifically on those factors that are result of conscious strategies of persuasion by individuals or groups. His interests include classical rhetorical theory, social media, the structure of news, religious rhetoric, modern propaganda, the rhetoric of science and technology, the power of aesthetics, the dynamics of social movements, and the history of philosophy.

Rosa A. Eberly is a free-range rhetorician who studies histories and theories of rhetoric, publics theory, public memory, and deliberation in higher education contexts and beyond. Eberly is the author of *Citizen Critics: Literary Public Spheres* and the coeditor of *A Laboratory for Public Scholarship and Democracy* and *The SAGE Handbook of Rhetoric*, and has written articles on publics theory, proto-public deliberation in rhetoric classrooms, public memory, sound studies, and rhetoric and identity.

Christopher J. Gilbert is an assistant professor of English at Assumption College. His work is published in journals such as the *Quarterly Journal of Speech*, *Rhetoric & Public Affairs*, *Philosophy & Rhetoric*, *Rhetoric Society Quarterly*, *Text and Performance Quarterly*, and *Critical Studies in Media Communication*. Chris's work traverses topics in rhetoric and cultural studies, but he has a particular interest in past (including ancient) and present politics of comic artworks and performances; wartime travesties of national identity; circulations of the peculiar and the ugly in public culture; and the combat logics of armed conflicts as well as rhetorical contests.

E. Johanna Hartelius is an associate professor in the Department of Communication at the University of Pittsburgh. Her research areas are the rhetoric of expertise, digital culture, immigration, and public memory. She is the author of *The Rhetoric of Expertise* and the editor of *The Rhetorics of US Immigration*. She is the recipient of the 2013 Janice Hocker Rushing Early Career Research Award, and her scholarship has appeared in *Argumentation and Advocacy*, *Communication and Critical/Cultural Studies*, *Critical Studies in Media Communication*, *Culture, Theory, and Critique*, *Management Communication Quarterly*, the *Quarterly Journal of Speech*, *Review of Communication*, *Rhetoric Society Quarterly*, and the *Southern Communication Journal*. Her current book project on the gifting *logos* theorizes rhetorical expertise in the digital commons.

Ekaterina V. Haskins is a professor of rhetoric in the Department of Communication and Media at Rensselaer Polytechnic Institute. She studies rhetoric as an intellectual and pedagogical tradition and as a practice that shapes individual and collective identities. Her research contributes to three distinct yet related areas of scholarship: the history of rhetoric, public memory, and rhetorics of display. She is the author of two books, *Logos and Power in Isocrates and Aristotle* and *Popular Memories: Commemoration, Participatory Culture, and Democratic Citizenship*. She has published numerous articles and book chapters on the history of rhetoric, public memory, and visual culture. Her current projects include studies of immersion and visuality in Greek rhetoric and philosophy, a book-in-progress on the rhetoric and politics of public memory in post-Communist Russia, and an investigation of the role of place and sensation in public life.

Gaines S. Hubbell is an assistant professor of English and director of composition at the University of Alabama in Huntsville. His research focuses on the rhetorical processes of invention and digital games. His work has appeared in *Game Studies*, and he is an editor for the *Journal of Games Criticism*.

Jeremy David Johnson is a doctoral candidate at Penn State University. Broadly concerned with media and rhetoric, his research focuses on the rhetorical dynamics of networked media, including social media and video games. His dissertation examines the rhetorical ramifications of algorithms in social media environments.

Michele Kennerly is an assistant professor in the Department of Communication Arts and Sciences and the director of the foundational communication course at Penn State University. She is the author of *Editorial Bodies: Perfection and Rejection in Ancient Rhetoric and Poetics* and the co-editor of another volume, *Information Keywords*. Her essays have appeared in such journals as *Rhetorica*, *Rhetoric Society Quarterly*, and *Advances in the History of Rhetoric*.

Arabella Lyon is a professor of English at the University at Buffalo. She is the recipient of the 2014 Rhetoric Society of America Book Award for *Deliberative Acts: Democracy, Rhetoric, and Rights* and the 1999 W. Ross Winterowd Award for Best Book in Composition Theory for *Intentions: Negotiated, Contested, and Ignored.* Her research interests include human rights, performance theory, and Chinese rhetoric.

Mari Lee Mifsud is a professor of rhetoric and the director of Women, Gender, and Sexuality Studies at the University of Richmond. Her work in ancient Greek rhetorical theory and contemporary critical and feminist theory focuses on the idea and practice of gift-giving in the orchestration of a free, just, and peaceful people. She is the author of *Rhetoric and the Gift: Ancient Rhetorical Theory and Contemporary Communication.* Her work with Jane Sutton theorizing rhetoric through difference culminated in *A Revolution in Tropes: Alloiostrophic Rhetoric.* Her work has appeared in *Philosophy & Rhetoric, Rhetoric Society Quarterly, Advances in the History of Rhetoric,* and *Informal Logic.*

Carolyn R. Miller is the SAS Institute Distinguished Professor of Rhetoric and Technical Communication, Emerita, at North Carolina State University. Her research interests include genre studies, digital rhetoric, rhetorical theory, and rhetoric of science and technology, and her publications have explored rhetorical concepts of invention, agency, *kairos*, *ēthos*, genre, and the *polis*. Her professional service includes terms as president of the Rhetoric Society of America and editor of its journal, *Rhetoric Society Quarterly.*

Damien Smith Pfister is an associate professor of communication at the University of Maryland. His work examines the confluence of digitally networked media, rhetorical practice, public deliberation, visual culture, and technology. His interest in how nascent genres of communication provide new opportunities for citizens to affect public deliberation is reflected in *Networked Media, Networked Rhetorics: Attention and Deliberation in the Early Blogosphere.* His essays have appeared in such journals as *Philosophy & Rhetoric, Argumentation and Advocacy, Environmental Communication, Rhetoric Review*, and *Social Epistemology.*

Scott R. Stroud is an associate professor in the Department of Communication Studies at the University of Texas at Austin. His research focuses on the intersection of philosophy and rhetoric. Many of his articles engage the relationship between pragmatism and rhetoric, as well as topics in comparative and non-Western rhetoric. He is the author of *John Dewey and the Artful Life: Pragmatism, Aesthetics, and Morality* and *Kant and the Promise of Rhetoric.*

Index